Navigating Technological Advancement in the VUCA and BANI World

G.S. Prakasha
https://orcid.org/0000-0002-1287-7606
Christ University, India

Maria Lapina
https://orcid.org/0000-0001-8117-9142
North-Caucasus Federal University, Russia

Francisco David Guillén Gámez
https://orcid.org/0000-0001-6470-526X
Universidad de Málaga, Spain

Juan J. Leiva-Olivencia
https://orcid.org/0000-0002-0082-1154
University of Málaga, Spain

IGI Global
Scientific Publishing
Publishing Tomorrow's Research Today

Published in the United States of America by
IGI Global Scientific Publishing
701 East Chocolate Avenue
Hershey, PA, 17033, USA
Tel: 717-533-8845 | Fax: 717-533-7115
Website: https://www.igi-global.com E-mail: cust@igi-global.com

Library of Congress Cataloging-in-Publication Data

Names: Prakasha, G. S. editor http://id.loc.gov/rwo/agents/n2023011515 | Lapina, Maria, 1981- editor http://id.loc.gov/rwo/agents/n2023071571 | Guillén Gámez, Francisco David, 1985- editor http://id.loc.gov/rwo/agents/n2025022716 | Leiva Olivencia, Juan José, 1979- editor http://id.loc.gov/rwo/agents/no2013143986
Title: Navigating technological advancement in the VUCA and BANI world / edited by: G S Prakasha, Maria Lapina, Francisco David Guillén Gámez, Juan Leiva-Olivencia.
Other titles: Navigating technological advancement in the volatile, uncertain, complex, and ambiguous and brittle, anxious, non-linear, incomprehensible world
Description: Hershey, PA : IGI Global Scientific Publishing, [2026] | Includes bibliographical references and index. | Summary: "This volume is unique as it discusses the education, industry, and business around the recent technological advancement with the advent of artificial intelligence. It is necessary for stakeholders to understand how to handle the technological advancement in the VUCA world and the BANI world"-- Provided by publisher.
Identifiers: LCCN 2025003749 (print) | LCCN 2025003750 (ebook) | ISBN 9798337322407 hardcover | ISBN 9798337322414 paperback | ISBN 9798337322421 ebook
Subjects: LCSH: Technological innovations http://id.loc.gov/authorities/subjects/sh85133143
Classification: LCC HD45 .N3784 2025 (print) | LCC HD45 (ebook) | DDC 658.4/063--dc23/eng/20250522
LC record available at https://lccn.loc.gov/2025003749
LC ebook record available at https://lccn.loc.gov/2025003750

British Cataloguing in Publication Data
A Cataloguing in Publication record for this book is available from the British Library.

Table of Contents

Preface xv

Chapter 1
A Novel Method as Manufacturing Technologies to Detect the Negative Effect of Atmospheric Agents in Micro Deterioration of Shock Abers of Aircrafts 1
Ciceron Gonzalez Toxqui, Universidad Politécnica de Baja California, Mexico
Manuel Guzmán Herrera, Universida Autónoma de Baja California, Mexico
Daniel Barrera Román, Universidad Politécnica de Baja California, Mexico
Israel Sauceda Meza, Universida Autónoma de Baja California, Mexico
Arilí Cardenas Robles, Universida Autónoma de Baja California, Mexico
Eddna Teresa Valenzuela Martinez, Universida Autónoma de Baja California, Mexico
Jesús Márquez-González, Universida Autónoma de Baja California, Mexico
Yidanes Alejandra Ibáñez Acevedo, Tecnologico Nacional de México, Instituto Tecnológico de Mexicali, Mexico
Gustavo G. L. López Badilla, Universidad Vasconcelos Campus Murua, Tijuana, Mexico

Chapter 2
VUCA in the Emerging Indian Healthcare Sector Post Pandemic 25
Swati Dwivedi, Graphic Era University, India
Ashulekha Gupta, Graphic Era University, India

Chapter 3
Post-Quantum Cryptography: Robust Solutions Against Quantum Computing Threats 39
Tatyana Peleshenko, North Caucasian Federal University, Russia
Diana Fedash, Faculty of Mathematics and Computer Science, Stavropol, Russia
Alexandr Gish, Faculty of Mathematics and Computer Science, Stavropol, Russia

Chapter 4
Research of Approaches to Assessing the Information Security Risks of an Organization: Qualitative Risk Assessment Methods 47
Julia Andrusenko, North Caucasus Federal University, Russia
G. S. Prakasha, Christ University, Bangalore, India
Ivan Golovin, North Caucasian Federal University, Russia
Vladimir Lapin, Gymnasium 25, Russia

Chapter 5
Machine Learning in Information Security 61
Alexander Bogdan, North Caucasian Federal University, Russia
Svetlana Govorova, North Caucasus Federal University, Russia
Sergey Melnikov, North Caucasian Federal University, Russia
Egor Govorov, North Caucasian Federal University, Russia

Chapter 6
Using Probabilistic Models to Assess and Enhance Information Security 81
Maria Lapina, North Caucasian Federal University, Russia
Alina Bagautdinova, North Caucasus Federal University, Russia
Irina Pykhtina, North Caucasian Federal University, Russia

Chapter 7
Enhancing Well-Being in the Digital Age in the VUCA and BANI World 97
Oindrilla Ghosh, Christ University, India
Binod Kumar, Radha Govind University, India

Chapter 8
Digitalization of Metropolis: Resources, Risks, Capitalization......................... 127
Ekaterina Vladimirovna Kargapolova, Plekhanov Russian University of Economics, Russia
Arina Sergeevna Dikanova, Plekhanov Russian University of Economics, Russia
Anna Jurievna Aryasova, Astrakhan State University of Architecture and Civil Engineering, Russia
Julia Victorovna Georgievskaya, Astrakhan State University of Architecture and Civil Engineering, Russia

Chapter 9
Research, Development, and Comparative Characterization of Methods for Invisible Embedding Digital Watermarks in Electronic Text Documents 139
Maxim Martemyanov, Faculty of Mathematics and Computer Sciences Named After Prof. Nikolay Chervyakov, Russia
Maria Lapina, Faculty of Mathematics and Computer Sciences Named After Prof. Nikolay Chervyakov, Russia
Mary Anita E. A., Independent Researcher, India

Chapter 10
Megalopolis as a Space of Challenges in the Context of Digital Transformation: Real vs. Virtual.. 157
Arina Dikanova, Plekhanov Russian University of Economics, Moscow, Russia
Ekaterina Kargapolova, Plekhanov Russian University of Economics, Moscow, Russia
Victor Mikhailovich Nemchinov, Institute of Oriental Studies, Russian Academy of Sciences, Russia
Irina Gavrilova, Moscow State University, Moscow, Russia

Chapter 11
Analysis of the Dynamics of the Multiplier Effect of Rural Tourism and Its Impact on the Integrated Rural Development 173
Natalia Sergeeva, Russian State Agrarian University - Moscow Timiryazev Agricultural Academy, Russia
Ludmila Evgrafova, Russian State Agrarian University - Moscow Timiryazev Agricultural Academy, Russia

Chapter 12
Cultural Adaptation of Digital Payment Systems With a Focus on Community-Based Financial Practices 189
Nisha K. Kumar, Independent Researcher, India
N. Giribabu, Sri Venkateswara College of Engineering and Technology, India
Meera Sharma, USCS, Uttaranchal University, India
M. Siva Krishna Raju, SRKR Engineering College, India
V. Sri Krishna Chaturya, SRKR Engineering College, India

Chapter 13
Ethical and Legal Dimensions of Drones and Robots in Smart Business: Navigating the Autonomous Innovations in Logistics 259
Ananya Pandey, Christ University, Bangalore, India
Jipson Joseph, Christ University, Bangalore, India

Chapter 14
A Technique for Collecting and Analyzing Physico-Chemical Parameters of Soils Based on Wireless Sensor Network Technology 287
Sergey Melnikov, North-Caucasus Federal University, Russia
Svetlana Govorova, North-Caucasus Federal University, Russia
Egor Govorov, North-Caucasus Federal University, Russia

Compilation of References 301

About the Contributors 353

Index 359

Detailed Table of Contents

Preface.. xv

Chapter 1
A Novel Method as Manufacturing Technologies to Detect the Negative Effect of Atmospheric Agents in Micro Deterioration of Shock Abers of Aircrafts 1
Ciceron Gonzalez Toxqui, Universidad Politécnica de Baja California, Mexico
Manuel Guzmán Herrera, Universida Autónoma de Baja California, Mexico
Daniel Barrera Román, Universidad Politécnica de Baja California, Mexico
Israel Sauceda Meza, Universida Autónoma de Baja California, Mexico
Arilí Cardenas Robles, Universida Autónoma de Baja California, Mexico
Eddna Teresa Valenzuela Martinez, Universida Autónoma de Baja California, Mexico
Jesús Márquez-González, Universida Autónoma de Baja California, Mexico
Yidanes Alejandra Ibáñez Acevedo, Tecnologico Nacional de México, Instituto Tecnológico de Mexicali, Mexico
Gustavo G. L. López Badilla, Universidad Vasconcelos Campus Murua, Tijuana, Mexico

An investigation was made in an aerospace industry located in the Tijuana city, which manufactures shock absorbers used in the landing gear of aircrafts, where an inefficient operation of the shock absorbers was detected, was caused by the activation action of the bar not being generated. expansion of the evaluated shock absorbers. This investigation was made from 2022 to 2023. With the SEM technique showed the atmospheric agents with the most effect in the microdeterioration of the metallic surface of the shock absorbers manufactured and with the automatized system was improved the air quality in indoors of the aerospace where was made this scientific study.

Chapter 2
VUCA in the Emerging Indian Healthcare Sector Post Pandemic 25
Swati Dwivedi, Graphic Era University, India
Ashulekha Gupta, Graphic Era University, India

The VUCA paradigm representing Volatility, Uncertainty, Complexity, and Ambiguity originally conceptualized by the U.S. Army War College to define ambiguous military situations, has achieved constant rising relevance in the modern-day in commercial, business and policy environment. The Indian healthcare sector, categorized by ever-changing market subtleties, intricate supply chains, and volatile global healthcare demands, reflects the spirit of a VUCA world. The COVID-19 pandemic additionally increased these challenges while instantaneously activating invention, resilience, and transformation on a structural basis within the sector. This chapter explores the VUCA framework's ability of how it can be applied to analyse the post-pandemic progression of India's healthcare industry. Its highpoints moves towards the strategic approaches fostering sustainability, adaptability, and competitive strength—accentuating India's place as a noteworthy global benefactor of affordable pharmaceuticals and vaccines.

Chapter 3
Post-Quantum Cryptography: Robust Solutions Against Quantum Computing Threats.. 39
Tatyana Peleshenko, North Caucasian Federal University, Russia
Diana Fedash, Faculty of Mathematics and Computer Science, Stavropol, Russia
Alexandr Gish, Faculty of Mathematics and Computer Science, Stavropol, Russia

Quantum computers are a revolutionary technology based on the principles of quantum mechanics. Unlike conventional computers that work with binary bits (0 or 1), quantum devices use qubits that can exist in multiple states ("superpositions") and interact with each other through the phenomenon of "quantum entanglement." Thanks to this, quanta can greatly accelerate the solution of a number of important problems that were previously considered insurmountable for classical machines. This review describes the principles of operation of quantum computers, considers their unique advantages, and demonstrates the potential for their impact on the future of computer science and the technology industry.

Chapter 4
Research of Approaches to Assessing the Information Security Risks of an Organization: Qualitative Risk Assessment Methods.. 47
Julia Andrusenko, North Caucasus Federal University, Russia
G. S. Prakasha, Christ University, Bangalore, India
Ivan Golovin, North Caucasian Federal University, Russia
Vladimir Lapin, Gymnasium 25, Russia

This article explores modern approaches to assessing information security risks within organizations, focusing on qualitative, quantitative, and hybrid risk management methodologies. The study examines the core principles of each approach, their practical applicability in different organizational contexts, the tools and frameworks commonly employed, and key implementation challenges. A comparative analysis of these methods highlights their respective strengths and limitations, providing insights for selecting the most suitable risk assessment strategy based on organizational needs, industry requirements, and available resources.

Chapter 5
Machine Learning in Information Security 61
Alexander Bogdan, North Caucasian Federal University, Russia
Svetlana Govorova, North Caucasus Federal University, Russia
Sergey Melnikov, North Caucasian Federal University, Russia
Egor Govorov, North Caucasian Federal University, Russia

The features of the application of machine learning methods in information security tasks and development prospects and future trends are considered. The article analyzes the application of a machine learning model to detect attacks in the field of information security using the RandomForestClassifier classifier. Various aspects of data processing are considered, including balancing classes and selecting the most significant features for building a model. This approach can be useful for developing real-time monitoring and protection systems.

Chapter 6
Using Probabilistic Models to Assess and Enhance Information Security 81
Maria Lapina, North Caucasian Federal University, Russia
Alina Bagautdinova, North Caucasus Federal University, Russia
Irina Pykhtina, North Caucasian Federal University, Russia

The article explores the possibilities of using probabilistic models to assess and improve information security. The theoretical foundations of the probabilistic approach are considered, including methods of risk analysis, threat modeling and prediction of potential vulnerabilities. Special attention is paid to the use of models such as Bayesian networks, Markov chains, and Monte Carlo simulations, which allow quantifying the likelihood of incidents and their consequences. The paper presents practical examples of the use of probabilistic models for analyzing the security of information systems, as well as recommendations for their integration into information security management processes.

Chapter 7
Enhancing Well-Being in the Digital Age in the VUCA and BANI World 97
Oindrilla Ghosh, Christ University, India
Binod Kumar, Radha Govind University, India

In today's VUCA (Volatile, Uncertain, Complex, Ambiguous) and BANI (Brittle, Anxious, Nonlinear, Incomprehensible) reality, enhancing well-being in the digital age is crucial. Digital technology can both support and hinder mental health, making it essential to cultivate a balanced approach. Encouraging digital detoxes and mindfulness practices can reduce stress and promote emotional resilience. Leveraging online communities for support fosters connection, but awareness of the potential for digital addiction remains essential. Tools like mental health apps can provide resources, yet they should complement real-life interactions to ensure authenticity. Emphasizing digital literacy empowers individuals to navigate information critically, enhancing decision-making regarding their mental and emotional health. Finally, organizations must prioritize employee well-being by implementing flexible work arrangements, promoting open communication, and providing mental health resources. By consciously shaping our digital environments, we can enhance well-being and thrive in this complex world.

Chapter 8
Digitalization of Metropolis: Resources, Risks, Capitalization......................... 127
Ekaterina Vladimirovna Kargapolova, Plekhanov Russian University of Economics, Russia
Arina Sergeevna Dikanova, Plekhanov Russian University of Economics, Russia
Anna Jurievna Aryasova, Astrakhan State University of Architecture and Civil Engineering, Russia
Julia Victorovna Georgievskaya, Astrakhan State University of Architecture and Civil Engineering, Russia

Digitalization of metropolis transforms social connections and human capital structures. Relevance of the study on the resource base, content and negative consequences of using Internet resources by residents of the metropolis is obvious. This survey presents results of a sociological study of Moscow residents, conducted according to the Standard Methodology of the All-Russian Scientific Research Program "Sociocultural Evolution of Russia and Its Regions" in December 2023. The assessments of the Internet resources used by respondents for entertainment, work, interaction with local authorities, opportunities for saving money and time are analyzed, risks and negative consequences are considered. Respondents use capabilities of online stores, various platforms for online communication, with a predominance of young and adult respondents, as well as respondents from the wealthy category. Low level of benefit from using Internet resources is observed among elderly and low-income respondents. These categories also need adapted digital services and accessible opportunities for their needs.

Chapter 9
Research, Development, and Comparative Characterization of Methods for Invisible Embedding Digital Watermarks in Electronic Text Documents 139
Maxim Martemyanov, Faculty of Mathematics and Computer Sciences Named After Prof. Nikolay Chervyakov, Russia
Maria Lapina, Faculty of Mathematics and Computer Sciences Named After Prof. Nikolay Chervyakov, Russia
Mary Anita E. A., Independent Researcher, India

This article discusses the concept of digital watermark (DWC), its classification, the main file formats for storing textual information electronically, types of attacks on DWC, and methods of embedding digital watermarks in electronic text documents based on formatting changes, using metadata, structural and linguistic changes. The practical development of each method is carried out, their advantages, disadvantages and resistance to various attacks are identified, and their comparative characterization is presented. On the basis of the analysis, we experimentally developed our own methods of introducing DWC into text documents, including the approach using zero-width spaces (ZWSP) (U+200B), which provides high resistance to various attacks and imperceptibility for the user. The results of the work demonstrate the effectiveness of each implementation and provide recommendations for their application depending on the requirements for copyright protection and authentication of text files.

Chapter 10
Megalopolis as a Space of Challenges in the Context of Digital Transformation: Real vs. Virtual.. 157
Arina Dikanova, Plekhanov Russian University of Economics, Moscow, Russia
Ekaterina Kargapolova, Plekhanov Russian University of Economics, Moscow, Russia
Victor Mikhailovich Nemchinov, Institute of Oriental Studies, Russian Academy of Sciences, Russia
Irina Gavrilova, Moscow State University, Moscow, Russia

In modern conditions of very high change rates socio-cultural space of megacities is often characterized by VUCA-risks. Hence research relevance of citizens' adaptation to everyday life challenges. This study presents results of risks investigation of Moscow residents, conducted in December, 2023. Respondents are most protected from harassment by gender, age, nationality and religion, and are least protected from crime, poverty, job loss, arbitrariness of officials and of the law enforcement agencies. The study has revealed that respondents feel insecurity from the Internet risks and online scammers being much higher than dangers existing in real life. This conclusion requires further verification with identical questions and relevant answers. But the obtained results make us think about paradoxes of transitive society and culture, in which real world with all its problems is more understandable and closer to residents of a megalopolis than the challenges of online services and the cyberspace attractions.

Chapter 11
Analysis of the Dynamics of the Multiplier Effect of Rural Tourism and Its Impact on the Integrated Rural Development 173

Natalia Sergeeva, Russian State Agrarian University - Moscow Timiryazev Agricultural Academy, Russia
Ludmila Evgrafova, Russian State Agrarian University - Moscow Timiryazev Agricultural Academy, Russia

Rural tourism, including agritourism, is a growing branch of the industry in rural areas. According to experts, in the near future the flow of rural tourists will increase and reach about seven million per year, while agricultural production enterprises will receive additional income from rural tourism in the amount of 5% of the total income of farms. Tools for the rural tourism development are also included in the state program "Tourism Development". With the growing demand, there is a challenge to ensure the supply of rural tourism services. Currently, the share of rural tourism in the total volume of tourism in Russia is estimated at about 2.0%. However, the situation will change as rural tourism is recognized as a priority trend of state policy in the near future. One of the priorities in promoting the rural tourism development is to assess its multiplier effect on related industries. This will make it possible to determine the trends of the sustainable rural development and the main areas of the rural tourism development.

Chapter 12
Cultural Adaptation of Digital Payment Systems With a Focus on Community-Based Financial Practices 189
Nisha K. Kumar, Independent Researcher, India
N. Giribabu, Sri Venkateswara College of Engineering and Technology, India
Meera Sharma, USCS, Uttaranchal University, India
M. Siva Krishna Raju, SRKR Engineering College, India
V. Sri Krishna Chaturya, SRKR Engineering College, India

The emergence of digital payments is changing how financial transactions are carried out; however, cultural adoption of these practices within varying socio-economic frameworks is critical for success. This chapter analyzes the digital payments landscape using VUCA and BANI models and centers discussion on aspects like cultural pluralism, deficits in digital literacy, and trust gaps. M-Pesa, WeChat Pay, and Paytm serve as case studies for this investigation. The results suggest that culturally adaptive systems of digital payment enhance adoption, trust, and financial resilience. The framework argues for the necessity of truism-driven, inclusive FinTech. Further investigations are needed on AI, blockchain, and the regulatory components of these issues in order to build resilient ecosystems for digital finance.

Chapter 13
Ethical and Legal Dimensions of Drones and Robots in Smart Business: Navigating the Autonomous Innovations in Logistics 259
Ananya Pandey, Christ University, Bangalore, India
Jipson Joseph, Christ University, Bangalore, India

Use of drones and robots are slowly but surely making their way into people's personal and professional lives. International companies are spending millions of dollars on drone research and development for logistics. Companies like Google, Amazon, DHL, and Walmart are actively experimenting on autonomous systems and robotics. Drones and robots are used in the business strategies of these organizations to ensure last-mile delivery, which involves transporting packages straight from a depot to the ultimate customer, maintaining inventory, loading and unloading of trucks, order processing, and the like. But the usage of these advancements poses some critical ethical and legal issues. Authorities must pay attention to these challenges and difficulties in order to ensure their meaningful usage. This chapter, in this perspective, overviews the most significant legal and ethical concerns pertaining to the use of autonomous vehicles in smart businesses and logistics and provides recommendations for their effective utilization in future.

Chapter 14
A Technique for Collecting and Analyzing Physico-Chemical Parameters of Soils Based on Wireless Sensor Network Technology 287
Sergey Melnikov, North-Caucasus Federal University, Russia
Svetlana Govorova, North-Caucasus Federal University, Russia
Egor Govorov, North-Caucasus Federal University, Russia

Taking into account soil and climatic conditions, based on measuring the physico-chemical parameters of the soil, meteorological parameters and processing statistical data on climatic conditions, makes it possible to ensure the completeness of the implementation of plant biopotential due to the rational choice of parameters and timing of agrotechnical measures. The controlled parameters vary significantly in the frequency of measurements and the accuracy of the representation. The results of the analysis of the physico-chemical parameters of soils and climatic conditions are summarized in a summary table, the contents of which allow calculations based on the volume of transmitted traffic and the speed of the wireless sensor network channels. Various standards of low-power wireless networks were considered. Based on the specifications of this standard, the effective data transfer rate, current consumption and operating time of the devices were determined at the declared and calculated data transfer rates.

Compilation of References 301

About the Contributors 353

Index 359

Preface

The twenty-first century has ushered in an era of unprecedented complexity, where education is challenged to prepare learners for rapidly changing social, technological, and economic landscapes. In this context, smart education has emerged as both a necessity and a transformative force, offering innovative pathways for teaching, learning, and professional development. This book, *Navigating Technological Advancement in the VUCA and BANI World*, seeks to critically examine the intersections of pedagogy, technology, and human development, and to present diverse insights on how education can be reshaped to meet the demands of our times.

The fourteen chapters compiled in this volume reflect a breadth of scholarship and practice. They begin with explorations of STEM approaches and practice-oriented methodologies that employ smart tools to enhance student engagement and professional preparation. Subsequent contributions investigate the power of cyberspace as augmented reality, the evolving significance of digital credentials, and the development of teacher competencies for effective use of artificial intelligence in classrooms. Perspectives from India and beyond provide a nuanced understanding of how students and teachers perceive AI integration, while other studies highlight critical issues such as video gaming addiction, socio-emotional learning, and inclusive practices in special education.

CHAPTER OVERVIEW

Chapter 1: A novel method as manufacturing technologies to detect the negative effect of atmospheric agents in micro deterioration of shock abers of aircrafts: Novel method to detect microdeterioration in shock absorbers of aircrafts

An investigation was made in an aerospace industry located in the Tijuana city, which manufactures shock absorbers used in the landing gear of aircrafts, where an inefficient operation of the shock absorbers was detected, was caused by the activation action of the bar not being generated. expansion of the evaluated shock absorbers. This investigation was made from 2022 to 2023. With the SEM technique showed the atmospheric agents with the most effect in the microdeterioration of the metallic surface of the shock absorbers manufactured and with the automatized system was improved the air quality in indoors of the aerospace where was made this scientific study.

Chapter 2: VUCA in Emerging Indian Healthcare Sector Post-Pandemic: VUCA and Healthcare

The VUCA paradigm representing Volatility, Uncertainty, Complexity, and Ambiguity originally conceptualized by the U.S. Army War College to define ambiguous military situations, has achieved constant rising relevance in the modern-day in commercial, business and policy environment. The Indian healthcare sector, categorized by ever-changing market subtleties, intricate supply chains, and volatile global healthcare demands, reflects the spirit of a VUCA world. The COVID-19 pandemic additionally increased these challenges while instantaneously activating invention, resilience, and transformation on a structural basis within the sector. This chapter explores the VUCA framework's ability of how it can be applied to analyse the post-pandemic progression of India's healthcare industry. Its highpoints moves towards the strategic approaches fostering sustainability, adaptability, and competitive strength—accentuating India's place as a noteworthy global benefactor of affordable pharmaceuticals and vaccines.

Chapter 3: Post-Quantum Cryptography Robust Solutions Against Quantum Computing Threats

Quantum computers are a revolutionary technology based on the principles of quantum mechanics. Unlike conventional computers that work with binary bits (0 or 1), quantum devices use qubits that can exist in multiple states ("superpositions") and interact with each other through the phenomenon of "quantum entanglement." Thanks to this, quanta can greatly accelerate the solution of a number of important problems that were previously considered insurmountable for classical machines. This review describes the principles of operation of quantum computers, considers their unique advantages, and demonstrates the potential for their impact on the future of computer science and the technology industry.

Chapter 4: Research Of Approaches To Assessing Information Security Risks Of An Organization: Qualitative Risk Assessment Methods

This article explores modern approaches to assessing information security risks within organizations, focusing on qualitative, quantitative, and hybrid risk management methodologies. The study examines the core principles of each approach, their practical applicability in different organizational contexts, the tools and frameworks commonly employed, and key implementation challenges. A comparative analysis of these methods highlights their respective strengths and limitations, providing insights for selecting the most suitable risk assessment strategy based on organizational needs, industry requirements, and available resources.

Chapter 5: Machine learning in information security

The features of the application of machine learning methods in information security tasks and development prospects and future trends are considered. The article analyzes the application of a machine learning model to detect attacks in the field of information security using the RandomForestClassifier classifier. Various aspects of data processing are considered, including balancing classes and selecting the most significant features for building a model. This approach can be useful for developing real-time monitoring and protection systems.

Chapter 6: Using probabilistic models to assess and enhance information security

The article explores the possibilities of using probabilistic models to assess and improve information security. The theoretical foundations of the probabilistic approach are considered, including methods of risk analysis, threat modeling and prediction of potential vulnerabilities. Special attention is paid to the use of models such as Bayesian networks, Markov chains, and Monte Carlo simulations, which allow quantifying the likelihood of incidents and their consequences. The paper presents practical examples of the use of probabilistic models for analyzing the security of information systems, as well as recommendations for their integration into information security management processes.

Chapter 7: Enhancing Well-Being in the Digital Age in the VUCA & BANI world

In today's VUCA (Volatile, Uncertain, Complex, Ambiguous) and BANI (Brittle, Anxious, Nonlinear, Incomprehensible) reality, enhancing well-being in the digital age is crucial. Digital technology can both support and hinder mental health, making it essential to cultivate a balanced approach. Encouraging digital detoxes and mindfulness practices can reduce stress and promote emotional resilience. Leveraging online communities for support fosters connection, but awareness of the potential for digital addiction remains essential. Tools like mental health apps can provide resources, yet they should complement real-life interactions to ensure authenticity. Emphasizing digital literacy empowers individuals to navigate information critically, enhancing decision-making regarding their mental and emotional health. Finally, organizations must prioritize employee well-being by implementing flexible work arrangements, promoting open communication, and providing mental health resources. By consciously shaping our digital environments, we can enhance well-being and thrive in this complex world.

Chapter 8: Digitalization Of Metropolis - Resources, Risks, Capitalization

Digitalization of metropolis transforms social connections and human capital structures. Relevance of the study on the resource base, content and negative consequences of using Internet resources by residents of the metropolis is obvious. This survey presents results of a sociological study of Moscow residents, conducted according to the Standard Methodology of the All-Russian Scientific Research Program "Sociocultural Evolution of Russia and Its Regions" in December 2023. The assessments of the Internet resources used by respondents for entertainment, work, interaction with local authorities, opportunities for saving money and time are analyzed, risks and negative consequences are considered. Respondents use capabilities of online stores, various platforms for online communication, with a predominance of young and adult respondents, as well as respondents from the wealthy category. Low level of benefit from using Internet resources is observed among elderly and low-income respondents. These categories also need adapted digital services and accessible opportunities for their needs.

Chapter 9: Research, Development And Comparative Characterization Of Methods For Invisible Embedding Digital Watermarks In Electronic Text Documents

This article discusses the concept of digital watermark (DW?), its classification, the main file formats for storing textual information electronically, types of attacks on DWC, and methods of embedding digital watermarks in electronic text documents based on formatting changes, using metadata, structural and linguistic changes. The practical development of each method is carried out, their advantages, disadvantages and resistance to various attacks are identified, and their comparative characterization is presented. On the basis of the analysis, we experimentally developed our own methods of introducing DWC into text documents, including the approach using zero-width spaces (ZWSP) (U+200B), which provides high resistance to various attacks and imperceptibility for the user. The results of the work demonstrate the effectiveness of each implementation and provide recommendations for their application depending on the requirements for copyright protection and authentication of text files.

Chapter 10: Megalopolis as a Space of Challenges in the Context of Digital Transformation (Real vs Virtual)

In modern conditions of very high change rates socio-cultural space of megacities is often characterized by VUCA-risks. Hence research relevance of citizens' adaptation to everyday life challenges. This study presents results of risks investigation of Moscow residents, conducted in December, 2023. Respondents are most protected from harassment by gender, age, nationality and religion, and are least protected from crime, poverty, job loss, arbitrariness of officials and of the law enforcement agencies. The study has revealed that respondents feel insecurity from the Internet risks and online scammers being much higher than dangers existing in real life. This conclusion requires further verification with identical questions and relevant answers. But the obtained results make us think about paradoxes of transitive society and culture, in which real world with all its problems is more understandable and closer to residents of a megalopolis than the challenges of online services and the cyberspace attractions.

Chapter 11: Analysis of the Dynamics of the Multiplier Effect of Rural Tourism and Its Impact on the Integrated Rural Development

Rural tourism, including agritourism, is a growing branch of the industry in rural areas. According to experts, in the near future the flow of rural tourists will increase and reach about seven million per year, while agricultural production enterprises will receive additional income from rural tourism in the amount of 5% of the total income of farms. Tools for the rural tourism development are also included in the state program "Tourism Development". With the growing demand, there is a challenge to ensure the supply of rural tourism services. Currently, the share of rural tourism in the total volume of tourism in Russia is estimated at about 2.0%. However, the situation will change as rural tourism is recognized as a priority trend of state policy in the near future. One of the priorities in promoting the rural tourism development is to assess its multiplier effect on related industries. This will make it possible to determine the trends of the sustainable rural development and the main areas of the rural tourism development.

Chapter 12: Cultural Adaptation of Digital Payment Systems with a Focus on Community-Based Financial Practices

The emergence of digital payments is changing how financial transactions are carried out; however, cultural adoption of these practices within varying socioeconomic frameworks is critical for success. This chapter analyzes the digital payments landscape using VUCA and BANI models and centers discussion on aspects like cultural pluralism, deficits in digital literacy, and trust gaps. M-Pesa, WeChat Pay, and Paytm serve as case studies for this investigation. The results suggest that culturally adaptive systems of digital payment enhance adoption, trust, and financial resilience. The framework argues for the necessity of truism-driven, inclusive FinTech. Further investigations are needed on AI, blockchain, and the regulatory components of these issues in order to build resilient ecosystems for digital finance.

Chapter 13: Ethical and Legal Dimensions of Drones and Robots in Smart Business: Navigating the Autonomous Innovations in Logistics

Use of drones and robots are slowly but surely making their way into people's personal and professional lives. International companies are spending millions of dollars on drone research and development for logistics. Companies like Google, Amazon, DHL, and Walmart are actively experimenting on autonomous systems and robotics. Drones and robots are used in the business strategies of these organizations to ensure last-mile delivery, which involves transporting packages straight from a depot to the ultimate customer, maintaining inventory, loading and unloading of trucks, order processing, and the like. But the usage of these advancements poses some critical ethical and legal issues. Authorities must pay attention to these challenges and difficulties in order to ensure their meaningful usage. This chapter, in this perspective, overviews the most significant legal and ethical concerns pertaining to the use of autonomous vehicles in smart businesses and logistics and provides recommendations for their effective utilization in future.

Chapter 14: A technique for collecting and analyzing physico-chemical parameters of soils based on wireless sensor network technology

Taking into account soil and climatic conditions, based on measuring the physico-chemical parameters of the soil, meteorological parameters and processing statistical data on climatic conditions, makes it possible to ensure the completeness of the implementation of plant biopotential due to the rational choice of parameters and timing of agrotechnical measures. The controlled parameters vary significantly in the frequency of measurements and the accuracy of the representation. The results of the analysis of the physico-chemical parameters of soils and climatic conditions are summarized in a summary table, the contents of which allow calculations based on the volume of transmitted traffic and the speed of the wireless sensor network channels. Various standards of low-power wireless networks were considered. Based on the specifications of this standard, the effective data transfer rate, current consumption and operating time of the devices were determined at the declared and calculated data transfer rates.

Further sections of the book emphasize the transformative potential of personalized learning, interdisciplinary collaboration through digital platforms, and the innovative application of gamification across different educational levels. Analyses of virtual reality learning environments, digital teaching competencies, and systematic reviews of responsible AI use underscore both the promises and challenges of these technologies. The concluding chapters point toward the future of teacher education and smart universities, underscoring the values of equity, innovation, and inclusivity.

It is our sincere hope that this volume will serve as a valuable resource for educators, researchers, and policymakers committed to advancing education in ways that are responsive, ethical, and future-oriented in an increasingly complex world.

Chapter 1
A Novel Method as Manufacturing Technologies to Detect the Negative Effect of Atmospheric Agents in Micro Deterioration of Shock Abers of Aircrafts

Ciceron Gonzalez Toxqui
Universidad Politécnica de Baja California, Mexico

Arilí Cardenas Robles
Universida Autónoma de Baja California, Mexico

Manuel Guzmán Herrera
Universida Autónoma de Baja California, Mexico

Eddna Teresa Valenzuela Martinez
Universida Autónoma de Baja California, Mexico

Daniel Barrera Román
Universidad Politécnica de Baja California, Mexico

Jesús Márquez-González
Universida Autónoma de Baja California, Mexico

Israel Sauceda Meza
Universida Autónoma de Baja California, Mexico

Yidanes Alejandra Ibáñez Acevedo
Tecnologico Nacional de México, Instituto Tecnológico de Mexicali, Mexico

DOI: 10.4018/979-8-3373-2240-7.ch001

Gustavo G. L. López Badilla
https://orcid.org/0009-0004-0193-4806
Universidad Vasconcelos Campus Murua, Tijuana, Mexico

ABSTRACT

An investigation was made in an aerospace industry located in the Tijuana city, which manufactures shock absorbers used in the landing gear of aircrafts, where an inefficient operation of the shock absorbers was detected, was caused by the activation action of the bar not being generated. expansion of the evaluated shock absorbers. This investigation was made from 2022 to 2023. With the SEM technique showed the atmospheric agents with the most effect in the microdeterioration of the metallic surface of the shock absorbers manufactured and with the automatized system was improved the air quality in indoors of the aerospace where was made this scientific study.

INTRODUCTION

The investigation consisted of two steps. The first step was an analysis of the manufacturing activities using the VEGAM (Ver-Guiar-Analizar-Mejorar, in the Spanish version) as a novel to evaluate the industrial process about the productivity and quality indices. The second step was the evaluation of the atmospheric agents corelated with the productivity and quality levels, which originated the micro deterioration in the shock absorbers of the aircraft damping system. The three step was the microevaluation of metallic surfaces of shock absorbers correlated with the principal atmospheric agents that were generating the micro deterioration, using the Scanning Electron Microscopy (SEM) technique. Aircraft damping systems are of great importance for takeoffs and especially in landing actions, because when mechanical failures occur during aircraft landings, they can cause critical situations (Najmon et al, 2019). This could cause possible damage to the health of passengers and sometimes even deaths, due to cases of stress in people with cardiovascular systems or the possibility of aircraft collapsing and causing relevant catastrophes. This is why specialists (design engineers, manufacturing and quality engineers) in the aerospace industries, who manufacture damping systems (aircraft landing system dampers) (Ford et al, 2016). This relevant aspect was caused concern among the directive, managerial, supervisory and specialized personnel of the aerospace industry where the scientific study was made. One of the factors of mechanical failure of aircraft landing system shock absorbers is the deterioration of the metallic materials, which, although essentially made of 316 steels, suffer from corrosion actions. Steel contains molybdenum, which helps improve the physical property of corrosion resistance, but as is known, aircraft circulate in various types of environments, the most common being arid, marine, tropical and tundra (Amaro et al, 2019). This generates, together with atmospheric pollutants, mainly derivatives of sulfur (sulfides), nitrogen (nitrogen oxides), carbon (carbon oxides) and sea salts (chlorides); the deterioration of the metallic materials of the cushioning equipment of the landing systems. Figure 1, shows two images of a damping system of a light aircraft (figure 1a and 1b) illustrated in the upper section, together with microphotographs of deterioration of this type of system, where the parts can be seen in the two images. of the landing systems damping equipment. In the lower section, two photomicrographs of different scales are observed (figure 1c at 100 microns and figure 1d at one micron). In the photomicrographs, the deterioration represented by the white areas is illustrated, where white dots are shown in Figure 1c at the level of 100 microns, and in Figure 1d, the deterioration is represented with greater visibility. This indicates, in the white areas, those with the greatest metallic damage to the evaluated shock absorber, which was removed from a light aircraft, to avoid some critical situation of a possible and terrible catastrophe and damage to the health of the passengers of this aircraft (Wohlers et al, 2016; Gustavo López Badilla et al, 2017).

Figure 1. Analysis of damping equipment of the aircraft damping system: (a) image 1 of damper (https://www.aeroexpo.online/es/prod/airframes-alaska/product-181588-42640.html), (b) image 2 of shock absorber (https://www.aeroexpo.online/es/prod/airframes-alaska/product-181588-42637.html), (c) microdeterioration of shock absorber at 100 X, (d) microdeterioration of damper at 1X.

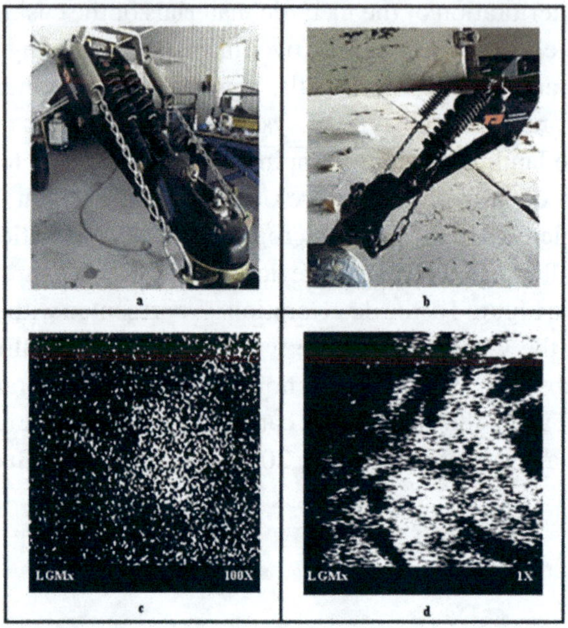

Source. Analysis of the investigation.

The microscopy analysis of metallic surfaces is very important in the evaluation of deterioration of metallic parts of airplanes, which are exposed to aggressive environments in the outdoor activities as in the flight, with exposition of air pollutants as sulfur dioxide, nitrogen oxides, carbon monoxide and ion chlorides in some places of the world with coastal zones of some countries.

Manufacturing Technologies

This relevant thematic represent the ability to evaluate industrial processes and apply specialized strategies with manual and automated activities with sophisticated equipment and machines, which manage to develop efficient and optimal actions (George, 2017). This is of great relevance in obtaining efficient productivity and quality levels, which can indicate the generation of added value to manufactured products, and with it the economic profits of each industry anywhere in the world.

Figure 2 represents the benefits of applying manufacturing technology in any type of industry (Ardolino et al, 2018).

Figure 2. Benefits of use the manufacturing technologies in each industry at the worldwide.

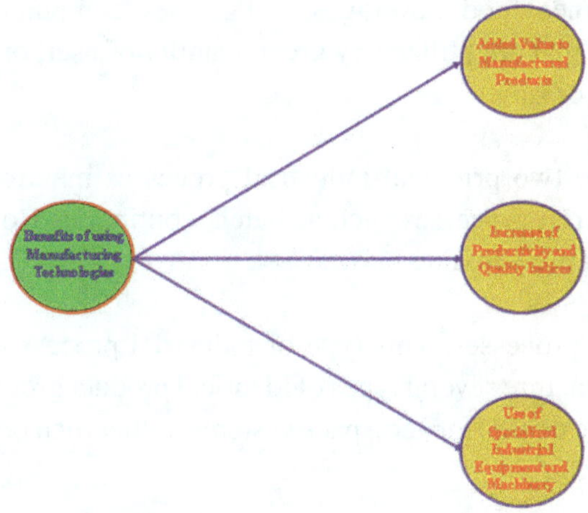

Source. Analysis of the investigation.

Last figure illustrates the principal advantages of the use of the manufacturing technologies in the industrial process of any type of industry in any place of the world, being very important in the growth of the industrial companies of any type of manufactured products (Bienhaus et al, 2018), having high technology in its industrial processes, to obtain efficient fabricated products and the major satisfaction of customers to follow buying the products of any industrial company (APQC, 2019).

With the manufacturing line tools, the raw material is transformed in finished products with optimal productivity and quality levels to satisfice to customers to desire new, secure and technological products to his daily activities. Is very important know about the product to fabricates and the raw material, to develop efficient industrial processes, which have the capacity with the workers of the manufacturing areas and the industrial specialized equipment and machinery. Exists principally two types of industrial processes:

a) Manual activities. Are developed by workers with manual handling in industrial operations where is necessary use this type of actions, and where some industrial equipment and machinery can't make the industrial operations. In this type of industrial process, the activities need be realized with strict actions, but only can be developed in small quantities, as the soldering action in some electronic boards of aerospace systems, which debit be elaborated by womens, essentially, because have more capacity of man, and make the industrial operations with care than man.

b) Automatized activities. Are elaborated by specialized industrial equipment and machinery in industrial activities where is necessary fabricates a lot industrial products in small periods of time. This is for example, when is necessary manufactures medicines, food, beverages, cell phones, computers, tablets or other type of products that are utilized by great quantity of user, or debit be replaced in a few periods of time.

In addition of the two principal industrial processes mentioned above, exists other types of industrial processes such as batch, continuous flow, job-based, and mass production that are explained now:

a) Batch industrial processes. This type of industrial process uses a specialized method to manufacture several types of identical product groups, called batches, such as electronic boards for aerospace systems, rather than being manufactured one at a time.

Relevant aspects

- This type of industrial process helps manufacture products in such a way that they can only pass through a manufacturing area with similar activities.
- Products are manufactured simultaneously and in a constant quantity, in the same sequence.
- Products can be manufactured with great flexibility and in large quantities.
- It is possible to manufacture several types of identical products, with different sizes and a wide variety of similar products, on the same production line with similar industrial activities, experienced or inexperienced operating personnel, and the same industrial equipment and machinery.

The principal advantages include the following actions:

1) It is possible to reduce production and customer delivery times.
2) Fashionable or up-to-date products can be manufactured quickly to compete in the market.

The disadvantages can show the following information:

1) Manufacturing in large quantities and sometimes in short periods of time could result in downtime for industrial process workers at certain times of the production day in industries around the world.
2) Sometimes, when manufacturing large quantities of a product, storage is necessary, which requires additional costs.
3) If errors occur, they could occur in a large number of manufactured products.

The main industries worldwide that use this type of industrial process are food and beverages, pharmaceuticals, personal care (cosmetics), and fashion and up-to-date products, such as electronics and textiles.

b) Continuous flow industrial process. It is developed with specialized methods that support the manufacturing of products, where the activities of industrial processes are uninterrupted (i.e., various shifts (morning, afternoon, and night)). That is, they are only changed every certain period of time (8- or 12-hour shifts on different days of the week). This industrial process can be applied to the manufacture of products in large quantities and for quick use by consumers, such as medicines, personal hygiene products (toilet paper, razors, certain cosmetics, food and beverages, plastics for food storage, and some other everyday products).

Relevant Aspects

This type of industrial process must receive raw materials consistently and be processed quickly to avoid storage.

Among the advantages are the following:

1) Standardized production processes are available.
2) The First In, First Out (FIFO) method is applied.
3) Greater efficiency, quality, and productivity can be achieved, as well as less waste.
4) It is possible to minimize production and customer delivery times, reducing production downtime.

This industrial process may present certain disadvantages, which are listed below:

1) It requires well-trained operating personnel in manufacturing areas.
2) Specialized equipment and machinery must be available.
3) It is necessary to develop automation of industrial manufacturing operations.

Some industries that use this type of industrial process are petroleum refining and the production of its derivatives (mainly gasoline and oils), as well as sugar and fertilizer manufacturing, and the extraction of gas for domestic, commercial, and industrial activities. To realize this manufacturing activities with this industrial process, an optimal production flow is required, where everything operates properly, with ongoing training for the workers operating the industrial processes, and always having standardized industrial processes.

c) Job-based industrial process. This is applied in the development of activities focused on a single product, when the product is very large, such as the manufacture of complete aircraft for the aerospace industry, ships, submarines, automobiles with special characteristics, or any other large product for another type of industrial company. Sometimes these products can be manufactured in a single location, or sometimes in different areas of an industrial city around the world, or sometimes in different cities or areas around the world, depending on the industrial company. This is why this type of industrial process is known as an intermittent industrial process.

Relevant aspects

This type of industrial process involves certain types of actions, which are represented below:

1) The parts of the product must be produced separately and through the batch production process.
2) It has a single process flow with a single production cycle.
3) It could represent a slightly more flexible process than other types of production processes, which can sometimes lead to long production times due to safety concerns.

This type of industrial process has some advantages, which are listed below:

1) Few products are manufactured with strict safety requirements.
2) Since it is a slow process, errors are rare.
3) They are produced with strict quality standards.

The job-based industrial process often has certain disadvantages:

1) It is possible to have highly trained personnel in manufacturing areas.
2) It requires specialized and expensive industrial equipment and machinery.
3) It is necessary to constantly train industrial process workers.

This type of industrial process requires strict receipt and inspection of raw materials, thorough manufacturing activities, and an effort to streamline all industrial operations for proper material transformation and efficient product development for on-time delivery to the customer. Furthermore, this type of industrial process can generate direct jobs (those developed in industries) and indirect jobs (those requiring consulting and services external to industries), as well as constantly promote technological development, innovation, and sustainability. All of this is intended to produce highly efficient products, improving users' quality of life.

d) The mass industrial process. It is developed when they are required to develop products manufacturing activities that have a required standard, such as handbags for women and backpacks for school actions, essentially. It is a type of industrial process that requires the manufacture of large amounts with fashion products, which must constantly be in improvement and update actions. It contains specialized assembly operations and actions with manufacturing cell activities (annexes to the production lines), where industrial operations are carried out that cannot be developed in these, due to their complex action, such as in some industries, where it is necessary to place large or very small parts of the products.

Relevant aspects.

In this industrial process it is possible to develop the design, manufacture and testing of the products, developing all these activities in a single industrial company, without depending on other industries and sometimes generating dead times in industrial processes.

Among the advantages they have are the following:

1) It generates large volume production processes and at low cost.
2) It requires standardized processes, without having to prepare modifications that cause dead times in production activities.
3) It has control of processes optimized from the reception of the raw material to the delivery of the product to the customer.

Among the disadvantages can be presented:

1) You must have well -trained personnel in the design, development, product test and delivery to the customer.
2) Sometimes, there may be errors in industrial processes, if the design of the product to be manufactured is not adequate.
3) Sometimes, specialized marketing strategies are required.

4) Sometimes, it can generate large amounts of waste due to bad design or bad industrial process.
5) The manufactured product may not be well received by consumers.

In this type of industrial process, it is necessary to use specialized design softwares such as computer -assisted design type (CAD) or be necessary to use computer - assisted manufacturing (CAM).

Industry in Tijuana

This city has relevant industries in the northwest of the Mexican Republic, being an important zone of the commerce with the California State with the Free Treatment Agreement (FTA) between Mexico, United States of the America (USA) and Canada (IMT, 2023). In this city, are 350 industries from the Brazil, Canada, China, France, Germany, Italy, Japan, Netherlands, Spain and USA; essentially, which are presented in table 1, with the type of principal industrial companies and manufactured products (Sames et al, 2016).

Table 1. Essential industries and manufactured products in Tijuana (2023)

Industry Country	Aerospace	Automotive	Biomedical	Electronic	Metallic	Plastics	Textile	Wood
Brazil	Parts of aircraft				Stainless Steels bars to Buildings			
Canada			Catheters, syringes and pacemakers	Computer systems and Tablets		Bags to package of biomedical products		
China	Parts of aircraft		Catheters, syringes and pacemakers		Stainless Steels bars to Buildings			
France					Metallic products to gastronomy commerce		Clothes and Evening Clothes	
Germany		Parts and Electronic Sesnors to Cars			Aluminum to Construction Industry			Office Furniture
Italy							Clothes and Evening Clothes	

continued on following page

Table 1. Continued

Industry / Country	Aerospace	Automotive	Biomedical	Electronic	Metallic	Plastics	Textile	Wood
Japan	Parts of aircraft	Parts and Electronic Sesnors to Cars		Computer systems, Tablets and Televisions				
Netherlands								Guitars and Office Furniture
Spain					Aluminum to Construction Industry			Office Furniture
USA	Parts of aircraft	Parts and Electronic Sesnors to Cars	Catheters, syringes and pacemakers	Computer systems, Tablets and Televisions		Bags to package of biomedical products		

Source. Analysis of the investigation.

Table1 shows the most relevant industries located in the Tijuana city, where works a lot persons as operation workers, technics, professional people (aeronautic engineers, computer engineers, electronic engineers, electric engineers, mechanical engineers, mechatronic engineers, administrative as accouters, communication area, psychologists, doctors and nurses, essentially). This table shows the principal industrial companies of the countries mentioned above (illustrated in the light green color (Gustavo López et al, 2014).

Aerospace industry in Tijuana

This type of industry is located in this important border city of Mexico, where are installed around 40 aerospace industries, manufacturing parts of aircrafts, which the main industry of this type of industrial companies in the USA, where fabricates the complete aircrafts (Avalle et al, 2019). This industry, have strict regulations in its manufacturing areas, to obtain optimal functionability, and avoid any type of critical situations that can generates catastrophes and possibly hard health symptoms or deaths. In table 2 is showed the main fabricated products in this important city of the Mexican Republic.

Table 2. Principal manufactured products of the aerospace industry in Tijuana (2023)

Relevant Aspects / **Manufactured Products**	**Type of Industrial Flow**	**Type of Industrial Equipment and Machinery**	**Type of Materials**
Aircraft Textile Interiors	Lineal and U types	Cutters, Assembly, Inspection Equipments and Machineries	Fabric materials
Automatic Pilot	Lineal and G types	Assembly, Inspection Equipments and Machineries	Electronic and Metallic Materials
Components for Propulsion Systems	Lineal and T types	Assembly, Gluing Equipment Inspection Equipments and Machineries	Plastics Materials
Electrical and Electronic Systems of Aircraft	Lineal, T and U types	Assembly, Inspection Equipments and Machineries	Electronic and Metallic Materials
Landing system components (Shock Absorbers)	Lineal, G and U types	Assembly, Gluing Equipment, Inspection Equipments and Machineries	Metallic Materials

Source. Analysis of the investigation.

This interesting table shows the essential fabricated products of the aerospace industry installed in the Tijuana city, where are observed in addition with the relevant aspects mentioned in this table, which are very important in the manufacturing processes to obtain the productivity and quality levels for the strict regulations in this relevant industry (Gustavo López Badilla et al, 2017).

Atmospheric Agents

This interesting factor has important effect in the deterioration of metallic materials utilized in some industries at the worldwide, where the main atmospheric aspects are the air pollutants and climatic parameters (principally temperature and relative humidity). The deterioration of the metallic materials of manufactured products, can generates simplex or critical situations in the fabricated products. The simplex situations can be electrical failures in cell phones or computer systems, as the main products fabricated by the electronics industry and damaged by the atmospheric agents in its metallic connections as connectors to charge the electrical battery. The critical events can be catastrophic actions, as failures in the automatic pilot or shock absorbers of landing gear of aircrafts, which will be landing or crashing of aircraft and originates drastic health symptoms or deaths of passengers (Abidi et al, 2019). The main air pollutants are derivates from Sulphur, nitrogen, carbon emitted of industries and car vehicles, essentially, added of chlorides from marine salt. Also, was appears fine dust in arid environments and chlorides from marine ecosystems, as salt, which is involved of dust devils and are transported to some places of cities as the Tijuana city and can causes deterioration of metallic materials of some manufactured products. Table 3 shows the principal atmospheric factors and the main negative effect in the diverse fabricated products with metallic materials (Gustavo López Badilla et al, 2021).

Table 3. Main atmospheric factors and negative effects in metallic materials of manufactured products in the industries of Tijuana (2023)

Metallic Materials / Atmospheric Factors	Aluminum	Cooper	Gold	Silver	Steel
Climatic Parameters **Relative Humidity (RH), %** **Temperature, °C**	Can be damage by RH and levels higher than 50%, that is common in the majorly of the year in Tijuana. Also, can be damage at low temperatures as less than 20° C.	Will deteriorate at by RH and levels higher than 40%, that is common in the majorly in autumn and winter of the year in Tijuana. Also, can be damage at low temperatures as less than 15° C.	Can be damage by RH and levels higher than 35%, that is common in the majorly of the year in Tijuana. Also, can be damage at low temperatures as less than 25° C.	Will deteriorate damage by RH and levels higher than 45%, that is common in the majorly of the year in Tijuana. Also, can be damage at low temperatures as less than 20° C.	Can be damage by RH and levels higher than 50%, that is common in the majorly of the year in Tijuana. Also, can be damage at low temperatures as less than 25° C.
Air Pollutants **Carbon Oxides** **Chlorides**	Is deteriorate at levels higher than 3 mg/m^3/day of carbon monoxide (CO), which is common in Tijuana in the majorly of the periods of the year. Ion Chlorides is measure by Silver Nitrate solution, and have a little effect in this metallic material	Is damaged at levels higher than 5 mg/m^3/day of carbon monoxide (CO), which is common in Tijuana in the majorly of the periods of the year. Ion Chlorides is measure by Silver Nitrate solution, and have a little effect in this metallic material	Is deteriorate at levels higher than 5 mg/m^3/day of carbon monoxide (CO), which is common in Tijuana in the majorly of the periods of the year. Ion Chlorides is measure by Silver Nitrate solution, and have a little effect in this metallic material	Is damaged at levels higher than 5 mg/m^3/day of carbon monoxide (CO), which is common in Tijuana in the majorly of the periods of the year. Ion Chlorides is measure by Silver Nitrate solution, and have a little effect in this metallic material	Is deteriorate at levels higher than 7 mg/m^3/day of carbon monoxide (CO), which is common in Tijuana in the majorly of the periods of the year. Ion Chlorides is measure by Silver Nitrate solution, and have a little effect in this metallic material

continued on following page

Table 3. Continued

Metallic Materials / Atmospheric Factors	Aluminum	Cooper	Gold	Silver	Steel
Nitrogen Oxides	Can be damage at levels higher than 20 μg/m^3/day of NOx (Nitrogen Oxides) and, which is common in the majorly of the year in Tijuana.	Will deteriorate at levels higher than 15 μg/ m^3 m^3/day of NOx (Nitrogen Oxides), which is common in the majorly of the year in Tijuana.	Can be damage at levels higher than 25 μg/ m^3/ day of NOx (Nitrogen Oxides), which is common in the majorly of the year in Tijuana.	Will deteriorate at levels higher than 10 μg/m^3/ day of NOx (Nitrogen Oxides), which is common in the majorly of the year in Tijuana.	Can be damage at levels higher than 25 μg/ day of NOx (Nitrogen Oxides), which is common in the majorly of the year in Tijuana.
Sulfur Oxides	Is deteriorate at levels higher than 30 μg/ m^3/day of Sulphur dioxide (SO_2), which is common in Tijuana in the majorly of the periods of the year.	Is damaged at levels higher than 20 μg/ m^3/day of Sulphur dioxide (SO_2), which is common in Tijuana in the majorly of the periods of the year.	Is deteriorate at levels higher than 25 μg/ m^3/day of Sulphur dioxide (SO_2), which is common in Tijuana in the majorly of the periods of the year.	Is damaged at levels higher than 20 μg/ m^3 of Sulphur dioxide (SO_2) which is common in Tijuana in the majorly of the periods of the year.	Is deteriorate at levels higher than 40 μg/ m^3 of Sulphur dioxide (SO_2), which is common in Tijuana in the majorly of the periods of the year.

Source. Analysis of the investigation.

Air Quality standards:

https://www.who.int/news-room/fact-sheets/detail/ambient-(outdoor)-air-quality-and-health?gad_source=1&gclid=CjwKCAjwh4-wBhB3EiwAeJsppHLOM6Y9ov6pKLxGMp4FEF3bcrIfZq8Cdkd-1agsAZn0tCCCB5DFdBoCCpYQAvD_BwE

VEGAM Matrix to evaluate industrial processes

This novel method is utilized to evaluate industrial processes in a matrix-shaped guide to detect very easy and fast the possible errors of operative workers of industrial equipment and machinery, and to improve the distribution plant (DebRoy et al, 2019). In figure 3, is represented an analysis with this method, which indicates the flow processes and the places of the operative workers and industria equipment and machinery of each industrial company with any type of manufactured products (Gustavo López Badilla et al, 2014).

Figure 3. Matrix guide to evaluate industrial processes of manufacturing shock absorbers in an aerospace industry of Tijuana (2023)

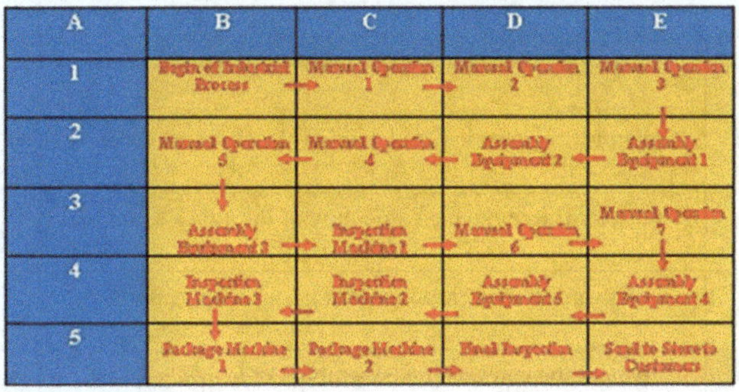

Source. Analysis of investigation

Figure 3, shows the industrial processes of the fabrication of the manufacturing of shock absorbers, with the manual operations, industrial equipments and machineries, final inspection and the action to send to store and to send to customers by land transportation from Tijuana, Baja California, Mexico to Dallas, Texas, United States of America, where was the main industrial company of this industry located in this important of the northwest zone of the Mexican Republic. An apology is offered for not being able to show the industrial process of this relevant product manufactured in this aerospace industry, for not having the required permits, but we explain each workstation stage of the process as best as possible in the following information in table 4 (Coykendall et al, 2014; Ngo et al, 2018).

Table 4. Industrial operations improved of manufactured products in the aerospace industry in Tijuana (2023)

Function	**Operation**	**Type of operation**	
Activities		**Simplex**	**Complex**
Begin the industrial processes	**Reception of the raw material**	**X**	
Manual Operation 1	**Installs the piston component of the shock absorber**		**X**
Manual Operation 2	**Connects the inner cylinder of the shock absorber**		**X**
Manual Operation 3	**Installs the piston red of the shock absorber**		**X**
Assembly Equipment 1	**Assembly the components of *a* and *b* operations**		**X**

continued on following page

Table 4. Continued

Function	Operation	Type of operation	
Activities		Simplex	Complex
Assembly Equipment 2	Assembly the component *c* with *a* and *b* operations		X
Manual Operation 4	Installs the metering tube part of the shock absorber.		X
Manual Operation 5	Connects the centering came part of the shock absorber		X
Assembly Equipment 3	Assembly the components of *g* and *h* operations		X
Inspection Machine 1	Inspect the components assembled of *b, c, d, g* and *h* operations, evaluating its functionability with strict aviation regulations.		X
Manual Operation 6	Installs the air valve of the shock absorber.		X
Manual Operation 7	Adjust the air valve of the shock absorber		X
Assembly Equipment 4	Assembly the components of k operation.		X
Assembly Equipment 5	Assembly the component *l* with *k* operation.		X
Inspection Machine 2	Inspect the components assembled of the *k* operation, evaluating its functionability with strict aviation regulations.		X
Inspection Machine 3.	Inspect the components assembled of the *k* operation, evaluating its functionability with strict aviation regulations.		X
Package Machine 1	Elaborates the first part of the package action of the manufactured product (shock absorber, evaluating its functionability with strict aviation regulations		X
Package Machine 2.	Realizes the second part of the package action of the manufactured product (shock absorber, evaluating its functionability with strict aviation regulations.		X
End of Industrial Process	Manufactured products are grouped to send to store.	X	
Store the manufactured product	Send to store to prepare the manufactured product to send to customers	X	

Source. Analysis of investigation

Table 4 shows the industrial processes in the manufacturing areas, where was illustrated each industrial action and its function, and also the type of operation (simplex or complex), observing that the majorly of the industrial operations are complex functions. This is was based on the diagram of figure 5, which is presented now, as the part of shock absorbers (Ang, et al, 2017).

Figure 4. Parts of a shock absorbers of landing gear of aircrafts.

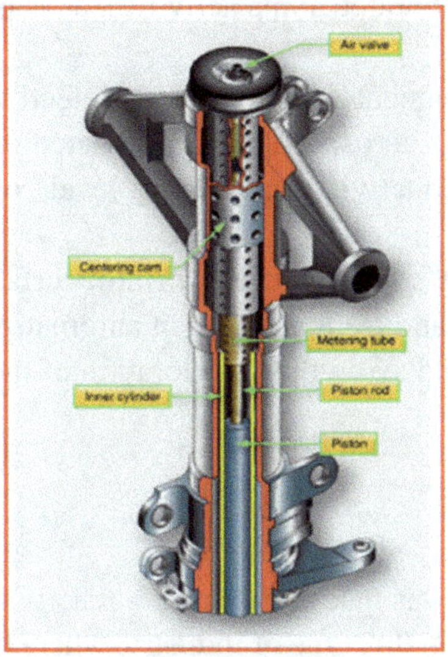

https://www.google.com/search?sca_esv=82cfab1f5a57e002&sca_upv=1&sxsrf=ACQVn09ta4sZdoO 6Y2uZBSCOTyuGRrXP7g:1711572393513&q=armado+de+un+amortiguador+del+tren+de+aterrizaje+de+un+avion&tbm=isch&source=lnms&prmd=ivsnbmtz&sa=X&ved=2ahUKEwj895aEqJWFAxUBI0QIHeOIAg8Q0pQJegQIDRAB&biw=1600&bih=773&dpr=1#imgrc=Ps2QQTi8zJJboM

Source. Analyses of investigation.

SEM analysis

Is an important specialized technique utilized to measure observe at microscopy scale some situations that are necessarily know, to determine if any factor, even if the negative or positive effect is on a smaller scale (Gustavo López Badilla et al, 2021). This is very important to take actions about any type of situation, which can be simplex or complex, and know the grade of effect in any type of operation. In this investigation, was used this relevant technique to determine the principal factors that it had on the deterioration of the metallic parts of shock absorbers, and take actions to avoid this electrochemical phenomenon called corrosion process (Brien et al, 2018). With this specialized technique, the degree of deterioration of the metallic materials of the analyzed shock absorbers could be observed, and collocate filters of dust and gases in the aerospace industry of Tijuana to avoid the presence of the corrosion action, where was made this scientific study. Some microphotographies are showed in a section mentioned above and in in a next section.

METHODOLOGY

This investigation was made to correlate some factors as atmospheric parameters with the deterioration and functionability of shock absorbers utilized in landing gear of aircrafts, where was elaborated some activities, as is presented next:

a) Analysis of the mechanical failures of the defective products manufactured (shock absorbers) in an aerospace industry located in the Tijuana city.
b) Evaluation of the productivity and quality levels of the industrial company analyzed.
c) Microanalysis of the deterioration of the metallic surfaces of the shock absorbers.
d) Desing, development and evaluation of an automatized system to improve the industrial processes and avoid the deterioration of the shock absorbers.

RESULTS

In this scientific study was made some relevant activities as is evaluated in next sections, being important in this type of industry, where have strict regulations and is inspected with high care to avoid any type of mechanical or electrical failures, which can generate any type of catastrophic event.

Evaluation of the mechanical failures

In this section of the investigation, was made an analysis of the principal mechanical failures of the shock absorbers, where was presented in this aerospace industry evaluated and is presented in table 5. In this table was observed that

Table 5. Analysis of the mechanical failures of shock absorbers in the aeroespacial industry evaluated (2023)

Causes / Mechanical Failures	Presence of Atmospheric Factors	Inadequate Industrial Process	Inadequate Raw Material
Metallic Deterioration	Presence of stains and small fractures due to the adhesion of contaminating particles (49%)		Raw material of poor air quality respect in the corrosion resistance (32%)
Bad Function of Piston	The piston remains stuck and does not open completely due to the presence of metallic deterioration of the piston (32%).	Presence of errors and defective manufactured products by not check the function of piston before install in the cylinder of shock absorber (30%)	
Stuck Air Valve	Due to the deterioration of the metallic material of the air valve, appear small metallic chips that originates the air valve not close correctly and not function adequately (39%).	Operation workers not installs adequately and piston rod not function correctly and causes errors and defective products of shock absorbers (29%).	Cylinder exceeds dimensions to insert into cylindrical structure (32%)
Stuck Piston Rod	Due to the deterioration of the metallic material of the piston, small metallic chips are generated, which cause the piston to become stuck (43%)	Operation workers not installs adequately the piston rod in the cylinder and originates errors and defective products of shock absorbers (24%).	
Failures in Centering Cam	Appears small metallic chips and holes of the cam generates bad function of the piston rod (34%)		Raw material of poor air quality respect in the corrosion resistance (31%)

Table 4 illustrates the bad circumstances of the industrial processes of manufacturing shock absorbers, observing that the major incidence was in the presence of atmospheric factors and with the major percentages of occurrence and presenting the essential causes of the defective shock absorbers in the aerospace industry evaluated. The main percentage was 49%, representing the metallic deterioration of cylinders, piston, piston rod and centering cam of the shock absorbers fabricated.

Analysis of productivity and quality indices

Once the macro-level analysis of the deterioration of aircraft landing gear shock absorbers was evaluated, we proceeded to analyze in this scientific study, the productivity and quality levels, to determine by periods of seasons (spring, summer, autumn and winter), in the period of this investigation from 2022 to 2023, which is mentioned above, which is presented in table 6.

Table 6. Evaluation of productivity and quality levels by seasons in the industry analyzed (2022-2023)

Industrial Parameters / Periods	Productivity, % before this investigation	Quality, % before this investigation	Productivity, % after this investigation	Quality, % after this investigation
Spring 2022	68	65	83	89
Summer 2022	72	69	86	86
Autumn 2022	80	80	87	88
Winter 2022	84	82	88	89
Spring 2023	82	80	92	91
Summer 2023	80	72	90	93
Autumn 2023	83	70	92	92
Winter 2023	84	75	94	94

Table 6 shows the productivity and quality indices in seasonably periods, being in this type of periods, because was monitoring the atmospheric parameters by days, weeks, month and seasons; and the seasons were where was reflected more changes about the industrial parameters evaluated (productivity and quality). This illustrates that conforms advances this scientific study was improved the productivity and quality indices, indicating that the specialized strategies of manufacturing technologies (use of the VEGAM matrix in the industrial process, production flow and inspections of manufactured products), was function adequately.

Microanalysis of metallic surfaces deteriorated

This part of the investigation supported relevantly to determine the impact of the influence of atmospheric parameters in the metallic surfaces of the metallic components of the shock absorbers used in the landing gear of aircrafts. This represents a microevaluation of different microscales, observing with major detail, the microdeterioration and the consequences of this dangerous actions, which can cause possible catastrophic events, essentially in the aircraft landing, with the possibility of severe damage to human health of passengers of possible deaths of passengers of aircrafts or persons who are in any type of the airports. Figure shows diverse types of microdeterioration of metallic surfaces of shock absorbers at different microscales (50X, 10X, 5X and 1X). With this microevaluation was observed the deterioration and detected with the zone of red circle the main evaluation, where is located connection of the piston rod with the cylinder, where was the piston sometimes got stuck and produced a defective manufactured product. The dark zones represent the Sulphur oxides (SO_2) and the light zones illustrates the chlorides ions; being the two more essential air pollutants, which caused the microdeterioration. Can observes in the fours microphotographies (a, b, c and d), the dark and light zones, representing this type of air pollutants with the SEM technique, the 70% of the air pollution agents, and the rest (30%), was representing the Nitrogen oxides (NO_X) and the Carbon oxides (CO), as the main air pollutants, which generates the microdeterioration and with this, the defective manufactured product.

Figure 5. Microanalysis of deterioration of metallic surfaces of a shock absorber (2023).

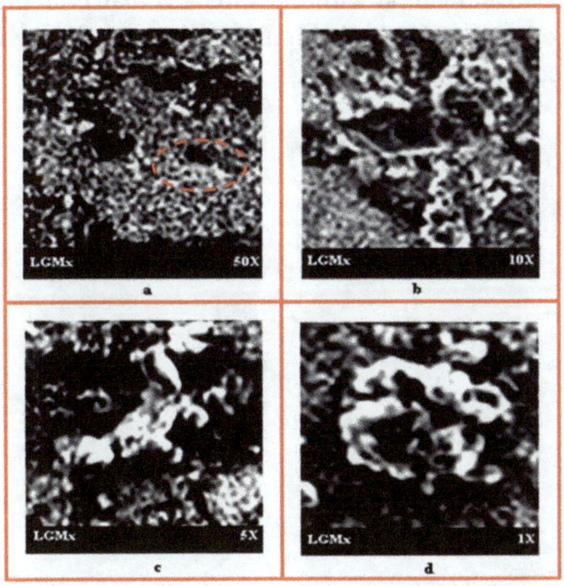

CONCLUSIONS

This scientific study was relevant to avoid the manufacturing of defective products as the shock absorbers of landing gear of aircrafts, which are very in the in-aircraft landing action to not permit any catastrophic event to can originate severe health symptoms in the passengers of aircrafts. With this study was can determined the presence of mechanical failures of shock absorbers and the principal causes of the microdeterioration of the metallic surfaces of this relevant components of the aircraft landing gear. The SEM technique supports with a great intensity to determine the principal air pollutants that were added to the metallic surface of the shock absorbers, which generating the drastic microdeterioration and with this the defective manufactured products. This scientific study could be interesting to avoid defective products in the aerospace industry, because with the microdeterioration, originating bad function of the shock absorbers and was necessary disintegrate this relevant type of fabricated products, losing time and delays in the delivery to customers. This aerospace caused delays in the next part of the fabrication of the aircrafts of the aerospace industry in United States of America. The use a VEGAM matrix a novel process to inspect the defective shock absorbers, redistribution plant with the adequate production floe and the identification of the adequate raw material in the correct place of the shock absorbers, supports to improve the productivity and quality levels. This novel method with the VEGAM matrix in this is investigation can be utilized in other type of production flow, industrial processes and in other type of industrial company in the world.

REFERENCES

Abidi, M., Al-Ahmari, A., Ahmad, A., Ameen, W., & Alkhalefah, H. (2019). Assessment of Virtual Reality-Based Manufacturing Assembly Training System. *International Journal of Advanced Manufacturing Technology*, *105*(9), 3743–3759. DOI: 10.1007/s00170-019-03801-3

Amaro, P., Anabela, C., & Rui, M. (2019). Lean Thinking: A Transversal and Global Management Philosophy to Achieve Sustainability Benefits. In Alves, A., Kahlen, F. J., Flumerfelt, S., & Siriban-Manalang, A. (Eds.), *Lean Engineering for Global Development* (pp. 1–31). Springer. DOI: 10.1007/978-3-030-13515-7_1

Ang, J., Goh, C., Saldivar, A., & Li, Y. (2017). Energy-Efficient Through-Life Smart Design, Manufacturing and Operation of Ships in an Industry 4.0 Environment. *Energies*, *10*(5), 610. DOI: 10.3390/en10050610

APQC. 2019. *Process Classification Framework*. Houston, TX: American Productivity & Quality Center. Accessed June 16, 2019. https://www.apqc.org/resource-library/resource-listing/apqc-process-classification-framework-pcf-cross-industry-excel-7

Ardolino, M., Rapaccini, M., Saccani, N., Gaiardelli, P., Crespi, G., & Ruggeri, C. (2018). The Role of Digital Technologies for the Service Transformation of Industrial Companies. *International Journal of Production Research*, *56*(6), 2116–2132. DOI: 10.1080/00207543.2017.1324224

Avalle, G., De Pace, F., Fornaro, C., Manuri, F., & Sanna, A. (2019). An Augmented Reality System to Support Fault Visualization in Industrial Robotic Tasks. *IEEE Access : Practical Innovations, Open Solutions*, *7*, 132343–132359. DOI: 10.1109/ACCESS.2019.2940887

Badilla, G. L., & Juan, M. T. G. (2021). "The Electronics Industry and Their Chemical Effects in the Environment of Arid Regions of Northwest of Mexico", Indian Journal of Advanced Chemistry (IJAC) ISSN: 2582-8975 (Online), Volume-1 Issue-2, October 2021.

Badilla, G. L., María, M. A. G., Samaniego, E. R., & Sandra, L. T. P. (2014). Microcorrosion and their effect in the operation in of industrial equipments in the electronics industry of Mexicali. In Aliofkhazraei, M. (Ed.), *Book Chapter INTECH, Full Book: Development of Corrosion Protection.*

Bienhaus, F., & Abubaker, H. (2018). Procurement 4.0: Factors Influencing the Digitisation of Procurement and Supply Chains. *Business Process Management Journal*, *24*(4), 965–984. DOI: 10.1108/BPMJ-06-2017-0139

Brien, M., & Alues, A. (2018). “Existing Standards as the Framework to Qualify Additive Manufacturing of Metals. In Proceedings of the 2018”, *IEEE Aerospace Conference*, Big Sky, MT, USA, 3–10 March 2018; pp. 1–10.

Coykendall, J., Cotteleer, M., Holdowsky, J., & Mahto, M. (2014). *3D Opportunity in Aerospace and Defense: Additive Manufacturing Takes Flight* (Vol. 1). Deloitte University Press.

DebRoy, T., Mukherjee, T., Milewski, J., Elmer, J., Ribic, B., Blecher, J., & Zhang, W. (2019). Scientific, technological and economic issues in metal printing and their solutions. *Nature Materials, 18*(10), 1026–1032. DOI: 10.1038/s41563-019-0408-2 PMID: 31263223

Ford, S., & Despeisse, M. (2016). Additive manufacturing and sustainability: An exploratory study of the advantages and challenges. *Journal of Cleaner Production, 13*, 1573–1587. DOI: 10.1016/j.jclepro.2016.04.150

George, C. (2017). *Specification for Control and Qualification of Laser Powder Bed Fusion Metallurgical Processes*. NASA Marshall Space Flight Center.

Najmon, J., Raeisi, S., & Tovar, A. (2019). In Froes, F., & Boyer, R. (Eds.), *Review of additive manufacturing technologies and applications in the aerospace industry. In Additive Manufacturing for the Aerospace Industry* (pp. 7–31). Elsevier Inc.

Ngo, T., Kashani, A., Imbalzano, G., Nguyen, K., & Hui, D. (2018). Additive manufacturing (3D printing): A review of materials, methods, applications and challenges. *Composites. Part B, Engineering, 143*, 172–196. DOI: 10.1016/j.compositesb.2018.02.012

Sames, W., List, F., Pannala, S., Dehoff, R., & Babu, S. (2016). The metallurgy and processing science of metal additive manufacturing. *International Materials Reviews, 61*(5), 315–360. DOI: 10.1080/09506608.2015.1116649

Wohlers, T., Campbell, R., & Caffrey, T. (2016). *3D Printing and Additive Manufacturing State of the Industry: Annual Worldwide Progress Report*. Wohlers Associates.

Chapter 2
VUCA in the Emerging Indian Healthcare Sector Post Pandemic

Swati Dwivedi
Graphic Era University, India

Ashulekha Gupta
https://orcid.org/0000-0001-5155-6090
Graphic Era University, India

ABSTRACT

The VUCA paradigm representing Volatility, Uncertainty, Complexity, and Ambiguity originally conceptualized by the U.S. Army War College to define ambiguous military situations, has achieved constant rising relevance in the modern-day in commercial, business and policy environment. The Indian healthcare sector, categorized by ever-changing market subtleties, intricate supply chains, and volatile global healthcare demands, reflects the spirit of a VUCA world. The COVID-19 pandemic additionally increased these challenges while instantaneously activating invention, resilience, and transformation on a structural basis within the sector. This chapter explores the VUCA framework's ability of how it can be applied to analyse the post-pandemic progression of India's healthcare industry. Its highpoints moves towards the strategic approaches fostering sustainability, adaptability, and competitive strength—accentuating India's place as a noteworthy global benefactor of affordable pharmaceuticals and vaccines.

DOI: 10.4018/979-8-3373-2240-7.ch002

INTRODUCTION

The term "VUCA" was originally introduced by the United States army to delineate the dynamic and perilous contexts in which commanders must function, sometimes lacking comprehensive information or a complete knowledge of the factors that may impact a given circumstance (Kaivo-oja & Lauraeus, 2018). The acronym VUCA represents the four key elements of volatility, uncertainty, complexity, and ambiguity. In contemporary corporate settings, the VUCA framework is employed as a conceptual tool for comprehending the intricacies and difficulties inherent in the operational landscape. VUCA pertains to the escalating rates of transformation and the distinct difficulties that initiatives encounter, encompassing volatility, uncertainty, complexity, and ambiguity. The concept of VUCA holds significance within the realm of corporate foresight and management systems as well. Leaders must possess agility in order to effectively respond to volatility, effectively manage information and knowledge amidst uncertainty, adapt organisational structures in response to complexity, and willingly engage in innovation when faced with ambiguity. Strategic human resource management is of paramount importance in enhancing organisational performance within the VUCA era (Karneli, 2022).

It is important to study VUCA in the healthcare sector because it enables organisations to comprehend and adjust to the unexpected and difficult nature of the sector (Mathew et al., 2023). The healthcare industry is currently experiencing significant transformations as a result of digitization and the presence of VUCA situations. These factors are causing disruptions to the established routines and stability within the sector (Steffen et al., 2023). The COVID-19 pandemic has underscored the necessity for healthcare organisations to exhibit resilience and efficacy in VUCA circumstances (Pandit, 2021). Moreover, the integration of artificial intelligence and digital technology assumes a pivotal position within the healthcare industry, as it offers essential services and contributes significantly to the contentment of healthcare professionals and those seeking medical treatment (Mijwil et al., 2022). It is crucial to comprehend and evaluate the multifaceted nature of performance under VUCA contexts. This understanding is essential for effectively adjusting to the dynamic and unpredictable work environments prevalent in the healthcare sector (Chakraborty, 2019). In the healthcare sector, the study of VUCA aids organisations in formulating strategies, fostering innovation, and effectively navigating the intricate dynamics of the industry, ultimately leading to the provision of enhanced patient-centered care.

The healthcare industry in India is now experiencing substantial transformations within the framework of VUCA circumstances. The industry is currently undergoing significant transformations as a result of personalised treatment, the presence of rich health data, and the imperative for data sharing (Mathew et al., 2023). Nevertheless, the majority of healthcare organisations in India are facing challenges in bridging the gap in digitization and adjusting to the emerging circumstances (Steffen et al., 2023). In order to effectively tackle this issue, it is imperative to have a mindset that values and encourages innovation, agility, and adaptation within the healthcare industry (Hass et al., 2012). Furthermore, the application of health technology assessment (HTA) approaches may be employed to distribute healthcare funds in a fair and effective manner, guaranteeing optimal utilisation of resources in healthcare expenditure (Ahmad Mir & Singh, 2022). The COVID-19 pandemic and the sustainable development objectives have underscored the imperative to reevaluate health policy and services in India (Pandit, 2021). The culture of an organisation, collaborative partnerships, and a clear strategic goal are essential elements that contribute to the enhanced resilience and effectiveness of some healthcare organisations in navigating VUCA circumstances.

The objective of this paper is to comprehensively analyze the Indian healthcare sector within the context of the VUCA framework. By examining the impact of VUCA factors on the healthcare industry, this paper aims to evaluate both the opportunities and challenges that the sector faces. Furthermore, it seeks to shed light on the sustainability of the healthcare sector in India, particularly in the post-COVID-19 era, where the industry has witnessed significant transformation and growth. Through a detailed exploration of VUCA elements specific to the healthcare sector, this review paper aims to provide insights that can inform strategic decision-making, enabling healthcare organizations to attain a substantial competitive advantage through distinctive competence and unique capabilities.

UNDERSTANDING VUCA

A. Volatility

The term "volatility" in the healthcare sector pertains to the inherent unpredictability and dynamic character of the industry (Mathew et al., 2023). The phenomenon is distinguished by abrupt changes, uncertainties, and intricacies that pose difficulties in the process of planning and strategizing (Pandit, 2021). The healthcare industry is now undergoing significant transformations as a result of personalised treatment, the increasing importance of health data, and the necessity for data exchange (Fuentealba et al., 2023). The aforementioned alterations necessitate the industry to engage in strategic reevaluation and adopt a mindset of innovation, agility, and adaptation (Steffen et al., 2023). The COVID-19 pandemic has underscored the necessity for healthcare organisations to exhibit resilience and proficiently navigate VUCA circumstances. The advent of digitization in the healthcare sector, along with patients' increasing demand for consumer rights, is further exacerbating the unstable circumstances within the business. In general, volatility within the healthcare industry pertains to the persistent and uncertain fluctuations that necessitate organisations to exhibit adaptability and promptness in response to the ever-changing environment.

The post-pandemic Indian healthcare industry is characterised by dynamic elements, such as the need for personal protective equipment (PPE) of superior quality and the availability of testing laboratories, facilities, centres, and kits (Gedam et al., 2022). Furthermore, the distribution of financial resources towards the National Health mission and its implications for the Human Development Index (HDI), Infant Mortality Rate (IMR), and Maternal Mortality Rate (MMR) are also matters of significant interest. The industry is influenced by the volatility of consumer behaviour and the growing demand for healthcare services, which can be attributed to increased income levels and a heightened awareness of health (Shetty, 2022). Additionally, it is crucial to take into account the mental health requirements of the populace and the burden placed on current mental health services during the ongoing epidemic (Tm. Kumar, 2020). In conclusion, the inadequacies in healthcare facilities, staff, and leadership, together with the advent of novel health issues such as the black fungus, all contribute to the instability observed within the healthcare system in India (Chakravorty, 2021).

B. Uncertainty

The idea of uncertainty has significant importance within the VUCA paradigm, particularly in relation to the healthcare sector. The healthcare sector is currently confronted with volatile and dynamic circumstances, posing difficulties in the formulation and implementation of plans and strategies (Mathew et al., 2023). The COVID-19 pandemic has underscored the imperative for healthcare organisations to proficiently navigate and address uncertainties (Pandit, 2021). The presence of uncertainty in the healthcare sector may be attributed to several causes, including the dynamic and evolving character of infectious illnesses, the intricate decision-making processes involved in implementing control measures, and the inherent ambiguity arising from limited information. In the current VUCA environment, it is imperative for healthcare organisations to actively adopt and integrate innovative practises, as well as cultivate agility and flexibility, in order to effectively manoeuvre through periods of unpredictability (Raja, 2021). The significance of organisational culture, collaborative partnerships, and a clear strategic goal cannot be overstated as crucial elements that contribute to resilience and successful leadership in situations characterised by uncertainty. The comprehension and resolution of uncertainty play a pivotal role in the ability of healthcare organisations to adeptly navigate crises and safeguard the welfare of patients and communities.

The presence of uncertainty has exerted a substantial influence on the process of decision-making within the healthcare industry (Dwivedi & Gupta, 2024). The healthcare industry is currently experiencing transformative shifts due to the process of digitization and the VUCA situations it introduces. These developments are causing disruptions to the established stability of traditional routines (Steffen et al., 2023). The complexities inherent in the contemporary corporate landscape, which is marked by VUCA, necessitate the adoption of several strategies for addressing these difficulties (Fridgeirsson et al., 2021). The process of making decisions in healthcare encompasses a range of uncertainty, and it is of utmost importance to consider and include all pertinent uncertainties in evaluations in order to arrive at judgements that can be justified. Decision-makers and analysts have seen a decline in their level of confidence about their capacity to accurately predict future events and gauge the preferences of stakeholders. As a result, decision-making processes have been characterised by a state of profound ambiguity (Grutters et al., 2015). Operations Research (OR) and stochastic programming methodologies are applicable in the modelling and resolution of healthcare decision-making issues in the presence of uncertainty. These approaches take into account several elements, including patient waiting times and operational expenses. In general, the presence of uncertainty has prompted the creation of several tools and methodologies to evaluate and manage the VUCA characteristics within healthcare decision-making processes.

C. Complexity

The healthcare systems in India exhibit a multifaceted nature and encounter a multitude of issues (Sayal et al., 2025). The healthcare system in India includes a combination of governmental and private institutions, with the private sector playing a prominent role in the dynamic medical landscape (Börner & Zohmann, 2022). The system exhibits notable attributes like inadequate financial resources, suboptimal administrative practises, and a convoluted framework for public healthcare (Fuentealba et al., 2023). The increasing focus on innovation is important in order to effectively respond to the VUCA nature of the contemporary world. It is crucial for Indian organisations to actively embrace innovation, agility, and flexibility as means to confront and overcome these difficulties (Rudrappa et al., 2019). The COVID-19 pandemic has underscored the imperative for transformation and advancement within the healthcare industry (Mehta, 2017). The scholarly literature mostly centres around the comprehension of the competencies necessary for effectively navigating a VUCA environment, as well as the responsibilities and influence of leaders in effectively managing organisational change.

The intricacy of the Indian healthcare system is shaped by a multitude of elements. The factors contributing to the challenges in the healthcare sector encompass deficiencies in infrastructure, a scarcity of healthcare personnel, discrepancies between urban and rural areas, restricted coverage of health insurance, poor allocation of public funds for healthcare, and a fragmented healthcare system (Mathew et al., 2023). Moreover, the healthcare system is confronted with a substantial difficulty in addressing the burden of non-communicable illnesses (A. Kumar, 2023; Gupta et al., 2023). The system is now facing challenges related to inadequate administration, insufficient financial resources, and limited accessibility and affordability of healthcare services. The increasing speed of change exacerbates the intricacy, since the healthcare landscape encompasses a diverse range of settings, spanning from advanced medical facilities in metropolitan locales to under-resourced health subcenters in rural places. The aforementioned variables contribute to the intricate nature of the healthcare system in India, necessitating continuous changes, investments in infrastructure and technology, and a concerted effort to enhance accessibility and affordability (Mehta, 2017).

D. Ambiguity

Ambiguity within the Indian healthcare sector pertains to the absence of lucidity or definitiveness in decision-making and comprehension, stemming from the unexpected and intricate characteristics inherent to the profession (D. The phenomenon is evident in several forms, including the imperative for healthcare institutions to recalibrate their strategies and confront the transformations in the contemporary VUCA landscape with enhanced lucidity (Mathew et al., 2023). The COVID-19 pandemic has underscored the difficulties encountered by the healthcare sector, resulting in unparalleled levels of uncertainty and ambiguity. Healthcare organisations are facing challenges in bridging the gap in digitization and adjusting to the changing circumstances, which highlights the urgency for a faster paradigm change (Steffen et al., 2023).

The healthcare industry's inherent unpredictability and intricate nature need the use of creative strategies and change management methodologies that are rooted in technological advancements (Mathew et al., 2023; Dwivedi, et al., 2022). In order to tackle these difficulties, it is imperative to transition from a healthcare system that is driven by supply to one that is centred around the needs of patients. This move necessitates that organisations adopt a mindset of innovation, agility, and flexibility (Fridgeirsson et al., 2021). The COVID-19 pandemic has underscored the importance of prompt interventions and the resolution of concerns pertaining to healthcare workers' working hours, mental well-being, and safety and security (Debbarma, 2023). Nevertheless, the majority of healthcare organisations in India are now facing challenges in bridging the digitalization divide and transforming into agile and adaptable VUCA service organisations (Steffen et al., 2023). In order to address these problems, it is essential to develop a comprehensive comprehension and decision-making approach, supplemented by information technology-based tools, which can facilitate the formulation of tailored digitalization strategies. Through the evaluation of projects from a VUCA viewpoint, project managers are able to efficiently deploy scarce resources in order to offset the effects of a VUCA environment.

STRATEGIES FOR NAVIGATING VUCA

a. Resilience and Adaptation

Healthcare organizations have the capacity to enhance their resilience within a VUCA environment through the adoption of innovative practices, fostering agility, and cultivating adaptation (Mathew et al., 2023). There is a need to transition from a healthcare system that is driven by supply to one that prioritizes the needs and preferences of patients (AFŞAR DOĞRUSÖZ et al., 2022). The enhancement of organizational resilience may be achieved by the implementation of various measures such as proactive planning, effective organization of emergency responses, and the development of policies aimed at ensuring company continuity (Sindila et al., 2023). In order to cultivate resilience, it is imperative for decision-makers to prioritize resourcefulness, accurate time interpretations, and entrepreneurial judgements (Angeler et al., 2022). In order to successfully manage healthcare systems during crises, it is imperative to implement adaptive and transformative adjustments (Velner et al., 2023). Various factors contribute to the resilience of an organization, such as its ethos, organizational culture, leadership, human capital, situational awareness, flexibility, organizational performance, and learning ability. Research has indicated that educational interventions, as well as factors such as gender and age groups, can serve as significant predictors of organizational resilience within healthcare organizations. Healthcare managers may effectively traverse the difficulties of a VUCA environment by consistently assessing the level of organizational resilience and placing emphasis on these characteristics.

b. Innovation and Technology

To effectively address the complexities of the VUCA environment, it is imperative to employ new approaches and solutions. One potential strategy involves the implementation of a hierarchical multi-expert learning framework, which draws inspiration from the structure and functioning of the central nervous system (CNS). The proposed architecture incorporates the concurrent processes of map exploration and route-planning in order to circumvent dead ends, analogous to the cerebrum in the central nervous system. Furthermore, it is possible to employ a localized adaptive model that effectively manages the trade-off between collision-avoidance and go-straight methods. This model may be seen as analogous to the cerebellum within the central nervous system (Kumar et al., 2023). The efficacy of this strategy has been demonstrated to surpass that of current methodologies in terms of accomplishing the goal, time effectiveness, and ensuring security (Holley et al., 2022). An additional approach involves the cultivation of leadership abilities that are important for effectively navigating the VUCA environment. The aforementioned talents encompass the ability to cultivate a collective objective, adaptability in acquiring knowledge, introspection, leadership via cooperation and persuasion, assurance in guiding amongst ambiguity, and possessing a mentality oriented towards progress. Through the implementation of these techniques, organizations have the capacity to surmount the obstacles presented by the VUCA age and guarantee the long-term viability of their operational procedures (Sinha & Sinha, 2020).

c. Collaboration and Partnerships

Collaborative efforts assume a pivotal role in effectively managing the VUCA environment within the healthcare sector. In order to provide complete and patient-centered treatment, it is imperative to engage in multidisciplinary teamwork due to the intricate nature of cancer care (Beth, 2023). Through engaging in interdisciplinary collaboration with experts in many fields, healthcare practitioners can enhance their comprehension of patients' requirements, hence resulting in improved results and the development of more efficacious treatment strategies. Furthermore, the utilization of collaboration technologies has the potential to enhance communication and ease the exchange of knowledge among healthcare professionals who are geographically separated, hence leading to an enhancement in the provision of patient care (Paul et al., 2013). The incorporation of innovation, agility, and flexibility into healthcare organizations is crucial in order to effectively traverse the VUCA conditions of the present day. Engaging in collaborative efforts with reliable colleagues and establishing professional relationships may significantly contribute to the optimization of healthcare practitioners' practice capabilities and the improvement of patient care outcomes (Gupta et al 2023). In general, partnerships provide healthcare professionals the opportunity to utilize a wide range of viewpoints and specialized knowledge, resulting in enhanced healthcare provision amidst VUCA issues.

CONCLUSION

The future outlook for the healthcare industry in India within a global environment appears to have promise, however it also presents significant challenges. The challenge of delivering high-quality healthcare services in emerging nations such as India is exacerbated by the combination of limited resources and a rapidly expanding population. The COVID-19 pandemic and the sustainable development objectives have brought attention to the necessity for policymakers to reassess healthcare policies and services, with the objective of achieving universal and sustainable healthcare. The healthcare industry in India has exhibited significant attributes during both pre-pandemic and peri-pandemic periods, and scholars have put forth predictive models for projecting stock values within this sector. The "Make in India" project presents prospects for private sector involvement in fostering the expansion of the healthcare sector. However, the extent to which it enhances accessibility to inexpensive and high-quality healthcare hinges upon the efficacy of government policies and their subsequent execution. In general, the healthcare industry in India exhibits promising prospects for expansion and enhancement. However, it necessitates the adoption of strategic planning, investment, and efficient policy execution to effectively address the complexities inherent in a VUCA environment.

The COVID-19 pandemic has exposed vulnerabilities in the Indian healthcare system, highlighting the need to bridge the gap between policy formulation and effective implementation at the grassroots level. In light of these insights, it is evident that the Indian healthcare sector's journey through the VUCA landscape is characterized by both challenges and opportunities. As the sector continues to grow and evolve, stakeholders must remain agile and adaptive, continually seeking innovative solutions to address emerging issues and capitalize on the potential for substantial development. By aligning policy, practice, and research, the Indian healthcare sector can chart a course towards greater resilience, sustainability, and improved healthcare services for its population.

REFERENCES

Afşar Doğrusöz, L., Yazici, S., & Gür Omay, E. G. (2022). Organizational Resilience in Healthcare Organizations: A Research in the Public and Private Sector. *Afet ve Risk Dergisi*, *5*(1), 318–329. DOI: 10.35341/afet.1083886

Ahmad Mir, T., & Singh, M. (2022). Indian Healthcare Sector and the Sustainable Development. *International Journal of Current Research and Review*, *14*(03), 43–47. DOI: 10.31782/IJCRR.2022.14308

Angeler, D. G., Eyre, H. A., Berk, M., Allen, C. R., Hynes, W., & Linkov, I. (2022). Adaptation, Transformation and Resilience in Healthcare Comment on "Government Actions and Their Relation to Resilience in Healthcare During the COVID-19 Pandemic in New South Wales, Australia and Ontario, Canada.". *International Journal of Health Policy and Management*. Advance online publication. DOI: 10.34172/ijhpm.2022.7043 PMID: 35247939

Beth, F. (2023). The Importance of Professional Communities and Collaborations. *Journal of the Advanced Practitioner in Oncology*, *14*(2), 115–116. DOI: 10.6004/jadpro.2023.14.2.1 PMID: 37009410

Börner, D., & Zohmann, A. (2022). VUCA. *Hands on - Manuelle Und Physikalische Therapien in Der Tiermedizin, 4*(04), 145–145. DOI: 10.1055/a-1942-6128

Chakraborty, D. (2019). Versatile Performance in Vuca World: A Case Study. *Ushus -. Journal of Business and Management*, *18*(4), 1–8. DOI: 10.12725/ujbm.49.1

Chakravorty, I. (2021). Key Factors Fuelling India's Second COVID Surge. *The Physician*, *7*(1), 1–6. DOI: 10.38192/1.7.1.9

Debbarma, S. (2023). COVID-19 Pandemic and Challenges Faced by Healthcare Professionals in India. *Cureus*. Advance online publication. DOI: 10.7759/cureus.38278 PMID: 37255890

Dwivedi, S., & Gupta, A. (2024). Strategically addressing skill gaps and imbalances among health employees. In *Contemporary Challenges in Social Science Management: Skills Gaps and Shortages in the Labour Market* (pp. 17-33). Emerald Publishing Limited. DOI: 10.1108/S1569-37592024000112A015

Dwivedi, S., Poddar, A., Gupta, A., Memoria, M., & Kumar, R. (2022, December). Analyzing the effectiveness of a health care waste reduction education and training interventions: A step towards green healthcare. In *2022 2nd International Conference on Innovative Sustainable Computational Technologies (CISCT)* (pp. 1-5). IEEE. DOI: 10.1109/CISCT55310.2022.10046456

Fridgeirsson, T. V., Kristjansdottir, B. H., & Ingason, H. T. (2021). An Alternative Risk Assessment Routine for Decision Making; Towards a VUCA Meter to Assess the Volatility, Uncertainty, Complexity and Ambiguity of Complex Projects. In *Research on Project* (pp. 41–54). Programme and Portfolio Management., DOI: 10.1007/978-3-030-60139-3_4

Fuentealba, D., Flores-Fernández, C., & Carrasco, R. (2023). Análisis bibliométrico y de contenido sobre VUCA. *Revista Española de Documentación Científica*, *46*(2), e354. DOI: 10.3989/redc.2023.2.1968

Gedam, V., Raut, R., Inamdar, Z., Narkhede, B., Dharaskar, S., & Narvane, V. (2022). <scp>COVID</scp> -19 critical success factors in Indian healthcare industry—A <scp>DEMATEL</scp> approach. *Journal of Multi-Criteria Decision Analysis*, *29*(1–2), 135–149. DOI: 10.1002/mcda.1763 PMID: 40477520

Grutters, J. P. C., van Asselt, M. B. A., Chalkidou, K., & Joore, M. A. (2015). Healthy Decisions: Towards Uncertainty Tolerance in Healthcare Policy. *PharmacoEconomics*, *33*(1), 1–4. DOI: 10.1007/s40273-014-0201-7 PMID: 25145801

Gupta, A., Chaithra, N., Jha, J., Sayal, A., Gupta, V., & Memoria, M. (2023, May). Machine learning algorithms for disease diagnosis using medical records: a comparative analysis. In *2023 4th international conference on intelligent engineering and management (ICIEM)* (pp. 1-6). IEEE. DOI: 10.1109/ICIEM59379.2023.10165850

Gupta, O., Joshi, P., & Gupta, A. (2023, May). Prioritizing the enablers for adoption of AI based health care wearables among elderly people. In *2023 4th International Conference on Intelligent Engineering and Management (ICIEM)* (pp. 1-4). IEEE. DOI: 10.1109/ICIEM59379.2023.10165740

Hass, B., Pooley, J., Feuring, M., Suvarna, V., & Harrington, A. (2012). Health technology assessment and its role in the future development of the Indian healthcare sector. *Perspectives in Clinical Research*, *3*(2), 66. DOI: 10.4103/2229-3485.96449 PMID: 22701823

Holley, D., Coulson, K., Buckley, C., & Corradini, E. (2022). Wellbeing in the workplace: Exploring the VUCA approach. *Journal of Learning Development in Higher Education*, *25*(25). Advance online publication. DOI: 10.47408/jldhe.vi25.954

Kaivo-oja, J. R. L., & Lauraeus, I. T. (2018). The VUCA approach as a solution concept to corporate foresight challenges and global technological disruption. *Foresight*, *20*(1), 27–49. DOI: 10.1108/FS-06-2017-0022

Karneli, O. (2022). Application of the Concept of Strategic Human Resources Management in Improving Organizational Performance in the Vuca Era. *Eduvest - Journal Of Universal Studies, 2*(4), 704–709. DOI: 10.36418/edv.v2i4.381

Kumar, A. (2023). The Transformation of The Indian Healthcare System. *Cureus*. Advance online publication. DOI: 10.7759/cureus.39079 PMID: 37378105

Kumar, K., Chaudhary, S., Anandaram, H., Kumar, R., Gupta, A., & Joshi, K. (2023, February). Industry 4.0 and health care system with special reference to mental health. In *2023 1st International Conference on Intelligent Computing and Research Trends (ICRT)* (pp. 1-5). IEEE.

Kumar, T. (2020). Community mental health services in India: The pandemic and beyond. *Indian Journal of Social Psychiatry*, *36*(5), 168. DOI: 10.4103/ijsp.ijsp_126_20

Mathew, J., Gupta, N., & Jagose, J. (2023). *Innovations In the Indian Healthcare Industry to Brave the Vuca Times*. Asia Pacific Journal of Health Management., DOI: 10.24083/apjhm.v18i1.1703

Mehta, P. (2017). Framework of Indian Healthcare System and its. *Challenges*, •••, 247–271. DOI: 10.4018/978-1-5225-0640-9.ch011

Mijwil, M. M., Faieq, A. K., & Al-Mistarehi, A.-H. (2022). The Significance of Digitalisation and Artificial Intelligence in The Healthcare Sector: A Review. *Asian Journal of Pharmacy. Nursing and Medical Sciences*, *10*(3). Advance online publication. DOI: 10.24203/ajpnms.v10i3.7065

Pandit, M. (2021). Critical factors for successful management of VUCA times. *BMJ Leader*, *5*(2), 121–123. DOI: 10.1136/leader-2020-000305 PMID: 36170532

Paul, S., Ramaprasad, A., & Wickramasinghe, N. (2013). Introduction to Technology Mediated Collaborations in Healthcare Minitrack. *2013 46th Hawaii International Conference on System Sciences*, 791–791. DOI: 10.1109/HICSS.2013.508

Raja, A. S. (2021). Business Research in the VUCA World. *Ushus Journal of Business Management*, *20*(1), v–xvi. DOI: 10.12725/ujbm.54.0

Rudrappa, S., Agarkhed, D. V., & Vaidya, S. S. (2019). Healthcare Systems: India. In *Quality Spine Care* (pp. 211–224). Springer International Publishing. DOI: 10.1007/978-3-319-97990-8_13

Sayal, A., Gupta, A., Jha, J., Chaithra, N., & Nithin, A. (2025). Application of Artificial Intelligence in Healthcare Sector: A Multidimensional Perspective. In *Driving Global Health and Sustainable Development Goals With Smart Technology* (pp. 115–138). IGI Global Scientific Publishing.

Shetty, K. P. (2022). A Study on the Transformation of the Indian Healthcare Sector during the COVID-19 Pandemic. *International Journal for Research in Applied Science and Engineering Technology*, *10*(1), 1057–1064. DOI: 10.22214/ijraset.2022.39986

Sindila, A., Foss, N. J., & Zhan, X. (2023). Building Resilience for Surviving and Thriving in a VUCA Context. *AIB Insights*, *23*(3). Advance online publication. DOI: 10.46697/001c.73812

Sinha, D., & Sinha, S. (2020). Managing in a VUCA World: Possibilities and Pitfalls. *Journal of Technology Management for Growing Economies*, *11*(1), 17–21. DOI: 10.15415/jtmge.2020.111003

Steffen, B., Braun von Reinersdorff, A., & Rasche, C. (2023). IT-Based Decision Support for Holistic Healthcare Management in Times of VUCA, Disorder, and Disruption. *Applied Sciences (Basel, Switzerland)*, *13*(10), 6008. DOI: 10.3390/app13106008

Velner, T., Rubinstein, Z., & Adini, B. (2023). Organizational Resilience Among Health Organizations in Israel. *Prehospital and Disaster Medicine*, *38*(S1), s143–s143. DOI: 10.1017/S1049023X23003758

Chapter 3

Post-Quantum Cryptography: Robust Solutions Against Quantum Computing Threats

Tatyana Peleshenko
https://orcid.org/0000-0003-2761-7847
North Caucasian Federal University, Russia

Diana Fedash
Faculty of Mathematics and Computer Science, Stavropol, Russia

Alexandr Gish
Faculty of Mathematics and Computer Science, Stavropol, Russia

ABSTRACT

Quantum computers are a revolutionary technology based on the principles of quantum mechanics. Unlike conventional computers that work with binary bits (0 or 1), quantum devices use qubits that can exist in multiple states ("superpositions") and interact with each other through the phenomenon of "quantum entanglement." Thanks to this, quanta can greatly accelerate the solution of a number of important problems that were previously considered insurmountable for classical machines. This review describes the principles of operation of quantum computers, considers their unique advantages, and demonstrates the potential for their impact on the future of computer science and the technology industry.

DOI: 10.4018/979-8-3373-2240-7.ch003

QUANTUM COMPUTERS

Quantum computers are a fundamentally new computing architecture based on the laws of quantum mechanics. Their capabilities are fundamentally different from classical computers due to the use of qubits - elements that can be in several states at the same time ("superposition") and interact with each other ("quantum entanglement"). These properties allow quantum computers to solve some problems much faster than traditional computers.

The Threat of Quantum Computers to Classical Cryptography

Classical cryptography is based on the complexity of certain mathematical problems, such as the factorization of large numbers (RSA), the discrete logarithm problem (DSA, ECDSA) and elliptic curves (ECC). However, it is precisely these problems that become vulnerable to the power of quantum computing. A quantum computer can be successfully used in various fields, including the use of quantum acceleration to factor a number using Shor's algorithm (Bogdanov & Kizhvatov, 2005)(Shor, 1994).

Shor's Algorithm

Peter Shor's algorithm allows efficient factorization of large numbers and finding discrete logarithms in polynomial time. This means that RSA, DSA, and ECC encryption algorithms can become easily cracked even on relatively small key sizes once sufficiently powerful universal quantum computers are available.

The performance of these algorithms was measured on both CPUs and GPUs using Nvidia's CUDA parallelization technology. The NTRU, RSA, and ECC-NIST-224 algorithms are presented in (Hermans, Vercauteren, & Preneel, 2010).

For example, for a classical computer, factorization of a large prime number hundreds of digits long is a virtually impossible task, taking millions of years. A quantum computer can solve such a task in just hours or days.

Thus, classical asymmetric cryptography methods such as RSA and elliptic curves will no longer provide the necessary security in the presence of powerful quantum devices.

Symmetric Cryptography

Symmetric systems such as AES are less susceptible to attack by quantum computers. Although Grover's algorithm significantly speeds up password brute-force, classical encryption remains secure as the key length increases. For example, AES-

256 provides sufficient protection even against attacks using a quantum computer. However, asymmetric encryption is widely used for key exchange of symmetric methods (e.g., TLS protocol), and it becomes the main target of quantum attacks. Grover's quantum algorithm (Zhukov, 2015) in the quantum computing model is an analogue of the method of complete enumeration of keys of encryption algorithms.

PREPARING FOR A POST-QUANTUM WORLD

To prepare for the advent of quantum computers, scientists are developing new types of cryptographic protocols that are resistant to such attacks. Among the promising areas are:

Lattice Cryptography

Lattice cryptography is one of the promising areas of modern cryptography, based on the theory of algebraic structures called lattices. The main idea is to create cryptographic primitives that are resistant even against attacks by quantum computers, which can easily break traditional encryption methods such as RSA or elliptic curve. There are two classes of lattice problems used in cryptography: the nearest vector problem (SVP) and the hidden nearest vector problem (CVP). The problem of finding the shortest non-zero vector in a given lattice is a computationally difficult problem. This problem underlies many lattice cryptography protocols.

The hidden nearest vector problem (CVP) is related to finding the nearest point on a lattice relative to a given arbitrary vector. The complexity of CVP is also used to build secure encryption schemes.

Table 1. The hidden nearest vector problem (CVP) is related to finding the nearest point on a lattice relative to a given arbitrary vector. The complexity of CVP is also used to build secure encryption schemes

NTRU encryption	GGH cryptosystem	LWE (Learning with Errors) Algorithm
NTRU (Nth Degree TRUncated polynomial ring) is a family of cryptographic algorithms developed by Joseph Silverman, Jeffrey Hoffstein, and Jill Pippenger in the mid-1990s. The main difference between NTRU and traditional cryptographic systems such as RSA or ElGamal is that it uses a special polynomial ring structure to provide security and high performance.	GGH (Goldreich–Goldwasser–Halevi cryptosystem) is a cryptographic system proposed by Oleg Goldreich, Shai Goldwasser, and Shamir Halevi in 1997. It was the first attempt to construct an encryption scheme based on the difficulty of lattice-theoretic problems such as the nearest vector problem (SVP) and the close vector problem (CVP).	The Learning with Errors (LWE) algorithm is a mathematical problem that underlies a number of modern cryptographic protocols aimed at ensuring security in the post-quantum era. First proposed by Øyster Rigvadtz in 2005, LWE has become the basis for a wide range of cryptographic schemes, ranging from encryption to homomorphic encryption constructions.

Hash Functions

Hash functions play a key role in information security and are one of the main tools for data protection. They allow you to transform input data of any size into a fixed short set of characters (hash) with certain properties.

Table 2. Hash functions are used to verify data integrity, store passwords, digital signatures, and cryptocurrency.

MD5 (Message Digest algorithm version 5)	SHA-1 (Secure Hash Algorithm 1)	SHA-2 (SHA-256, SHA-512)	BLAKE2/BLAKE3
Исторически использовалась повсеместно, однако была признана небезопасной из-за слабостей в защите от коллизий.	A more modern version of MD5, but also has shortcomings and is no longer recommended for critical applications.	Modern security standards recommend using the SHA-2 family, which includes algorithms with different output lengths (256 bits, 512 bits, etc.).	Fast-growing alternatives to SHA-2 that offer high performance and security.

Error Coding

Error coding is an important aspect of data transmission and storage aimed at detecting and correcting errors that occur due to interference or equipment failures. Errors can occur for various reasons: noise effects in communication channels, equipment failures, physical damage to data carriers, etc.

Table 3. The main types of codes used for error detection and correction are detection codes, correction codes, and combined approaches.

Detection codes			
Parity check	**CRC (Cyclic Redundancy Check)**		**Checksums**
The simplest type of detection code consists of adding an extra bit to ensure that the total number of ones in the data packet is even or odd. Changing even one bit will result in a violation of the parity condition, signaling the presence of an error.	**This is a technique of adding a special data field (checksum) calculated by a special polynomial. CRC is able to recognize large classes of errors, including multiple shifts and short bursts.**		**A simple method that involves calculating the sum of all bytes in a packet and comparing it to the expected result. A violation of consistency indicates a violation of data integrity.**
Correction codes			
Hamming code	RS codes	LDPC codes	Turbo codes
One of the oldest and simplest methods of correcting single errors. In this approach, additional check bits are placed in a special way, allowing the location of the error to be uniquely identified and the data to be corrected.	This class of codes has found its application in optical storage devices (CD, DVD) and data transmission networks. RS codes are effective in restoring continuous sections of damaged data.	A very popular modern class of codes, characterized by high quality error correction and the possibility of implementing high-performance hardware solutions.	Turbo codes emerged in the late 20th century and have shown excellent performance in combating noise and data loss in radio communications and mobile phones.
Combined approaches.			
Combination of detection and correction codes	Using multiple layers of protection	Integration of different architectures	Compatibility across platforms
One popular example is a combination of codes such as Reed-Solomon (to correct long consecutive errors) and BCH (to eliminate single errors). This combination improves the quality of data transmission in conditions of strong interference.	Data can pass through several stages of coding in sequence, starting with a simple method such as parity and continuing with more complex methods such as turbo codes or LDPC codes. Each layer complements the previous one, increasing the overall level of reliability.	The combination of distributed network architectures with redundant data transmission paths and error-correcting codes can guarantee the delivery of information even under conditions of partial failure of nodes or communication lines.	The creation of unified interfaces and protocols compatible with various operating systems and hardware architectures allows for the expansion of the scope of application of combined approaches.

Multivariate Cryptography

Multivariate Cryptography is a class of cryptographic methods based on the use of complex nonlinear functions of several variables to construct cryptographic primitives such as digital signatures and encryption schemes. This approach is distinguished by the use of open mathematical objects - multivalent (multi-valued) polynomials, which are extremely difficult to analyze analytically and disclose secret parameters. These approaches are actively researched by the US National Institute of Standards and Technology (NIST), which is holding a competition to standardize post-quantum cryptosystems.

Table 4. The main idea of the approach is built around the following principles: open polynomials, secret isomorphism and their application.

Properties of open polynomials		
Certainty	**Safety**	**Ease of verification**
The coefficients of open polynomials are formed in advance, according to established rules or procedures.	Public polynomials themselves do not reveal the internal structure of secret data.	Any participant can check the correctness of open polynomials in a simple way.
Advantages and features of isomorphism		
Increased security	Saving resources	Supporting a new era of cryptography
The outside world sees a complex object with no obvious way to recover secret data.	The internal structure of private keys can be processed much more simply and efficiently by a computer.	This approach supports the concept of post-quantum cryptography, where classical methods (RSA, elliptic curves) lose their power.

Application: the encryption and decryption process is reduced to solving the corresponding system of equations. By introducing additional variables and complicating the degrees of polynomials, high resistance to known crypto-attacks is achieved, including brute force attacks and differential analysis.

CONCLUSION

The development of quantum computers poses a serious threat to modern cryptographic infrastructure, especially to methods based on factorization and discrete logarithms. To maintain the security of information systems, it is necessary to switch to technologies capable of countering new threats using post-quantum cryptography methods.

REFERENCES

Bogdanov, A. Y. (2005). *Quantum algorithms and their impact on the security of modern classical cryptographic systems/A. Yu. Bogdanov, IS Kizhvatov*. RGGU.

Brassard, G., Hoyer, P., Mosca, M., & Tapp, A. (2000). Quantum amplitude amplification and estimation. arXiv preprint quant-ph/0005055.

Grover, L. K. (1996, July). A fast quantum mechanical algorithm for database search. In *Proceedings of the twenty-eighth annual ACM symposium on Theory of computing* (pp. 212-219).

Hermans, J., Vercauteren, F., & Preneel, B. (2010, March). Speed records for NTRU. In *Cryptographers' track at the RSA conference* (pp. 73–88). Springer Berlin Heidelberg.

Shor, P. W. (1994). Algorithms for quantum computers: Discrete logarithms and factoring. In *Proc. 35th Annual Symposium on Foundations of Computer Science*, IEEE Computer Society Press, Los Alamitos, CA (pp. 124-134).

Zhukov, A. (2015, September). Lightweight cryptography: modern development paradigms. In *Proceedings of the 8th International Conference on Security of Information and Networks* (pp. 7-7).

Chapter 4

Research of Approaches to Assessing the Information Security Risks of an Organization:
Qualitative Risk Assessment Methods

Julia Andrusenko
https://orcid.org/0009-0001-3426-2741
North Caucasus Federal University, Russia

G. S. Prakasha
https://orcid.org/0000-0002-1287-7606
Christ University, Bangalore, India

Ivan Golovin
North Caucasian Federal University, Russia

Vladimir Lapin
Gymnasium 25, Russia

ABSTRACT

This article explores modern approaches to assessing information security risks within organizations, focusing on qualitative, quantitative, and hybrid risk management methodologies. The study examines the core principles of each approach, their practical applicability in different organizational contexts, the tools and frameworks commonly employed, and key implementation challenges. A comparative analysis

DOI: 10.4018/979-8-3373-2240-7.ch004

of these methods highlights their respective strengths and limitations, providing insights for selecting the most suitable risk assessment strategy based on organizational needs, industry requirements, and available resources.

1. INTRODUCTION

The current stage of digital transformation, characterized by a rapid increase in the volume of processed data and the complexity of the architecture of information systems, makes information security issues relevant (Fenz et al., 2014; Karabacak & Sogukpinar, 2005). Recently, there has been an increase in both the frequency and falsity of information security incidents associated with the violation of key properties of data, confidentiality, integrity, and availability (Schneier, 2000; Peltier, 2005). In these conditions, the development of integrated information risk management systems is of particular importance (Chockalingam et al., 2017; Rahman et al., 2024).

According to international studies, the number of information security incidents continues to increase steadily, while a significant part of them is due to insufficient development of the processes of identification and assessment of potential threats (Shameli-Sendi et al., 2016; Wangen et al., 2017). Against the background of increased regulation by both the Russian regulatory authorities FSTEC, Roskomnadzor, and international standardization organizations, the creation of adaptable risk assessment methods that correspond to current threats to information security is of particular importance (ISO/IEC, 2022; ENISA, 2006). The key features of modern approaches to risk management are: the need to take into account the specifics of the IT infrastructure of a particular organization; adaptation to the constantly changing threat landscape; compliance with regulatory requirements; and ensuring the economic efficiency of protective measures (Stoneburner et al., 2002; The Open Group, 2013). At the same time, the existing risk assessment methodological apparatus is distinguished by significant diversity, and each approach has its own unique advantages and limitations depending on the application scenarios (Jones, 2006; Tran & Ross, 2021; Olaes, 2025). The purpose of this study is a systemic analysis of modern methods for assessing information security risks, with particular attention paid to the criteria for choosing the optimal method (Rosa et al., 2017; Kirvan & Irei, 2023).

2. THE PURPOSE OF INFORMATION SECURITY RISK ANALYSIS

The goal of any organization is to achieve certain indicators that characterize the results of its activities. For example, for commercial companies this is to make a profit, increase capitalization, market share, or turnover, and for government organizations—to provide public services to the population and solve management problems (Peltier, 2005; Schneier, 2000). In any case, regardless of the purpose of the organization, the achievement of this goal may be hampered by the implementation of information security risks (Fenz et al., 2014; Karabacak & Sogukpinar, 2005). At the same time, each organization assesses the risks and the possibility of investing in their reduction in its own way (Jones, 2006; The Open Group, 2013).

Thus, the goal of information security risk management is to maintain them at an acceptable level for the organization (ISO/IEC, 2022; ENISA, 2006). To solve this problem, organizations create integrated information security systems (ISS) (Chockalingam et al., 2017; Rahman et al., 2024).

Analysis of information security risks allows you to determine the necessary and sufficient set of information protection tools, as well as organizational measures aimed at reducing information security risks, and to develop an ISS architecture for an organization that is most effective for its specific activities and aimed at reducing its information security risks. When analyzing risks, the expected damage in the event of a threat being realized is compared with the costs of measures and means of protection, after which a decision is made regarding the assessed risk, which can be:

- reduced, for example, by introducing means and mechanisms of protection that reduce the probability of the threat being realized or the destructiveness coefficient;
- eliminated by refusing to use the resource exposed to the threat;
- transferred, for example, insured, as a result of which, in the event of a security threat being realized, the losses will be borne by the insurance company, and not the owner of the resource;
- accepted.

But in order to make a decision, it is first necessary to conduct a risk analysis using one of the existing methods, which we will consider below.

3. QUALITATIVE RISK ASSESSMENT METHODOLOGY

Qualitative risk assessment in information security is one of the most widely used and accessible approaches to threat analysis, especially in cases where quantitative data is not available or where the level of maturity of the organization's information security does not allow for accurate numerical calculations (Peltier, 2005; Shameli-Sendi et al., 2016). Within this approach, risks are described using verbal formulations or categorical scales, such as “low,” “medium,” and “high,” without referring to numerical values of probability or damage (ENISA, 2006; ISO/IEC, 2022). The main objective of qualitative assessment is to identify and structure the most significant threats and vulnerabilities, followed by their distribution by priority based on expert analysis (Fenz et al., 2014; Chockalingam et al., 2017).

The process of qualitative risk assessment begins with the identification of assets that are valuable to the organization: these can be information systems and data, as well as people or business processes (Stoneburner et al., 2002; Karabacak & Sogukpinar, 2005). Next, an analysis of potential threats and vulnerabilities is carried out, as well as an expert determination of the probability of an incident and the severity of its consequences (Wangen et al., 2017; Rosa et al., 2017). Based on this data, a risk matrix is formed in which the values of probability and damage intersect, allowing a clear determination of the risk level for each scenario (Jones, 2006; The Open Group, 2013). It is common to use scales with different gradations—for example, three- or five-level—for both probability and impact, after which the final risk categories are determined (Rahman et al., 2024; Tran & Ross, 2021). The resulting assessments are used to rank threats and form risk management priorities, in particular, to develop decisions on the implementation of protective measures, as well as for further monitoring of residual risks (Olaes, 2025; Kirvan & Irei, 2023).

The meaning and value of a qualitative risk assessment lies in the ability to identify and understand vulnerable areas in the protection system with minimal resource expenditure (Peltier, 2005; Fenz et al., 2014). This approach is especially useful in the early stages of building an information security management system, with regular risk reassessment, as well as in situations where visualization of the current state of information security is needed for management (Karabacak & Sogukpinar, 2005; ISO/IEC, 2022). In addition, a qualitative approach is often used as a preliminary stage before a more detailed quantitative analysis (Jones, 2006; The Open Group, 2013).

Practical implementation of qualitative analysis involves the use of risk matrices, SWOT analysis, threat and vulnerability directories, as well as regular interaction with key company employees (Stoneburner et al., 2002; Shameli-Sendi et al., 2016). Support can be provided both by means of simple tables and templates in Excel, and by means of specialized tools supporting visualization and documentation of risks (ENISA, 2006; Rahman et al., 2024).

Among qualitative methods of information assessment, ISO/IEC 27005, OCTAVE (SEI) and CRAMM (CCTA) are distinguished. Each of these methods reflects its own development context, maturity level, intended goals and characteristics of the organization for which it is applied.

The ISO/IEC 27005 methodology has been developed as a practical guide for implementing the risk management process in the context of the information security management system described in ISO/IEC 27001. It does not impose a rigid analysis structure, but, on the contrary, offers a flexible approach. A special feature of this methodology is its compatibility with other ISO standards and the possibility of integration into a broader IT risk management system. ISO/IEC 27005 covers the entire assessment lifecycle, from context definition to residual risk monitoring and review

OCTAVE (Operationally Critical Threat, Asset, and Vulnerability Evaluation), developed by the Software Engineering Institute (SEI) at Carnegie Mellon University, focuses primarily on the involvement of the organization's employees and the use of internal expertise to analyze risks. Unlike standard models, OCTAVE focuses on how the organization's business processes relate to information assets, and how internal practices affect resilience to threats.

The CRAMM (CCTA Risk Analysis and Management Method), developed by the UK Central Telecommunications Authority, is one of the most formalized approaches (ENISA, 2006; ISO/IEC, 2022). It is based on an extensive library of assets, threats, and countermeasures, on which a structured risk assessment model is built (Karabacak & Sogukpinar, 2005; Fenz et al., 2014). CRAMM relies on expert assessment of the impact of threats, the presence of vulnerabilities, and the effectiveness of protective measures, which allows obtaining a qualitative risk model with the possibility of moving to a semi-quantitative form (Shameli-Sendi et al., 2016; Wangen et al., 2017). The methodology has been widely used in the UK public sector and has been applied to automate analysis using specialized software (Chockalingam et al., 2017; Rahman et al., 2024).

Table 1. Comparison of qualitative risk assessment methods

Comparison criterion	ISO/IEC 27005	OCTAVE (SEI)	CRAMM (CCTA)
Main focus	Universal risk management model	Involvement of staff and business context	Formalized analysis with threat catalogs
Analysis approach	Flexible structure, ISO 27001 support	Scenario-based, interview-based and self-assessment	Formalized using reference books
Level of automation	Requires external tools	Suitable for manual analysis	CRAMM Tool software available
Assets and threats approach	Contextual, at the choice of the organization	Linked to critical business processes	Rigid cataloging and classification
Implementation flexibility	High, scalable to any organization	Medium, requires staff involvement	Low, targeted at large organizations

Thus, the choice of methodology depends on the organization's goals, the maturity level of its processes, and available resources (Fenz et al., 2014; Wangen et al., 2017). ISO/IEC 27005 is suitable for organizations seeking to build an integrated information security management system (ISO/IEC, 2022; ENISA, 2006). OCTAVE will be effective when it is necessary to analyze risks with the participation of internal teams and with an emphasis on practices related to behavior and processes (Stoneburner et al., 2002; The Open Group, 2013). CRAMM seems appropriate in cases where a formalized and repeatable approach is required, especially in large organizations or in the presence of regulatory requirements (Karabacak & Sogukpinar, 2005; Chockalingam et al., 2017).

4. QUANTITATIVE RISK ASSESSMENT METHODOLOGY

Quantitative risk assessment methods in information security are an analytical approach based on the numerical expression of the probability of occurrence of threats and potential losses. Unlike qualitative models, which use expert categorization, quantitative methods rely on exact or probabilistic data, providing the ability to reasonably calculate expected damage, compare risks between different systems and conduct an economic analysis of costs and benefits. The main goal of these approaches is to ensure objectivity and reproducibility of results, which is especially valuable when making management decisions in organizations with a high level of process maturity.

One of the most well-known quantitative risk analysis methods is the ISRAM method, which was developed in December 2003 at the National Research Institute of Electronics and Cryptology and the Gebze Institute of Technology. This risk analysis approach and its questionnaire-based model are used to assess risks in the field of information security. The method includes two main risk components - probability and consequences, for which two separate and independent surveys are conducted. The method is based on risk modeling as a combination of probability and consequences of a security breach.

The ISRAM method is designed to analyze risks in the field of information security (Karabacak & Sogukpinar, 2005). It is designed to assess risks in complex information systems with the possibility of participation of both management and employees (Fenz et al., 2014; Wangen et al., 2017). The method consists of seven stages, the first four of which represent a preparatory phase during which surveys are formed. At the fifth stage, a questionnaire is conducted, and at the sixth, a risk analysis is performed. The final, seventh stage is an assessment of the obtained results (Karabacak & Sogukpinar, 2005). An overview of the elements of the ISRAM operating principles is shown in Figure 1. The ontology defines a risk, which includes probability and consequences (Shameli-Sendi et al., 2016). There are certain factors that affect the probability and consequences of a risk. These factors are transformed into questions of a questionnaire, which is filled out by participants

(Rosa et al., 2017). Each question has a weight, with the help of which the answers are determined. The risk table scales the questionnaire answers, and with the help of a given risk formula, the answers are converted into a numerical risk value (Rahman et al., 2024; Tran & Ross, 2021).

Figure 1. The working principle of the ISRAM method

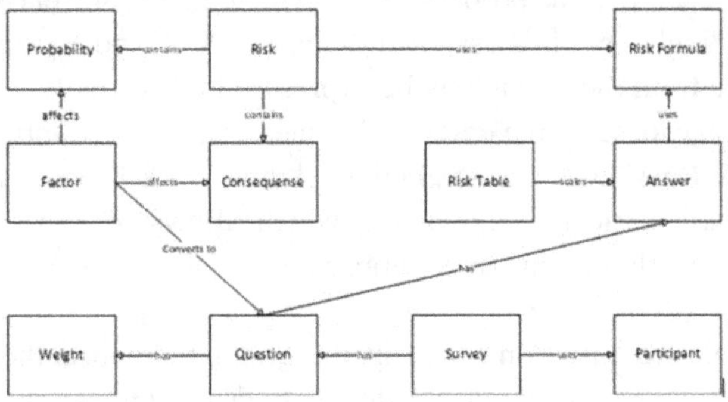

The ISRAM methodology is distinguished by its balanced approach between formal numerical assessment and the ability to involve experts at the business process level. This makes it especially valuable in conditions where precise statistical data is not available, but the organization seeks to move beyond subjective categorization. ISRAM is suitable both for the initial stage of risk management system development and for regular threat review as part of the security management cycle.

5. HYBRID RISK ASSESSMENT METHODOLOGY

Hybrid methods of information security risk assessment are an attempt to combine the strengths of qualitative and quantitative analysis, striving to ensure both the reliability of the assessment and the manageability of the process in conditions of limited resources (Chockalingam et al., 2017; Fenz et al., 2014). Unlike strictly qualitative models based on expert assessments and verbal scales, and strictly quantitative models that require precise statistical data and numerical calculations, hybrid approaches use semi-formalized methods that allow for the formation of an objective picture of risks even with incomplete data (Shameli-Sendi et al., 2016; Wangen et al., 2017). Their main purpose is to provide organizations with a tool for conducting a systematic and reproducible analysis of threats, without going beyond the available analytical capabilities (ISO/IEC, 2022; ENISA, 2006).

Hybrid methods are based on the idea of comparing expert assessments with conventional numerical scales, in which probabilities and consequences are formalized using categorical gradations, but at the same time, each gradation is assigned numerical values or weighting coefficients (Karabacak & Sogukpinar, 2005; Rosa et al., 2017). For example, the probability of an incident may be classified as "low," "medium," and "high," but it is assigned the values 1, 3, and 5 respectively. Similarly, the damage from the threat may be expressed both verbally and based on the ranges of monetary losses, time costs, or business process disruptions (Rahman et al., 2024; Tran & Ross, 2021). Then, in the risk matrix, these values are multiplied or otherwise mathematically aggregated, which allows obtaining the final risk index and subsequently ranking and comparing scenarios (Jones, 2006; The Open Group, 2013).

Hybrid models are based on formalized logic that supports the comparability of risks without the need for precise source data. This approach is widely used in corporate practice, especially in those organizations where it is necessary to justify the choice of protective measures, but there are no conditions for full-fledged quantitative measurements (Peltier, 2005; Karabacak & Sogukpinar, 2005). At the same time, hybrid methods are able to provide support for decision-making based on a balance between expertise and formalization (Fenz et al., 2014; Wangen et al., 2017).

One of the most famous examples of a hybrid model is the FAIR (Factor Analysis of Information Risk) method, developed by The Open Group (The Open Group, 2013). This methodology is based on the concept of factor analysis of risk and provides an analytical structure in which probability and impact are defined through several dependent parameters, each of which can be assessed both numerically and qualitatively-quantitatively (Jones, 2006; Kirvan & Irei, 2023). The FAIR method offers a more rigorous decomposition of risk, taking into account such components as threat frequency, asset vulnerability, business impact, and a probabilistic loss model (Olaes, 2025; Tran & Ross, 2021). At the same time, it is possible to use statistical data where they are available and expert assessments where they are not available, which makes the methodology flexible and applicable in various conditions of information security maturity (Rahman et al., 2024; Fenz et al., 2014).

A hybrid approach is often used in corporate Risk Management Frameworks (RMFs), which use multi-criteria scales that aggregate probabilities and consequences along various vector axes, including reputational, financial, legal, and operational consequences. At the same time, organizations develop their own assessment scales and correspondence tables to ensure standardized interpretation of data.

The idea behind hybrid methods is to increase the accuracy and reproducibility of risk assessment without excessive dependence on statistically accurate source data. The hybrid model allows for risk analysis to be performed on a regular basis, experts from various departments to be involved, documentation to be formalized, and audit support to be maintained. The advantage of such methods is that they combine the intuitive clarity of qualitative analysis with the formal strict logic of quantitative justification, allowing for protective measures to be justified to management or auditors.

Table 2. Components of the FAIR Risk Assessment Methodology

Component of the FAIR model	Description	Assessment methods	Notes
The trigger	The reason why the threat may occur. It includes an action or event that triggers a threat.	The assessment is based on expert judgments or statistical data.	The trigger can be external or internal.
Threat	The possibility that the events will have a harmful effect. It is an act that leads to loss.	Assessment of the likelihood of a threat (for example, through statistics or expert assessments).	It includes threat types such as external hackers, employee errors, etc.
Vulnerability	the degree of vulnerability of assets to the effects of threats. This is the probability that the threat will be used for attacks.	Asset vulnerability assessment with the help of experts or analysis of security systems.	It is important to take into account the protective mechanisms already in place in the organization.
Malicious impact	Losses that may occur if the threat is realized. It includes both direct and indirect consequences.	Impact assessment based on data on business assets and their value.	This may include financial losses, reputational damages, and legal consequences.
General risk	Assessment of the probability of losses and their impact on the organization. The total risk is calculated using a combination of the probability of threats and the impact on assets.	Risk is assessed as the product of the probability of threat, vulnerability, and impact.	It is summarized for each threat, which allows you to rank the risks.
Control measures	Measures aimed at reducing or minimizing risks. Determining the effectiveness of current or proposed protective measures.	Effectiveness assessment based on previous expertise or successful practices.	It is important to consider both current and future countermeasures.
Risk Level	The final score, which is expressed in numerical value, helps to make decisions about protection actions.	It is calculated using a model based on numerical estimates of probability and impact.	The FAIR model allows you to determine the level of risk that can be managed.

The introduction of hybrid techniques requires clearly defined scales, an asset valuation system, a description of threat scenarios, and a formalized settlement model. Analysis can be supported using software tools such as Karcher GRC, Risk

Less, or open-source solutions such as OpenFAIR. Many hybrid models include mechanisms for calibrating expert assessments, weighting mechanisms, and scenario analysis, which allows for a balanced assessment even in conditions of limited information. Such techniques play a key role in mature information security management systems, especially when it is necessary to demonstrate the cost-effectiveness of investments in protection.

6. COMPARISON OF RISK ASSESSMENT METHODS

To analyze various methods for assessing information security risks, it is important to take into account their features, advantages and limitations. Let's consider three main approaches: a qualitative methodology based on the example of ISO/IEC 27005, a quantitative methodology based on the example of ISRAM, and a hybrid methodology based on the example of FAIR. Each of them has unique characteristics that make them suitable for different types of organizations and environments.

The qualitative methodology presented by the ISO/IEC 27005 standard is focused on the use of expert assessments and subjective judgments to identify and analyze risks (ISO/IEC, 2022; ENISA, 2006). This allows an organization to assess threats and vulnerabilities using general categories and scales such as "high," "medium," and "low," which makes the approach accessible and flexible (Fenz et al., 2014; Stoneburner et al., 2002). The main advantages of this technique include its ease of use and low resource consumption. It is ideal for organizations that are just starting to implement a risk management system, or for those cases where a quick and general understanding of the situation is required (Peltier, 2005; Schneier, 2000). However, the methodology has a number of disadvantages, among which it is possible to single out the low accuracy of estimates, since they often depend on the subjective opinions of experts (Shameli-Sendi et al., 2016; Wangen et al., 2017). In addition, the lack of quantitative data makes it difficult to analyze in detail the extent of risks and make informed decisions to minimize them. This approach may be ineffective in the case of complex and multifaceted threats, where it is important to take into account the interaction of multiple factors (Chockalingam et al., 2017; Karabacak & Sogukpinar, 2005).

The ISRAM method is a survey-based quantitative approach proposed to analyze security risks of information technologies by allowing the participation of the manager and staff of the organization (Karabacak & Sogukpinar, 2005; Rahman et al., 2024). It is compliant with standards ISO/IEC 17799 and ISO/IEC 13335 (ISO/IEC, 2022; ENISA, 2006). The input of this method comes from discussions among the risk analysis team. The team decides relevant factors that may affect probability and consequence of the occurrence of a security breach along with weights of each factor (Shameli-Sendi et al., 2016; Rosa et al., 2017). ISRAM does not require any expert to carry out analysis, as the owners of the organization are adequate to assess risk factors. The outcome of the analysis is a numeric value of risk that is calculated using a formula. This method is scalable, as the inputs are not group- or scenario-specific and can be reused with minor adjustment (Tran & Ross, 2021; Wangen et al., 2017).

A hybrid technique, such as the FAIR model, is a combination of qualitative and quantitative approaches (The Open Group, 2013; Jones, 2006). It allows the use of expert assessments in combination with numerical values, which makes it possible to conduct a more accurate risk analysis without requiring a complete set of statistical data (Kirvan & Irei, 2023; Olaes, 2025). FAIR provides a framework for assessing threats, vulnerabilities, and asset impacts, which helps organizations determine not only the probabilistic characteristics of threats, but also their financial or operational impacts (Rahman et al., 2024; Tran & Ross, 2021). This makes the model more universal and suitable for a wide range of organizations, including those where risk data is insufficient. However, the technique is also not without drawbacks. For example, it requires knowledge and experience in building models and interpreting results, and it can also be time-consuming to implement, since both qualitative and quantitative data must be taken into account (Fenz et al., 2014; Chockalingam et al., 2017). Unlike ISRAM, which relies solely on statistics, FAIR provides flexibility but remains dependent on the quality of expert assessments (Peltier, 2005; Kirvan & Irei, 2023).

Each of these techniques has its own strengths and weaknesses, and the choice of the appropriate one depends on the specifics of the organization and its level of maturity in risk management. The high-quality ISO/IEC 27005 methodology is a good choice for the initial stages of risk management, as it is easy to apply and does not require complex calculations. ISRAM's quantitative methodology is more accurate and objective, but it requires high costs for data collection and analytics, which makes it suitable for large organizations with a developed infrastructure. The FAIR hybrid model combines the advantages of both previous techniques, ensuring the accuracy of quantitative analysis and the flexibility of a qualitative approach, but it also requires a high level of specialist training for its effective application.

REFERENCES

Chockalingam, S., Hadziosmanovic, D., Pieters, W., Teixeira, A., & van Gelder, P. (2017). Integrated safety and security risk assessment methods: A survey of key characteristics and applications. *arXiv*. https://arxiv.org/abs/1707.02140 DOI: 10.1007/978-3-319-71368-7_5

ENISA. (2006). *Risk management: Implementation principles and inventories for risk management/risk assessment methods and tools*. https://www.enisa.europa.eu/publications/risk-management-principles-and-inventories

Fenz, S., Heurix, J., Neubauer, T., & Pechstein, F. (2014). Current challenges in information security risk management. *Computers & Security*, *45*, 1–17. DOI: 10.1016/j.cose.2014.01.005

ISO/IEC. (2022). *ISO/IEC 27005:2022 – Information security, cybersecurity and privacy protection: Guidance on managing information security risks*. https://www.iso.org/standard/80585.html

Jones, J. (2006). *An introduction to factor analysis of information risk (FAIR)*. https://en.wikipedia.org/wiki/Factor_analysis_of_information_risk

Karabacak, B., & Sogukpinar, I. (2005). ISRAM: Information security risk analysis method. *Computers & Security*, *24*(6), 428–438. DOI: 10.1016/j.cose.2004.07.004

Kirvan, P., & Irei, A. (2023). Using the FAIR model to quantify cyber-risk. *TechTarget*. https://www.techtarget.com/searchsecurity/tip/Using-the-FAIR-model-to-quantify-cyber-risk

Olaes, T. (2025). Using the FAIR model for cyber risk quantification. *Balbix*. https://www.balbix.com/insights/fair-model-for-risk-quantification-pros-and-cons/

Peltier, T. R. (2005). *Information security risk analysis*. CRC Press. https://www.crcpress.com/Information-Security-Risk-Analysis/Peltier/p/book/9780849333466

Rahman, M. M., Kshetri, N., Sayeed, S. A., & Rana, M. M. (2024). AssessITS: Integrating procedural guidelines and practical evaluation metrics for organizational IT and cybersecurity risk assessment. *arXiv*. https://arxiv.org/abs/2410.01750

Rosa, F. D. F., Bonacin, R., & Jino, M. (2017). The security assessment domain: A survey of taxonomies and ontologies. *arXiv*. https://arxiv.org/abs/1706.09772

Schneier, B. (2000). *Secrets and lies: Digital security in a networked world*. Wiley. https://www.schneier.com/books/secrets-and-lies/

Shameli-Sendi, A., Aghababaei-Barzegar, R., & Cheriet, M. (2016). Taxonomy of information security risk assessment (ISRA). *Computers & Security*, *59*, 32–62. DOI: 10.1016/j.cose.2015.11.001

Stoneburner, G., Goguen, A., & Feringa, A. (2002). *Risk management guide for information technology systems* (NIST SP 800-30). https://csrc.nist.gov/publications/detail/sp/800-30/rev-1/final

The Open Group. (2013). *Open FAIR™ risk analysis (O-RA) standard.* https://publications.opengroup.org/c13g

Tran, C., & Ross, T. (2021). Quantitative risk assessment with the FAIR model. *LogicGate.* https://www.logicgate.com/blog/the-fair-model-an-objective-approach-to-risk-measurement/

Wangen, G., Hallstensen, C., & Snekkenes, E. (2017). A framework for estimating information security risk assessment method completeness. *Journal of Computer Security*, *25*(5), 517–544. https://link.springer.com/article/10.1007/s10207-017-0382-0

Chapter 5

Machine Learning in Information Security

Alexander Bogdan
North Caucasian Federal University, Russia

Svetlana Govorova
https://orcid.org/0000-0002-3225-1088
North Caucasus Federal University, Russia

Sergey Melnikov
https://orcid.org/0000-0002-1669-8947
North Caucasian Federal University, Russia

Egor Govorov
https://orcid.org/0009-0006-7046-8217
North Caucasian Federal University, Russia

ABSTRACT

The features of the application of machine learning methods in information security tasks and development prospects and future trends are considered. The article analyzes the application of a machine learning model to detect attacks in the field of information security using the RandomForestClassifier classifier. Various aspects of data processing are considered, including balancing classes and selecting the most significant features for building a model. This approach can be useful for developing real-time monitoring and protection systems.

DOI: 10.4018/979-8-3373-2240-7.ch005

1. INTRODUCTION

The modern world is characterized by the rapid development of information technology, which leads to an increase in the volume of processed data and an increase in the number of cyber threats. In the context of digital transformation, traditional methods of protection are no longer always able to effectively withstand new, complex attacks. That is why the use of machine learning (ML) methods in information security (IS) is becoming one of the most promising areas for improving the level of protection of information systems. Machine learning offers innovative approaches for analyzing large amounts of data, allowing you to identify hidden patterns and anomalies, which is crucial for timely threat detection. The use of ML in information security allows not only to speed up the incident detection process, but also to significantly increase the accuracy of predicting potential attacks. The relevance of this topic is due to the need to find and implement new technologies capable of providing reliable protection in a dynamic environment.

2. OVERVIEW OF CURRENT TRENDS IN INFORMATION SECURITY AND MACHINE LEARNING

In recent years, there has been an active development of a number of areas in the field of machine learning in information security.

Anomaly detection. Systems that use deep learning to analyze user behavior and network traffic can identify not only known threats, but also previously undetected ones. Such approaches significantly improve the speed and accuracy of cyber incident detection (Apruzzese, Andreolini, & Ferretti, 2022).

Big data analysis. The use of machine learning methods to process huge arrays of logs and network data contributes to the rapid identification of patterns characteristic of cyber attacks, which facilitates the adoption of timely measures (Wiranda & Sadikin, 2021)(Das & Nene, 2017).

Automation of response. Using reinforcement learning algorithms to develop systems capable of making real-time decisions to block suspicious activity and minimize damage from attacks.

The security of the ML systems themselves. With the increasing use of machine learning, attention is being paid to protecting models from attacks aimed at their disorientation and manipulation, which stimulates the development of methods to increase the stability of algorithms (Sun, Bashir, Tariq, & Xiao, 2021).

Integration with traditional security systems: Hybrid approaches combining traditional analysis methods with modern ML technologies make it possible to create multi-level protection systems, providing a high level of adaptability and reliability (Paracha, Arshad, Farah, et al., 2024).

These trends reflect the desire to create intelligent and adaptive protection systems capable of effectively countering modern cyber threats (Adebayo, Chukwurah, & Ajayi, 2025).

In modern machine learning practice, algorithms are conventionally divided into three main categories (Paracha, Arshad, Farah, et al., 2024):

1. Supervised learning. Algorithms that learn from labeled data, where each input example corresponds to the correct answer. Such methods are widely used for classification tasks (for example, spam detection in e-mail) and regression (forecasting financial indicators) (Adebayo, Chukwurah, & Ajayi, 2025). Examples: logistic regression, decision trees, random forests, neural networks.
2. unsupervised learning. Methods that work with unlabeled data, the purpose of which is to identify hidden structures and patterns in the data. Typical tasks are clustering (dividing data into groups based on similarity) and dimensionality reduction (simplifying data representation without losing significant information). Examples: k-means algorithm, principal component method (PCA), self-organization algorithms.
3. Reinforcement learning. An approach in which an agent learns by interacting with the environment, receiving rewards or punishments for his actions. This method is especially effective for tasks that require consistent decision—making, where the goal is to maximize the total reward. Examples: Q-learning, policy-based algorithms, deep reinforcement learning (Deep Q-Networks) (Paracha, Arshad, Farah, et al., 2024).

3. FEATURES OF THE APPLICATION OF MACHINE LEARNING METHODS IN SECURITY TASKS

The use of machine learning in the field of information security has a number of specific features (Stamp, 2022).

Analysis of large and heterogeneous data. Modern information security systems generate huge amounts of logs, network traffic, and events. ML algorithms allow you to process this data in real time, identifying patterns characteristic of cyber attacks or abnormal behavior.

Anomaly detection. One of the key aspects of security is the detection of deviations from normal behavior. Unsupervised methods such as clustering or outlier detection algorithms can effectively identify suspicious activities that do not correspond to the usual operating model of the system (Sun, Bashir, Tariq, & Xiao, 2021).

Dynamism and adaptability. Cyber threats are constantly evolving, which requires information security systems to quickly adapt to new types of attacks. Machine learning provides flexibility by allowing models to learn from new data and adjust their predictions over time.

Interpretability issues. When using complex models (for example, deep neural networks), there is a need to ensure transparency and interpretability of solutions. This is especially important in a security context where mistakes can lead to serious consequences.

Vulnerability to attacks on models. Machine learning algorithms can become a target for special attacks, such as data integrity attacks (adversarial attacks), when an attacker intentionally modifies input data to obtain a false result (Sun, Ding, & Jiang, 2023). This requires the development of additional protective measures and regular review of trained models.

These features emphasize the importance of an integrated approach when integrating ML methods into information security systems, where success depends on data quality, regular adaptation of algorithms, and an integrated protection strategy.

Let's consider the application of machine learning in information security.

1.1. Detection of anomalies and incidents

One of the key tasks in information security is the timely detection of anomalies in the behavior of network devices, users and systems. Machine learning allows you to analyze huge amounts of data in real time, identifying deviations from "normal" behavior. Both supervised and unsupervised methods are used here. Unsupervised methods such as clustering algorithms (for example, k-means) or autoencoders are trained on normal data, after which any deviations are recorded as anomalies. Supervised methods, such as models trained on pre-tagged data (incident/normal behavior), can classify events, increasing detection accuracy. Hybrid approaches are also known, which allow not only to detect known patterns, but also to warn about previously unknown threats. Examples of real-world applications include monitoring systems that analyze server logs, network traffic, and user activity, promptly signaling suspicious events. This approach helps to significantly reduce incident response time and minimize damage from attacks.

1.2. Spam filtering and phishing attacks detection

Spam filtering and protection against phishing attacks are among the earliest and most successful applications of machine learning methods in information security. Spam filters use algorithms such as naive Bayes classifier, decision trees, or modern neural networks to analyze the content of emails, their headers, and metadata.

These methods allow you to effectively separate legitimate messages from spam messages. Phishing Attacks - These models analyze URL characteristics, message content, and behavioral patterns to identify suspicious links and patterns typical of phishing. Additionally, ML allows you to quickly update models based on new data on phishing attacks, which increases the adaptability of the system. Companies in the field of e-mail and online banking are already using similar systems to reduce the level of threats associated with online fraud.

1.3. Analysis of network traffic and logs

Analysis of network traffic and logs is fundamental to identify hidden threats and correlate events within the infrastructure. Modern security systems generate huge arrays of logs and network data. ML algorithms are able to automatically collect, aggregate and analyze this data, revealing hidden patterns and relationships (Stamp, 2022). Clustering algorithms help group events by similarity, which makes it easier to detect anomalies in data streams. Dimensionality reduction methods, such as PCA, allow us to identify the main components that characterize attacks. Automated analysis systems allow you to quickly detect and respond to threats in real time, improving network security and preventing the spread of attacks. The practical application of such methods helps organizations not only detect incidents, but also conduct an in-depth analysis of the causes and consequences of attacks, which contributes to the development of more effective protection strategies.

1.4. Detection of malicious software

Traditional signature-based malware detection methods often fail to deal with new or modified threats (Sun, Bashir, Tariq, & Xiao, 2021). Machine learning offers an alternative approach. Machine learning models allow for static file analysis. The models analyze file characteristics (for example, code structure, metadata) to identify signs of malware. This allows you to detect even polymorphic or metamorphic software that changes its structure to bypass traditional methods. The systems monitor the behavior of programs in a controlled environment (sandboxing), analyzing the actions of the application in real time, which allows for dynamic analysis. ML helps to identify anomalies in behavior, such as unusual access to system resources or attempts to hide activity. Ensamble methods, which include combining various algorithms (for example, decision trees, neural networks, and SVM), improve the accuracy and stability of the detection system. Effective use of these methods can significantly reduce the risk of infection of corporate systems and protect critical data from malware attacks.

4. EXAMPLES OF THE USE OF MACHINE LEARNING IN THE FIELD OF INFORMATION SECURITY

To date, machine learning has been successfully applied in the field of information security. The following examples of its use are known (Paracha, Arshad, Farah, et al., 2024).

IDS/IPS systems based on machine learning.

Detection (IDS) and intrusion prevention (IPS) systems have traditionally relied on signature analysis, but the dynamics of modern threats require more flexible approaches (Zhang et al., 2019). Machine learning allows you to:

1. Detect anomalies. Unsupervised methods such as clustering and autoencoder-based algorithms are used to identify uncharacteristic behavior in network traffic and logs.
2. Integrate contextual information. Supervised models trained on historical attack data help classify events and reduce the number of false positives.
3. Adapt to the changes. Hybrid solutions combining ML with traditional methods allow the system to dynamically update the knowledge base and respond to previously unknown threats. For example, commercial solutions such as Dark trace or Vectra AI actively use ML to analyze the behavior of network devices and automatically correlate suspicious incidents.

Protection against DDoS attacks. DDoS attacks remain one of the most common threats to infrastructures. Machine learning makes it possible to increase the effectiveness of protection by: analyzing traffic anomalies: models trained on normal and abnormal network traffic patterns help to quickly identify sudden load surges or unusual request distribution patterns; responding in real time by using reinforcement learning algorithms that can dynamically adjust filters and routing policies, reducing the impact of the attack; integration with CDN and cloud solutions. Many providers, such as Cloudflare and Akamai, already use ML for automated detection and minimization of DDoS attacks, which ensures the continuity of services.

User Behavior Analysis (UEBA). UEBA (User and Entity Behavior Analytics) is a field aimed at identifying anomalies in the behavior of users and entities within an organization. The main ML features in UEBA include (Sharafaldin, Lashkari, & Ghorbani, 2018):

1. Building behavioral profiles. The systems analyze user activity data (logins, resource movements, data actions) to form a "normal" profile.

2. Identification of deviations. If inconsistencies with standard behavior patterns are detected, the system can trigger an alarm, which is especially important to prevent insider threats and account compromise.
3. Integration with SIEM systems. Modern solutions such as Splunk or IBM QRadar use ML algorithms to correlate events and provide recommendations for responding to threats.
4. Automation of vulnerability and threat detection. Automated systems based on machine learning are able not only to identify existing threats, but also to predict the emergence of new vulnerabilities.

Models that analyze data from various sources (for example, CVE databases, event logs, forums) allow you to identify potential vulnerabilities before they are exploited, which allows you to identify vulnerabilities in real time.

Classification algorithms can be used to prioritize vulnerabilities, assess the risk for a specific infrastructure, and focus efforts on critical issues.

Integration with security management systems: Many automated vulnerability management platforms use ML to generate recommendations for patches and updates, which significantly reduces the response time to threats.

5. TECHNOLOGICAL TOOLS AND PLATFORMS USED TO IMPLEMENT MACHINE LEARNING IN INFORMATION SECURITY

Machine learning in information security requires powerful tools for data processing, model building, and integration into operational systems (Barreno, Nelson, Joseph, et al., 2010). Let's look at the main libraries used in this field.

TensorFlow is one of the most popular frameworks for deep learning. It is used to create complex neural network models, including for detecting anomalies in traffic and analyzing user behavior.

PyTorch is a flexible framework for creating and training neural networks. It is well suited for malware and cyber threat analysis tasks.

Scikit-Learn is a library with a large set of machine learning algorithms. It is often used in classification, regression, and clustering tasks when analyzing network traffic and logs.

XGBoost is a powerful gradient boosting library used in predictive analytics and phishing detection systems.

ELK Stack (Elasticsearch, Logstash, Kibana) is a toolkit for log analysis and visualization that is actively used to detect anomalies with ML algorithms.

OpenAI Gym and Reinforcement Learning frameworks are used to train models operating in a dynamically changing cybersecurity environment.

These tools allow you to develop customized solutions for information security and automated threat analysis (Sun, Ding, & Jiang, 2023).

In addition to open-source solutions, there are commercial products on the market in which machine learning is actively used to ensure cybersecurity.

Darktrace uses ML to analyze network behavior and identify anomalies in real time. Vectra AI is an artificial intelligence-based solution for detecting attacks within a network. Cylance (BlackBerry Protect) is an antivirus based on predictive machine learning models that detects malware without signatures. Another cloud platform that uses ML to protect end devices from CrowdStrike Falcon attacks.

For vulnerability analysis and security monitoring, the following are used Tenable.io and Qualys– platforms that use ML to automatically search for vulnerabilities in the IT infrastructure. Splunk Enterprise Security is a SIEM system with an ML engine for event correlation and threat detection.

6. PROBLEMS AND CHALLENGES OF IMPLEMENTING MACHINE LEARNING IN INFORMATION SECURITY

One of the most common problems when using machine learning methods in information security is false positives and algorithm errors. When analyzing large amounts of data, systems often generate an excessive number of warnings, which leads to "fatigue" of analysts and a decrease in the responsiveness to real threats ("A comprehensive study on CIC-IDS2017 dataset," 2024). False alarms can occur due to inaccuracies in training samples, incorrect settings of model hyperparameters, or changing operating conditions of information systems.

Also, machine learning does not guarantee absolute accuracy. Errors can occur due to insufficient training of the model, inadequate selection of features, or even as a result of deliberate attacks, for example, using adversarial examples. This requires constant monitoring of the quality of the models, regular data updates, and algorithm adjustments to reduce errors.

In terms of privacy and data security, the use of machine learning in information security requires access to large amounts of data, often containing sensitive or confidential information. Data related to user behavior, network traffic, and other aspects of information systems are used to train models. This raises questions about compliance with privacy standards such as GDPR and the need to implement additional data protection measures at all stages of processing (Barreno, Nelson, Joseph, et al., 2010).

Models can also become the target of attacks such as poisoning attack, in which attackers intentionally introduce distortions into the training data. Data security breaches can lead to leaks and compromise not only of the model, but also of the entire infrastructure.

There are also certain difficulties in integrating into the existing infrastructure. The introduction of ML technologies into the security infrastructure involves a number of technical and organizational challenges:

1. Compatibility with legacy systems: Many organizations operate on legacy or specialized systems, which makes it difficult to integrate new ML solutions that require modern computing resources and specific APIs for data processing.
2. The need for specialized competencies: The development and maintenance of ML models requires the participation of specialists with deep knowledge in the field of data analysis, statistics and programming, which is not always available within existing IT departments.
3. Resource-intensive and scalable: The implementation of machine learning systems, especially for real-time analysis, requires significant computing resources and investments in infrastructure. This can be an obstacle for companies with limited budgets or weak technical base.

7. DEVELOPMENT PROSPECTS AND FUTURE TRENDS

In recent years, there has been an active development of research areas aimed at improving the efficiency of ML systems in information security.

The development of methods to protect against attacks on models (adversarial attacks) is becoming critically important. Researchers are working on algorithms capable of detecting and neutralizing input data manipulation, which improves the reliability of threat detection systems. Federated Learning and distributed learning are used to preserve data confidentiality, enterprises are moving to models where learning takes place on local devices, and aggregated results are used to build a common model. This approach reduces the risk of confidential information leaks and ensures data sharing between organizations without direct exchange. Another urgent task is to increase the interpretability of decisions made by complex ML models. This is necessary so that security specialists can understand the logic of the system and quickly make informed decisions when threats arise (Das & Nene, 2017).

The combination of traditional (rule-based) methods with ML algorithms makes it possible to create adaptive systems that can learn from new data and independently adjust their forecasts. Such systems promise to significantly reduce the number of false positives and increase responsiveness to new types of attacks.

The future of information security will be determined by the dynamic development of threats and the improvement of counteraction mechanisms. Cybercriminals are increasingly using automation and even artificial intelligence elements to carry out attacks. Attacks such as targeted phishing campaigns, machine learning attacks, and the emergence of new types of threats related to the expansion of IoT and cloud services are expected to increase. In the future, security systems will rely on threat forecasting, analyzing historical data and identifying patterns that allow risk assessment in advance. This will allow you to move from reactive to proactive security management, minimizing the consequences of attacks. With the dynamic change in threats, security systems will become more flexible, able to adjust their rules and algorithms in real time in response to detected anomalies. This approach will significantly increase the resilience of the infrastructure to new attacks (Sun, Bashir, Tariq, & Xiao, 2021).

Artificial intelligence (AI) will continue to have a significant impact on information security, transforming traditional approaches to protection.

AI-based systems are capable not only of detecting threats, but also of independently making decisions to neutralize them. This is especially important in the context of the need for rapid response to incidents, when delays can lead to serious consequences.

AI helps create models that analyze the behavior of users, networks, and applications, identifying even the most subtle deviations from the norm. Such systems can predict potential threats even before they are implemented, which significantly reduces the response time.

With the growing number of connected devices, the development of 5G and IoT, AI is becoming an integral part of integrated protection systems. It is capable of processing huge amounts of data, providing real-time monitoring and decision support based on analytics.

The development of AI in information security also requires attention to issues of ethics and transparency. The development of standards and regulations for the use of AI will help minimize the risks of misuse and protect the rights of users.

8. SOLVING THE PRACTICAL PROBLEM OF USING MACHINE LEARNING MODELS TO DETECT ANOMALIES IN NETWORK TRAFFIC

The purpose of this work was to use machine learning to protect against cyber attacks, such as detecting anomalies in network traffic, filtering spam, detecting phishing attacks, and protecting against malware.

For the practical part of the study, the CIC-IDS2017 dataset was selected, which is one of the most advanced datasets for network security attack detection tasks. Unlike older datasets such as KDD'99 and KDD+, CIC-IDS2017 presents data obtained under more realistic conditions, including interactions between different devices such as servers, workstations, and various operating systems. This dataset features both external and internal networks with a variety of devices. Among them are various versions of Windows (7, 8.1, 10), Ubuntu (versions 12, 14, 16), as well as MAC and Kali Linux, which allows you to analyze the interaction between devices with different operating systems and architectures. This significantly increases the representativeness of the data, as it reflects a wider range of real devices used in modern network infrastructures.

The dataset includes various types of attacks such as DDoS, port scanning, phishing and SQL injection, which makes it very diverse and suitable for testing machine learning methods. Unlike older datasets such as KDD'99, which was assembled in the 90s, CIC-IDS2017 is relatively new and reflects current threats in modern information systems.

This dataset was taken for the purpose of conducting practical research using relevant and diverse data, which allows us to obtain more accurate and applicable results for real-world operating conditions of network systems.

To improve the accuracy and effectiveness of machine learning models, it is important to select only those features that actually affect predictions. In the process of data analysis, it often turns out that not all features are equally significant, and some of them may only make noise in the model. Assessing the importance of features helps eliminate unnecessary data, which simplifies the model and speeds up its learning.

The Random Forest algorithm was used to determine the most significant features. This method is effective for assessing the importance of features, as it builds multiple decision trees and analyzes how each feature affects the results. The advantage of Random Forest lies in its ability to account for interactions between features and avoid overfitting.

As a result of the model, the 20 most significant features that have the greatest impact on the classification were identified:

1. Fwd IAT Min (0.1367)
2. Init Bwd Win Bytes (0.0836)
3. Flow IAT Min (0.0697)
4. Fwd IAT Std (0.0563)
5. Fwd IAT Mean (0.0450)
6. Flow IAT Std (0.0412)
7. Flow Duration (0.0384)
8. Init Fwd Win Bytes (0.0359)
9. Flow IAT Mean (0.0356)
10. Fwd IAT Max (0.0331)
11. Fwd IAT Total (0.0299)
12. Fwd Packets/s (0.0284)
13. Flow Packets/s (0.0283)
14. Bwd Packets/s (0.0272)
15. Fwd Header Length (0.0271)
16. Avg Packet Size (0.0248)
17. Flow Bytes/s (0.0222)
18. Flow IAT Max (0.0220)
19. Fwd Packets Length Total (0.0216)
20. Packet Length Max (0.0210)

For further analysis of the significance of features in the machine learning process, it is necessary to assess the degree of relationship between them. Let's construct a correlation matrix for the 20 most significant features identified based on their importance to the model. The construction of a correlation matrix allows us to assess how various features affect the model, identify possible redundant data and improve the overall interpretability of the model.

Using the correlation matrix it is necessary to:

1. Identify the relationships between features: The correlation matrix helps to explore the relationships between features, which is important for understanding their contribution to the model. If the features are strongly correlated, this can affect the training of the model and its accuracy.
2. Detect multicollinearity. Multicollinearity is a situation where several features strongly correlate with each other. This can lead to distorted results, especially in linear models. The construction of the correlation matrix makes it possible to identify such features, which makes it possible to decide which of them should be removed or combined.

3. Optimize the feature set. Correlation analysis helps to reduce the dimension of data by eliminating highly correlated features that may not add new information to the model, while increasing its efficiency.

Figure 1 shows the correlation matrix for the 20 most significant features.

Figure 1. Correlation matrix

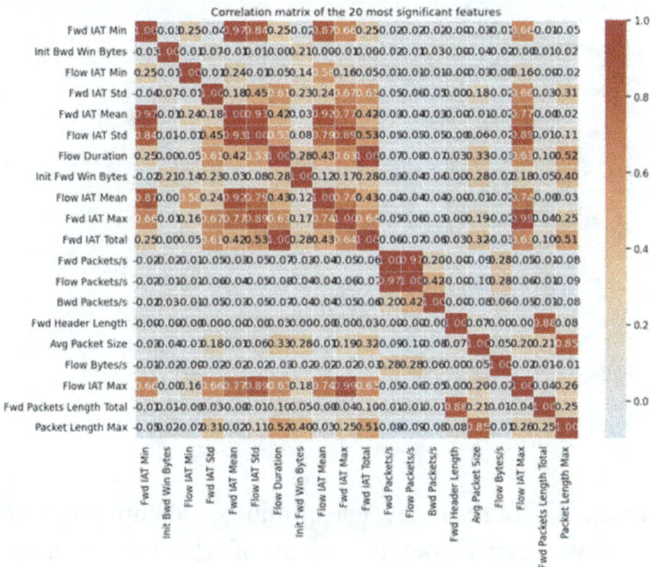

The analysis showed that there is a high correlation between a number of features. For example, Avg Packet Size and Packet Length Max have a correlation of 0.85, which indicates a strong relationship between packet size and maximum packet length.

Fwd IAT Max and Flow IAT Max show an almost perfect correlation (0.99), which indicates the similarity of the behavior of these features during data transmission.

The high correlation between these features allows us to consider the possibility of combining or excluding them to simplify the model without losing significant information.

At this stage of the work, features with a high correlation exceeding 0.8 were removed in order to avoid multicollinearity and improve the accuracy of the model. Features such as Avg Packet Size, Flow IAT Max, Flow IAT Std, and Flow IAT Min were excluded, as their correlations with other features were too high, which could lead to information redundancy in the model. The remaining signs will be used for further analysis.

There is a significant class imbalance in the presented dataset, which may affect the effectiveness of the machine learning model (see Figure 2).

Figure 2. Class distribution before balancing

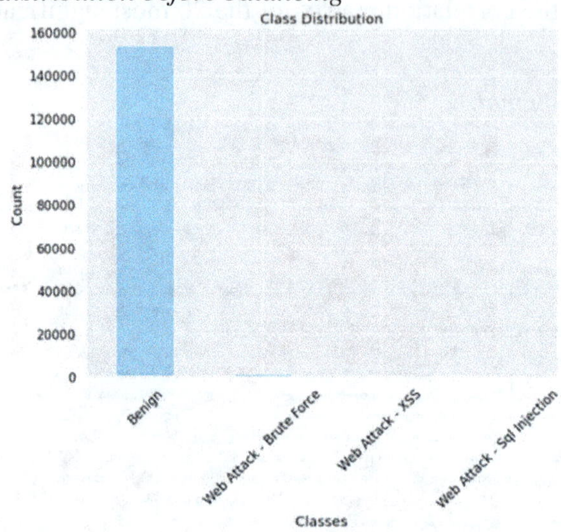

To improve the quality of the model and eliminate the impact of class imbalance, a balancing process was carried out, as a result of which the number of examples for the Benign class was reduced to 5,000 (Figure 3). This number was chosen taking into account the need to preserve the diversity of training data, but at the same time reduce the dominance of the Benign class in training. The classes Web Attack - Brute Force, Web Attack - XSS and Web Attack - Sql Injection remained unchanged, as their initial number was already quite small.

Figure 3. Class distribution after balancing

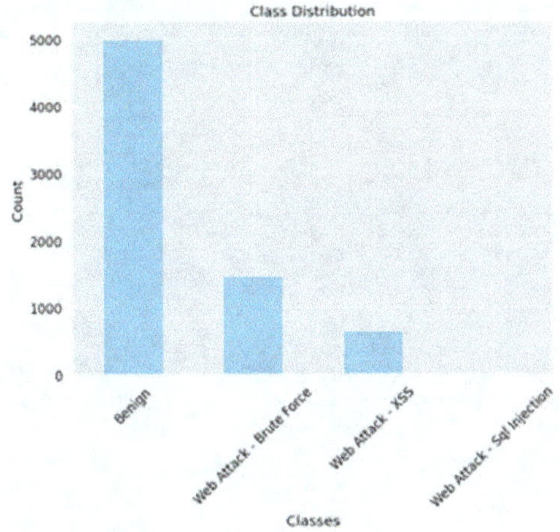

Class balancing helps to improve the model's ability to recognize less represented classes, providing more equitable training and increasing its overall effectiveness in classifying rare attacks.

To evaluate the effectiveness of the model and prevent overfitting, the initial data set was divided into training and test samples in a ratio of 80:20. This approach allows an objective assessment of the model's performance based on an independent sample, which helps to obtain reliable results and increase the validity of conclusions.

An analysis of existing research has shown that RandomForestClassifier provides high execution speed with a relatively low computational load, which is critical for attack detection systems. These characteristics make the model an effective choice for tasks that require rapid analysis of big data in real time. The training schedules of the model are shown in Figure 4.

Figure 4. Model training schedules

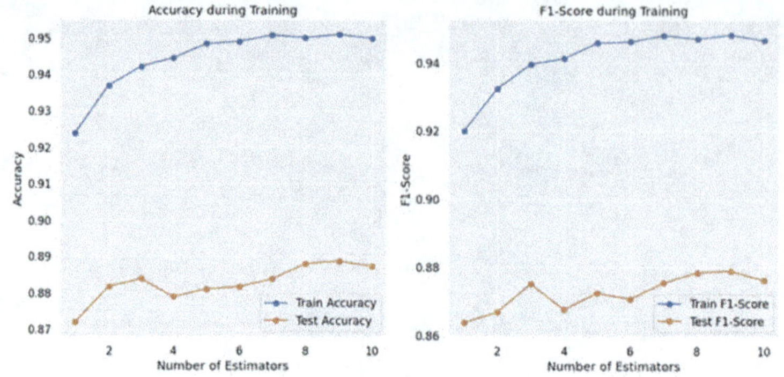

After training the RandomForestClassifier model on a test sample, the following results were obtained:

For the Benign (normal traffic) class, precision was 0.99, completeness (recall) was 1.00, and F1 measures were 0.99. This indicates an almost perfect classification of normal data.

For the Web Attack - Brute Force class, the accuracy was 0.66, the completeness was 0.90, and the F1 measures were 0.77, which shows a good balance between accuracy and completeness for this type of attack.

For the Web Attack - Sql Injection class, all metrics were 1.00, which means an error-free classification of SQL injection attacks.

For the Web Attack - XSS class, the accuracy was 0.58, the completeness was 0.22, and the F1-measures were 0.32. These results indicate the difficulty of correctly classifying XSS attacks, especially due to the low completeness.

The overall accuracy of the model in the test sample was 0.89, which indicates a good overall classification level. At the same time, indicators for less represented classes, such as XSS, require further model improvements.

For further analysis, the "Benign" class was removed, which allowed us to focus solely on the characteristics inherent in malicious activity and exclude samples related to normal traffic. This approach contributes to a more accurate assessment of methods and algorithms aimed at detecting attacks, allowing you to minimize the impact of normal traffic and focus on modeling the behavior of attackers. Figure 4 shows the classification of attacks.

Figure 5. Classification of attacks

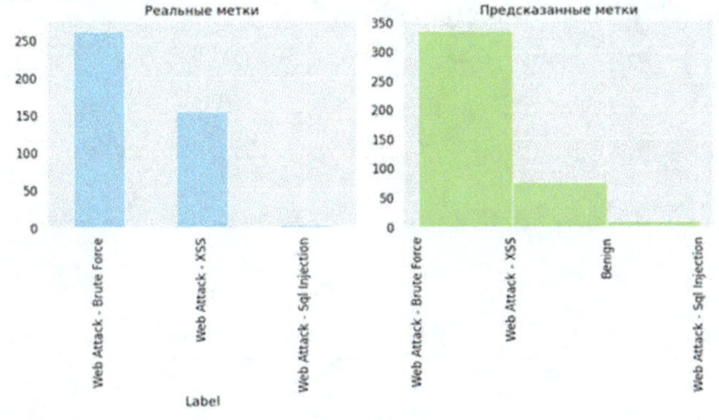

The graph (Figure 5) shows that the model, in some cases, mistakenly classifies XSS-type attacks as Bruteforce-type attacks. However, despite this, the model shows good potential for further improvement, as identifying this kind of confusion provides valuable information for fine-tuning the algorithm. This indicates the possibility of improving classification accuracy through deeper feature analysis and optimization of model parameters, which, in turn, will improve its ability to distinguish even similar types of attacks.

9. CONCLUSION

The study analyzed the application of a machine learning model to detect attacks in the field of information security using the RandomForestClassifier classifier. In the course of the work, various aspects of data processing were considered, including balancing classes and selecting the most significant features for building a model. Despite the high class imbalance in the original dataset, significant results were achieved in classification accuracy.

The model showed good performance in detecting various types of attacks, with high accuracy for classes such as “Web Attack - Sql Injection”, and with lower accuracy for more complex attacks, for example, XSS. The revealed confusion between XSS and Bruteforce attacks opens up opportunities for further optimization, which will improve classification accuracy in real conditions.

This approach can be useful for developing real-time monitoring and protection systems, as it combines high execution speed and efficiency in conditions of changing data. In the future, we can expect to improve the model through the use of more sophisticated data preprocessing methods, hyperparameter settings, and the use of more powerful machine learning algorithms.

REFERENCES

Adebayo, A. S., Chukwurah, N., & Ajayi, O. O. (2025). Artificial intelligence and machine learning algorithms for advanced threat detection and cybersecurity risk mitigation strategies. *Engineering and Technology Journal*, *10*(3), 4080–4094. DOI: 10.47191/etj/v10i03.18

Apruzzese, G., Andreolini, M., Ferretti, L., Marchetti, M., & Colajanni, M. (2022). Modeling realistic adversarial attacks against network intrusion detection systems. *Digital Threats : Research and Practice*, *25*(4), 46. Advance online publication. DOI: 10.1145/3469659

Barreno, M., Nelson, B., Joseph, A. D., & Tygar, J. D. (2010). The security of machine learning. *Machine Learning*, *81*(2), 121–148. DOI: 10.1007/s10994-010-5188-5

Das, S., & Nene, M. J. (2017). A survey on types of machine learning techniques in intrusion prevention systems. *2017 2nd International Conference on Wireless Systems and Networks (WiSPNET)*, 1–5. DOI: 10.1109/WiSPNET.2017.8300169

Paracha, A., Arshad, J., Farah, M., & Ismail, K. (2024). Machine learning security and privacy: A review of threats and countermeasures. *EURASIP Journal on Information Security*, *10*(1), 10. Advance online publication. DOI: 10.1186/s13635-024-00158-3

Sharafaldin, I., Lashkari, A. H., & Ghorbani, A. (2018). Toward generating a new intrusion detection dataset and intrusion traffic characterization. *Proceedings of the 10th International Conference on Information Systems Security and Privacy*, 108–116. DOI: 10.5220/0006639801080116

Stamp, M. (2022). *Introduction to machine learning with applications in information security* (2nd ed.). Chapman and Hall/CRC., DOI: 10.1201/9781003264873

Sun, N., Ding, M., Jiang, J., Xu, W., Mo, X., Tai, Y., & Zhang, J. (2023). Cyber threat intelligence mining for proactive cybersecurity defense: A survey and new perspectives. *IEEE Communications Surveys and Tutorials*, *25*(4), 3273282. DOI: 10.1109/COMST.2023.3273282

Sun, Y., Bashir, A. K., Tariq, U., & Xiao, F. (2021). Effective malware detection scheme based on classified behavior graph in IIoT. *Ad Hoc Networks*, *120*, 102558. DOI: 10.1016/j.adhoc.2021.102558

Wiranda, N., & Sadikin, F. (2021). Machine learning for security and security for machine learning: A literature review. *2021 International Conference on Information and Communications Technology (ICOIACT)*, 1–7. DOI: 10.1109/ICOIACT53268.2021.9563985

Zhang, H., Zhao, B., Yuan, H., Zhao, J., Yan, X., & Li, F. (2019). SQL injection detection based on deep belief network. *ACM Transactions on Internet Technology*, *19*(4), 46. Advance online publication. DOI: 10.1145/3331453.3361280

Chapter 6
Using Probabilistic Models to Assess and Enhance Information Security

Maria Lapina
https://orcid.org/0000-0001-8117-9142
North Caucasian Federal University, Russia

Alina Bagautdinova
https://orcid.org/0009-0003-4952-4497
North Caucasus Federal University, Russia

Irina Pykhtina
https://orcid.org/0009-0009-6635-2387
North Caucasian Federal University, Russia

ABSTRACT

The article explores the possibilities of using probabilistic models to assess and improve information security. The theoretical foundations of the probabilistic approach are considered, including methods of risk analysis, threat modeling and prediction of potential vulnerabilities. Special attention is paid to the use of models such as Bayesian networks, Markov chains, and Monte Carlo simulations, which allow quantifying the likelihood of incidents and their consequences. The paper presents practical examples of the use of probabilistic models for analyzing the security of information systems, as well as recommendations for their integration into information security management processes.

DOI: 10.4018/979-8-3373-2240-7.ch006

INTRODUCTION

Modern cyber threats are developing rapidly, becoming more complex and dangerous, which is why information protection requires new approaches (Alali et al., 2018; Figueira, Bravo, & López, 2020). Outdated methods, such as signature analysis, which does not recognize unknown attacks, or strict firewall rules, do not have time to adapt to new threats (Feng & Li, 2011; Wang, 2005). Manual incident handling is time-consuming, and risk assessment based on past data does not take into account the variability of attacking tactics (Mo, Beling, & Crowther, 2009; Kondakci, 2007). Rigid access control systems based on fixed roles and reactive measures aimed at eliminating consequences rather than preventing them are also inferior in effectiveness (ElSalamouny, 2011; Lee, 2011). In addition, disparate defenses without mutual integration reduce the overall resilience of the system to attacks (Odunola, Olawumi, & Adesola, 2015; Yermalovich & Mejri, 2020). In such conditions, traditional approaches can no longer cope, requiring more flexible and intelligent solutions (Carter, Idika, & Streilein, 2014; Shaikh et al., 2023).

Table 1. Traditional methods of ensuring information security

The traditional method	Description	Advantages	Disadvantages
Signature analysis	Threat detection based on pre -known patterns (signatures)	The ability to correctly identify and block malware or attacks that have already been studied and described in the signature database; ease of implementation	Ineffective against new, unknown attacks (zero-day)
Static rules and policies	Strict firewall and access control settings based on fixed rules	Predictability of work	Inflexibility in the face of changing threats
Manual incident management	Analyzing logs and responding to threats manually	The possibility of in-depth analysis and decision-making based on expert assessment	Requires significant resources; slow response; dependence on the human factor
Deterministic risk models	Risk assessment based on historical data and static parameters	clarity and transparency of the model	Ignores the dynamics of changes in threats and vulnerabilities; low adaptability

continued on following page

Table 1. Continued

The traditional method	Description	Advantages	Disadvantages
Fixed access systems	Assigning access rights based on static fields	separation of rights	Does not adapt to a change in context (for example, temporary access in emergency cases)
The reactive approach	The focus is on eliminating the effects of attacks, not preventing them.	Allows you to eliminate the consequences of the attack and restore system performance	Delayed response; vulnerability before the problem is fixed; high recovery costs
Isolated protection systems	Lack of integration between different security systems	Ease of learning how to manage components	Reduced overall protection effectiveness due to lack of interaction;

The table shows that classical approaches to information security have a number of advantages — they are understandable, stable and time-tested. However, in the context of the rapid growth of cybercrime, their effectiveness drops sharply if they are not combined with innovative technologies.

Modern data protection is based on fundamentally different solutions. For example, behavioral analysis tracks deviations in user actions or system performance, helping to identify previously unknown threats such as zero-day attacks or hidden insider activities. Unlike static rules, this method dynamically adapts to changing conditions through continuous data collection and processing.

Machine learning plays an equally important role. It not only automates malware detection, but also predicts potential attacks by analyzing huge amounts of information and identifying hidden patterns (Alali et al., 2018; Figueira, Bravo, & López, 2020; Shaikh et al., 2023). This significantly speeds up the response and reduces dependence on manual labor of specialists (Mo, Beling, & Crowther, 2009; Kondakci, 2007).

DLP systems are actively used to protect confidential data, which monitor all possible leakage channels—from corporate mail to cloud storage and removable media (Feng & Li, 2011; Odunola, Olawumi, & Adesola, 2015). This is especially critical for companies operating under strict regulations, such as the GDPR (Kondakci, 2015; Yermalovich & Mejri, 2020).

Another powerful tool is Big Data Analytics, which is capable of processing data from multiple sources: network traffic, event logs, and security sensors (Carter, Idika, & Streilein, 2014; Nicol, Sanders, & Trivedi, 2004). It can be used to identify complex targeted attacks (APT) that go unnoticed when using traditional methods (Lee, 2011; ElSalamouny, 2011).

Cloud technologies also make a significant contribution to information security by offering scalable and flexible solutions (Wang, 2005; Shaikh et al., 2023). Built-in encryption, access control, and threat monitoring mechanisms allow you to quickly adapt protection to new challenges (Alali et al., 2018; Figueira, Bravo, & López, 2020).

Predictive analytics complements this arsenal by using historical data and ML algorithms to prevent attacks before they occur (Mo, Beling, & Crowther, 2009; Feng & Li, 2011). As a result, companies are moving from a reactive to a proactive security model (Kondakci, 2007; Odunola, Olawumi, & Adesola, 2015).

The main advantage of modern approaches is not only enhanced protection, but also resource savings due to automation of routine operations and intelligent risk analysis (Shaikh et al., 2023; Carter, Idika, & Streilein, 2014).

Table 2. Modern methods of machine learning support

Method	Description	Application examples	Advantages	Disadvantages
Behavioral analysis	Detecting anomalies in the behavior of users or systems	Attack Detection zero-day, phishing prevention, internal threat detection	Adaptability to new threats	Requires large amounts of data for training
Machine Learning (ML)	Using algorithms for data analysis and threat forecasting	Malware classification, optimization of firewall rules, attack prediction	Process automation, high precision	Dependence on data quality and algorithms
DLP Systems	Control and prevention of confidential information leaks	Personal data protection, prevention of intellectual property leaks	Compliance with regulatory requirements	The difficulty of finding staff to set up and manage
Big Data Analysis	Processing large amounts of data to identify patterns and threats	Detection APT attacks, network traffic analysis, security optimization	The ability to process huge amounts of data	It requires powerful computing resources
Automation (SOAR)	Integration of security tools for process automation	Automatic incident response, threat analysis, attack blocking	Reducing response time, reducing the burden on specialists	High cost of implementation
The Blockchain	Ensuring data immutability and transparency	Log protection, digital identity management, integrity assurance	High resistance to falsification	Limited scalability, high complexity of implementation
Cloud solutions	Using cloud platforms to protect data and applications	Data encryption, access control, threat monitoring	Flexibility, scalability, built-in security tools	Dependence on a cloud service provider

continued on following page

Table 2. Continued

Method	Description	Application examples	Advantages	Disadvantages
Predictive analytics	Threat forecasting based on historical data and ML	Prevention attacks before their implementation, risk analysis	Early detection of threats, reducing the likelihood of successful attacks	It requires accurate data and high-quality models.
Artificial Intelligence (AI)	Using AI to analyze complex scenarios and automate security	Network traffic analysis, detection of suspicious activity, automation of protection	The ability to process complex data	High the cost of integration

Today's information security realities require comprehensive solutions combining time-tested approaches with innovative technologies (Alali et al., 2018; Sommestad, Ekstedt, & Johnson, 2010). Classic security tools such as antiviruses and firewalls are still an important first line of defense, providing a basic level of security (Wang, 2005; Kondakci, 2007). However, in the context of the rapid evolution of cyber attacks, these measures are no longer sufficient (Feng & Li, 2011; Mo, Beling, & Crowther, 2009).

Modern technologies, including neural network algorithms, real-time data processing, and intelligent automation systems, significantly expand security capabilities (Shaikh et al., 2023; Figueira, Bravo, & López, 2020). They allow not only to quickly identify new types of threats, but also to predict potential attacks, minimizing the human factor (ElSalamouny, 2011; Lee, 2011). For example, machine learning-based systems are able to analyze behavioral patterns and detect anomalies that go unnoticed when using traditional signature methods (Carter, Idika, & Streilein, 2014; Odunola, Olawumi, & Adesola, 2015).

The integration of different approaches is of particular value. The combination of reliable classical solutions with flexible modern technologies creates a multi-layered protection system capable of resisting both massive automated attacks and targeted complex threats (Yermalovich & Mejri, 2020; Kondakci, 2015). This synthesis allows organizations not only to respond promptly to incidents, but also to prevent them, significantly increasing the overall stability of the infrastructure (Nicol, Sanders, & Trivedi, 2004; Shaikh et al., 2023).

At the same time, it is important to understand that the introduction of advanced technologies does not eliminate the need for fundamental security measures, but rather complements them by creating a more reliable and adaptive defense mechanism (Alali et al., 2018; Feng & Li, 2011). It is precisely this balanced approach that is becoming a key factor in successfully countering modern cyber threats (Mo, Beling, & Crowther, 2009; Carter, Idika, & Streilein, 2014).

THEORETICAL FOUNDATIONS OF THE PROBABILISTIC APPROACH IN INFORMATION SECURITY

In the field of information security, probabilistic models serve as a powerful mathematical tool for dealing with uncertainty. Unlike deterministic approaches, they do not use exact values, but statistical patterns, which is especially valuable when assessing cyber threats. The bottom line is that events such as break-ins, information leaks, or equipment failures are random in nature - they cannot be predicted with absolute certainty, but the probability of occurrence can be calculated based on three key factors: archived incident data, expert opinions, and mathematical analysis of distributions.

The practical significance of these models is manifested in several aspects. First, they translate abstract threats into specific numerical indicators, allowing an objective comparison of different types of risks among themselves. Secondly, they make it possible to make forecasts - to assess not only the chances of an attack, but also the scale of possible damage. Finally, such an analysis becomes the basis for the rational allocation of the security budget, helping to allocate resources to those countermeasures that will maximize risk reduction at minimal cost. This is especially true for organizations with limited capacity to finance protective mechanisms.

Basic principles of working with probabilistic models

1. Assessment of the likelihood of incidents

One of the key tasks in information security is to assess the likelihood of incidents such as cyber attacks, data leaks, or system failures. The following approaches are used for this:

Historical data: An analysis of past incidents allows you to estimate the frequency of their occurrence. For example, if there have been 10 attacks on the system over the past year, we can assume that the probability of an attack next month is about 0.83 (10/12).

Expert estimates: In cases where historical data is insufficient, experts are involved who, based on their experience, assess the likelihood of certain events.

Statistical methods: Using probability distributions (for example, normal, Poisson, or exponential) to model random events.

2. Predicting impacts and vulnerabilities

Probabilistic models are also used to predict the consequences of incidents and identify potential vulnerabilities. For example, Bayesian networks can be used to model the relationship between various factors, such as the presence of vulnerabilities, the level of system protection, and the likelihood of a successful attack.

An example of predicting the consequences:

If the probability of a successful attack on the system is 20%, and the potential damage is estimated at $100,000, then the expected losses can be calculated as:

Expected losses=0.20×100,000=20,000 dollars.

Advantages of the probabilistic approach over traditional methods

Traditional risk analysis methods are often based on qualitative assessments such as “high”, “medium” or “low” risk. The probabilistic approach offers more accurate and objective tools for dealing with risks. Its advantages include:

Quantification: Probabilistic models allow you to express risks in numerical form, which simplifies their analysis and comparison.

Flexibility: Models can be adapted to different scenarios and types of threats.

Forecasting: The ability to predict the probability and consequences of incidents before they occur.

Integration with other methods: Probabilistic models can be combined with machine learning and artificial intelligence methods to improve the accuracy of analysis.

An example of using a probabilistic model

Let's look at an example of using a Bayesian network to analyze data leakage risks.

Network nodes

Vulnerability of the system (yes/no).

The presence of an attack (yes/no).

Data leakage (yes/no).

Probabilities

Vulnerability probability: 30%.

The probability of an attack in the presence of vulnerability: 70%.

The probability of data leakage during a successful attack: 50%.

Calculation

The probability of data leakage:

$P(Leak) - P(Vulnerability) \times P(Attack \mid Vulnerability) \times P(Leakage \mid The\ attack) =$

$=0.3 \times 0.7 \times 0.5 = 0.105(10.5\%)$.

The probabilistic approach provides powerful tools for information security risk analysis and management. It allows you to quantify the likelihood of incidents, predict their consequences, and identify vulnerabilities. Compared to traditional methods, probabilistic models are more accurate, flexible, and can be integrated with modern technologies.

Basic probabilistic models used in information security

Bayesian networks

Principles of construction and use

Bayesian networks are graphical models that represent probabilistic relationships between random variables. They consist of nodes (probability variables) and edges (dependencies between variables). The basis for working with Bayesian networks is Bayes' theorem, which allows you to update the probabilities of events based on new data. The Bayes formula looks like this:

$$P(A \mid B) = P(B)P(B \mid A) \cdot P(A),$$

where:

P(A|B) is the a posteriori probability of event A under the condition of event B;

P(B|A) is the probability of an event

P(A) and P(B) are a priori probabilities of events

Application examples for analyzing dependencies between threats and vulnerabilities

Let's look at an example of using a Bayesian network to analyze data leakage risks.

Network nodes:

Vulnerability of the system (yes/no).

The presence of an attack (yes/no).

Data leakage (yes/no).

Probabilities:

The likelihood of vulnerability:

$P(Vulnerability) = 0.3$.

The probability of an attack in the presence of a vulnerability:

$P(Attack \mid Vulnerability) = 0.7$.

The probability of data leakage during a successful attack:
$P(Leakage \mid Attack) = 0.5.$

Calculation:

The probability of data leakage:
$P(Leak) = P(Vulnerability) \times P(Attack \mid Vulnerability) \times P(Leakage \mid The\,attack) =$
Markov chains

Modeling transitions between security system states

Markov chains are mathematical models that describe a sequence of events where the probability of each event depends only on the current state. In the context of information security, Markov chains are used to model transitions between system states such as "secure state", "attack has begun", "attack is successful" and "system is compromised".

An example of a Markov chain for attack simulation:
States:
S1: Safe condition.
S2: The attack has begun.
S3: The attack is successful.
S4: The system is compromised.
The transition matrix:
where
Pij is the probability of transition from the Si state to the Sj state.

Use to predict the development of attacks

The Markov chain can be used to predict the development of an attack. For example, if the system is in state

S1, then the probability that after two steps it will be in state S3 is calculated as:
The

Monte Carlo simulations method

Simulation of random events and risk assessment

The Monte Carlo method is a numerical method that uses random samples to model complex systems and assess risks. In information security, it is used to analyze scenarios where an accurate analytical solution is impossible due to the complexity of the system.

Usage examples for analyzing complex scenarios

Let's look at an example of assessing the risk of a DDoS attack on a server.
Model parameters:
Probability of an attack per day: p=0.01
Attack damage: L=100,000
L=100,000 dollars.
Number of days: N=365

Simulation:

We generate random numbers
ri (from 0 to 1) for each day.
If ri<pi, we believe that an attack has occurred.
Let's sum up the damage for all the attacks.

Results:

After 10,000 simulations, the average damage was $365,000.

Bayesian networks, Markov chains, and the Monte Carlo method are powerful tools for analyzing information security risks. Bayesian networks can model the relationships between threats and vulnerabilities, Markov chains can predict the development of attacks, and the Monte Carlo method can assess risks in complex scenarios. These models help information security professionals make informed decisions and effectively manage risks.

Practical application of probabilistic models

Probabilistic models are an important tool for analysis and forecasting in various fields, including cybersecurity. Let's look at three examples of their practical application: assessing the risks of data leakage using Bayesian networks, predicting the time until the next attack using Markov chains, and analyzing the system's resistance to DDoS attacks using the Monte Carlo method.

Example 1: Assessing the risks of data leakage using Bayesian networks

Bayesian networks are graphical models that allow you to take into account the probabilistic relationships between various factors. They are especially useful for assessing the risks of data leakage, where many interrelated variables need to be considered.

Model:

1. Network nodes:

   V: Presence of vulnerabilities in the system (yes/no).
   A: The probability of an attacker attack (high/low).
   E: Employee errors (yes/no).
   P: Effectiveness of protection measures (high/low).
   L: Data leakage risk (high/low).

2. A priori probabilities:

   P(V = yes) = 0.2.
   P(A = high) = 0.1.
   P(E = yes) = 0.15.
   P(P = high) = 0.8.

3. Conditional probabilities:

   If V = yes, A = high, E = yes, P = low, then P(L = high) = 0.9.
   If P = high, then P(L = high) = 0.1.

Calculation example:

Let's assume that vulnerabilities have been discovered in the system (V = yes), but security measures are ineffective (P = low). Then the probability of data leakage will be:

P(L = high | V = yes, P = low) = 0.5.

If the protection measures are strengthened (P = high), the probability of leakage is reduced to 10%.

Conclusion:

Bayesian networks allow quantifying the impact of various factors on the risk of data leakage and making informed decisions to minimize it.

Example 2: Predicting the time until the next attack using Markov chains

Markov chains are mathematical models that describe a sequence of events where the probability of each event depends only on the current state. They are useful for predicting the time before the next attack on the system.

Model:

1. System status:

   S0: No attack.
   S1: The attack has begun.

2. The transition matrix:

   Where:
   p00 = 0.9: The probability of remaining in the S0 state.
   p01 = 0.1: Probability of moving from S0 to S1.
   p10 = 0.5: Probability of returning to S0 after an attack.
   p11 = 0.5: The probability of continuing the attack.

3. Predicting the time until the next attack:

   The average time until the transition from S0 to S1 is calculated as:
   T = 1p01 = 10.1 = 10 units of time

Conclusion:

The model allows you to predict the frequency of attacks and plan protection measures. For example, if attacks occur on average every 10 units of time, you can strengthen system monitoring during these periods.

Example 3: Analysis of the system's resistance to DDoS attacks using the Monte Carlo method

The Monte Carlo method is a numerical method that uses random samples to analyze complex systems. It is used to assess the system's resilience to DDoS attacks, where it is important to take into account the random nature of the load on servers.

Model:

1. System parameters:

   N: Number of requests per second (normal distribution, $\mu = 1000$, $\sigma = 200$).
   C: Server bandwidth (1500 requests per second).

2. Simulation:

10,000 random values of N are generated.

For each value, it is checked whether N exceeds the bandwidth of C.

3. Assessment of the probability of failure:

For example, if in 500 cases N > C, then P(failure) = 0.05.

Conclusion:

The Monte Carlo method allows you to estimate the probability of a system failure during a DDoS attack. For example, if the probability of failure is 5%, you can take measures to increase throughput or implement mechanisms to repel attacks.

Model	Advantages	Application areas
Bayesian networks	Consideration of complex relationships between factors.	Assessment of data leakage risks.
	The ability to update probabilities based on new data.	Troubleshooting the systems.
	Visual graphical representation of dependencies.	Medical diagnostics (assessment of the probability of diseases).
	Quantification of the influence of factors on the result.	Financial analysis (assessment of investment risks).
Markov Chains	Ease of modeling sequential events.	Predicting the time until the next attack.
	The ability to analyze the long-term behavior of the system.	Modeling the operation of network protocols.
	Ease of interpretation of the results.	Reliability analysis of technical systems.
	Efficiency for systems with discrete states.	Forecasting weather conditions.
The Monte Carlo method	The ability to analyze complex systems with random parameters.	Analysis of resistance to DDoS attacks.
	High accuracy with a large number of simulations.	Assessment of risks in the financial sector (for example, the cost of options).
	Flexibility in modeling different scenarios.	Experiment planning and hypothesis testing.
	The ability to account for uncertainties and random factors.	Optimization of production processes.

Each of these models has its own advantages and applications. Combining them allows you to create comprehensive strategies for managing risks and increasing the resilience of systems in the face of cyber threats.

The study confirms that probabilistic models provide a viable approach for transitioning from subjective assessments to quantitative risk analysis in information security. Techniques such as Bayesian networks and Markov chains enable not only threat forecasting but also dynamic evaluation of defense mechanisms' effectiveness. Case studies demonstrate that integrating these models into security management systems helps optimize resource allocation by focusing on the most critical vulnerabilities.

However, implementing probabilistic approaches requires substantial computational resources and high-quality training data. Additionally, the complexity of interpreting results may hinder widespread adoption among organizations lacking specialized expertise. Nevertheless, advances in machine learning and automation are gradually reducing these barriers.

Looking ahead, further refinement of probabilistic models will enable more adaptive security systems capable of predicting both known threats and emerging attack vectors. This positions probabilistic modeling as a key methodology in modern cybersecurity frameworks.

The findings suggest that despite existing challenges, probabilistic methods offer significant potential for developing proactive, data-driven security solutions that can evolve alongside the threat landscape. Future research should focus on improving model interpretability and scalability to facilitate broader industry adoption.

REFERENCES

Alali, M., Almogren, A., Hassan, M. M., Rassan, I. A., & Bhuiyan, M. Z. A. (2018). Improving risk assessment model of cyber security using fuzzy logic inference system. *Computers & Security*, *74*, 323–339. DOI: 10.1016/j.cose.2017.09.011

Carter, K. M., Idika, N., & Streilein, W. W. (2014). Probabilistic threat propagation for network security. *IEEE Transactions on Information Forensics and Security*, *9*(9), 1394–1405. DOI: 10.1109/TIFS.2014.2334272

ElSalamouny, E. (2011). *Probabilistic trust models in network security* (Doctoral dissertation, University of Southampton).

Feng, N., & Li, M. (2011). An information systems security risk assessment model under uncertain environment. *Applied Soft Computing*, *11*(7), 4332–4340. DOI: 10.1016/j.asoc.2010.06.005

Figueira, P. T., Bravo, C. L., & López, J. L. R. (2020). Improving information security risk analysis by including threat-occurrence predictive models. *Computers & Security*, *88*, 101609. DOI: 10.1016/j.cose.2019.101609

Kondakci, S. (2007). A new assessment and improvement model of risk propagation in information security. *International Journal of Information and Computer Security*, *1*(3), 341–366. DOI: 10.1504/IJICS.2007.013959

Kondakci, S. (2015). Analysis of information security reliability: A tutorial. *Reliability Engineering & System Safety*, *133*, 275–299. DOI: 10.1016/j.ress.2014.09.021

Lee, S. W. (2011, June). Probabilistic risk assessment for security requirements: A preliminary study. In *2011 Fifth International Conference on Secure Software Integration and Reliability Improvement* (pp. 11-20). IEEE. DOI: 10.1109/SSIRI.2011.12

Mo, S. Y. K., Beling, P. A., & Crowther, K. G. (2009, April). Quantitative assessment of cyber security risk using Bayesian Network-based model. In *2009 Systems and Information Engineering Design Symposium* (pp. 183-187). IEEE. DOI: 10.1109/SIEDS.2009.5166177

Nicol, D. M., Sanders, W. H., & Trivedi, K. S. (2004). Model-based evaluation: From dependability to security. *IEEE Transactions on Dependable and Secure Computing*, *1*(1), 48–65. DOI: 10.1109/TDSC.2004.11

Odunola, A. B., Olawumi, A. A., & Adesola, A. G. (2015). Performance prediction model for network security risk management. *Communications on Applied Electronics*, *2*(8), 1–7. DOI: 10.5120/cae2015651816

Shaikh, E., Al-Ali, A. R., Muhammad, S., Mohammad, N., & Aloul, F. (2023). Security analysis of a digital twin framework using probabilistic model checking. *IEEE Access : Practical Innovations, Open Solutions*, *11*, 26358–26374. DOI: 10.1109/ACCESS.2023.3257171

Sommestad, T., Ekstedt, M., & Johnson, P. (2010). A probabilistic relational model for security risk analysis. *Computers & Security*, *29*(6), 659–679. DOI: 10.1016/j.cose.2010.02.002

Wang, A. J. A. (2005, March). Information security models and metrics. In *Proceedings of the 43rd annual ACM Southeast Conference-Volume 2* (pp. 178-184).

Yermalovich, P., & Mejri, M. (2020, October). Information security risk assessment based on decomposition probability via bayesian network. In *2020 International Symposium on Networks, Computers and Communications (ISNCC)* (pp. 1-8). IEEE. DOI: 10.1109/ISNCC49221.2020.9297330

Chapter 7
Enhancing Well-Being in the Digital Age in the VUCA and BANI World

Oindrilla Ghosh
https://orcid.org/0000-0001-9609-0781
Christ University, India

Binod Kumar
https://orcid.org/0009-0002-9721-1367
Radha Govind University, India

ABSTRACT

In today's VUCA (Volatile, Uncertain, Complex, Ambiguous) and BANI (Brittle, Anxious, Nonlinear, Incomprehensible) reality, enhancing well-being in the digital age is crucial. Digital technology can both support and hinder mental health, making it essential to cultivate a balanced approach. Encouraging digital detoxes and mindfulness practices can reduce stress and promote emotional resilience. Leveraging online communities for support fosters connection, but awareness of the potential for digital addiction remains essential. Tools like mental health apps can provide resources, yet they should complement real-life interactions to ensure authenticity. Emphasizing digital literacy empowers individuals to navigate information critically, enhancing decision-making regarding their mental and emotional health. Finally, organizations must prioritize employee well-being by implementing flexible work arrangements, promoting open communication, and providing mental health resources. By consciously shaping our digital environments, we can enhance well-being and thrive in this complex world.

DOI: 10.4018/979-8-3373-2240-7.ch007

INTRODUCTION

In today's fast-paced, interconnected society, individuals are faced with unprecedented challenges and opportunities that significantly affect their well-being (Döner & Efeoğlu, 2023). As advances in technology proliferate, the digital landscape has become both a catalyst and a barrier to human flourishing (Juncos, 2018). This paradox is particularly pronounced in the context of the VUCA (Volatile, Uncertain, Complex, Ambiguous) and BANI (Brittle, Anxious, Nonlinear, Incomprehensible) frameworks, which provide lenses through which we can better understand the contemporary world (Zhang et al, 2024). These frameworks highlight how the rapid pace of change and the unpredictable nature of our environment can create stressors that undermine mental and emotional health (Evseeva et al., 2022).

Understanding well-being in the digital age requires a multi-faceted approach that takes into consideration the nuances introduced by technology (Morales et al., 2019). Digital tools have the potential to enhance our lives, providing resources for learning, connection, and self-care (Harrison, 2017). Yet, they also risk leading to isolation, information overload, and a diminished capacity for genuine interpersonal relationships (Cheruiyot & Venter, 2024). In a digital-first world, where social media interactions often eclipse face-to-face engagements, the boundaries between personal and professional life blur, introducing new complexities that can exacerbate feelings of anxiety and inadequacy (Chaudhri et al., 2024).

Addressing these challenges necessitates a concerted effort to integrate well-being into our daily digital experiences (Long et al., 2022). This involves not only individual strategies for cultivating resilience and emotional intelligence but also systemic approaches that promote supportive environments online and offline (Schoemaker & Heaton, 2018). In a VUCA and BANI world, enhancing well-being becomes an adaptive skill, allowing individuals to navigate uncertainties while fostering a sense of purpose and connection (Morales et al., 2019).

The stakes are high: mental health crises are on the rise, fueled by the digital landscape's challenges (Hickman, 2023), yet the same technology that contributes to these issues can also provide tools for healing (Schell, 2023). From mindfulness applications to online support communities, the digital age offers novel approaches to mental health and well-being (Saputra et al., 2024). However, realizing this potential requires an intentional and informed strategy to harness technology's benefits while mitigating its drawbacks (Fernández-Pérez de la Lastra & Sánchez-Gardey, 2024). This introduction sets the stage for a deeper exploration of how we can enhance well-being in a world that is constantly shifting, characterized by volatility and anxiety, and fraught with complexities that often feel overwhelming (Winkler et al., 2024).

BACKGROUND

The emergence of digital technologies has transformed how we communicate, work, and engage with the world around us (Rani & Widyowati, 2021). In this digital revolution, we've witnessed both significant advancements and notable challenges (Bakar & Dorasamy, 2023). The VUCA framework, which originated in military strategic planning, has gained traction in the corporate and social realms as a way to describe the unpredictable and fast-evolving nature of modern life (Bellis et al., 2024). Similarly, the BANI framework articulates the specific anxieties that come from living in a world characterized by brittleness, anxiety, nonlinearity, and incomprehensibility (Roblek et al., 2024).

Within this backdrop, well-being emerges as a critical topic of study and practice (Mustofa & Tjahjono, 2024). The principles of well-being spanning mental, emotional, and physical health are increasingly recognized as essential components of a fulfilling life (Ramadan et al., 2023). However, the digital age complicates these traditional concepts, introducing factors such as online harassment, digital fatigue, and a pervasive culture of comparison that can undermine well-being (Long et al., 2022).

Social media, for instance, presents a double-edged sword (Niemeyer-Rens et al., 2022). It can provide support, community, and connection, yet it often fosters feelings of inadequacy and anxiety through constant exposure to curated versions of others' lives (Schoemaker & Heaton, 2018). Understanding the impact of these dynamics is essential in promoting well-being in a digital context (Baron et al., 2018). Moreover, the nonlinear and often unpredictable nature of our interactions in the digital space adds another layer of complexity, necessitating adaptive strategies for coping with stress and uncertainty (Evans & Bahrami, 2020).

Research indicates that the prevalence of mental health issues is rising, particularly among younger generations who have grown up immersed in digital culture (Majnoor & Vinayagam, 2023). This has catalyzed a broader conversation about digital well-being, prompting individuals, organizations, and policymakers to consider how to create environments that foster resilience, connection, and meaningful interactions (de Diego Ruiz et al., 2024). It has also led to the development of various initiatives designed to promote a healthier digital experience, such as digital literacy programs and mental health awareness campaigns (Ramadan et al., 2023).

Furthermore, organizations are beginning to recognize the importance of employee mental health and well-being as a cornerstone for productivity and creativity (Bellis et al., 2024). Businesses that prioritize well-being through flexible work arrangements, mental health resources, and supportive cultures are likely to thrive in a volatile world (Nguyen et al., 2024).

THEORETICAL FRAMEWORKS

In our fast-paced, interconnected world, individuals face unprecedented challenges and opportunities for growth (Siregar et al., 2023). To navigate this complexity, we must rely on theoretical frameworks that provide insight into enhancing well-being, particularly in the context of VUCA (Volatility, Uncertainty, Complexity, Ambiguity) and BANI (Brittle, Anxious, Nonlinear, Incomprehensible) environments (de Diego Ruiz et al., 2024). These frameworks help us understand the psychological, social, and technological influences that shape well-being in the digital era (Kaya, 2023).

Positive Psychology: Positive psychology is foundational in enhancing well-being in today's digital landscape (Fernandes et al., 2023). Developed by psychologist Martin Seligman, it emphasizes strengths, virtue, and factors that contribute to human flourishing (Mustofa & Tjahjono, 2024). In a VUCA world, individuals are often overwhelmed by uncertainty and rapid change (Amiri et al., 2024). Positive psychology offers tools and perspectives, such as the PERMA model (Positive Emotion, Engagement, Relationships, Meaning, and Accomplishment), which can foster resilience and enhance an individual's overall well-being (Awah et al., 2024). By focusing on what makes life meaningful and fulfilling, individuals can better navigate the complexities of the digital age (Mohamed Shaffril et al., 2021).

Systems Theory: In a VUCA and BANI context, systems theory provides a comprehensive lens for understanding how various elements within a digital ecosystem interact and impact well-being (Jun & Weilin, 2024). Systems thinking encourages individuals to see beyond linear cause-and-effect relationships and recognize the interconnectedness of psychological, emotional, social, and technological factors (Lyu et al., 2023). By applying systems theory, individuals can design interventions that account for the complexity of their environments rather than pursuing isolated solutions (Weng & Feng, 2022). This perspective can be vital when addressing issues related to mental health in digital spaces, where social media algorithms, online interactions, and societal expectations interplay (Gantman & Dabós, 2018).

Ecological Perspective: An ecological perspective, rooted in Urie Bronfenbrenner's ecological systems theory, emphasizes the role of various environments family, community, culture, and policy in shaping individual well-being (Gao et al., 2021). In the digital age, this framework helps us understand how technology influences our ecological systems (Sampson et al., 2006). The constant bombardment of digital information can lead to information overload, impacting mental health (Mariani et al., 2023). Conversely, the digital landscape can also serve as a resource for social connection and engagement (Kelly et al., 2016). Adopting an ecological perspective allows for the development of comprehensive strategies that harness the positive aspects of digital interaction while mitigating its negative effects (Krupinski, 2019).

Cognitive Behavioral Theory (CBT): Cognitive Behavioral Theory focuses on the interplay between thoughts, feelings, and behaviors. In the face of VUCA and BANI, individuals may experience heightened anxiety, fear, or confusion (Haddaway et al., 2022). CBT techniques, such as cognitive restructuring and mindfulness practices, can be useful in managing these feelings (Rethlefsen et al., 2021). By helping individuals identify and challenge irrational thoughts often exacerbated by the digital landscape CBT promotes healthier thinking patterns and coping strategies (Øby, 2024). This framework is particularly relevant in addressing the mental health challenges associated with constant connectivity and digital distractions (D. Wang et al., 2023).

Theoretical frameworks such as positive psychology, systems theory, ecological perspectives, and cognitive behavioral theory offer critical insights into enhancing well-being in the VUCA and BANI world (Dwivedi et al., 2023). These frameworks highlight the importance of understanding the complex interactions between personal, social, and technological factors in fostering resilience and improving well-being in the digital age (Bronk et al., 2023). By employing these frameworks, individuals can navigate their realities more effectively, promoting mental health and overall life satisfaction in a rapidly changing environment (Song et al., 2023).

FOUNDATIONS

The foundations for enhancing well-being in the digital age amid VUCA and BANI conditions require a multifaceted approach (Utomo & Latukismo, 2022). The dynamic interplay between technology, society, and individual psychology necessitates strategies that account for both the inherent challenges and opportunities present in these contexts (Maupin et al., 2024). The foundations for well-being include understanding the roles of technology, fostering digital literacy, promoting community and connection, and prioritizing mental health (Chen & Song, 2019).

The Role of Technology: Technology plays a pivotal role in shaping contemporary experiences and, therefore, well-being (Bakar & Dorasamy, 2023). In a VUCA world, technological solutions like mental health apps, online support communities, and remote work capabilities can enhance flexibility and accessibility (Eberz et al., 2023). However, acknowledging the darker sides of technology, such as cyberbullying, social isolation, and information fatigue, is crucial (Niemeyer-Rens et al., 2022). Developing platforms that promote positive engagement while safeguarding against negative experiences forms a critical foundation for enhancing well-being (Dhandapani & Kaur, 2024). Efforts to create inclusive digital environments can help combat feelings of inadequacy and loneliness (Reyes et al., 2021).

Digital Literacy: Digital literacy is becoming increasingly important for individual well-being in the digital age (Mäkinen, 2018). In a BANI world, misinformation, biased content, and online harassment can adversely affect mental health (Mitchell, 2016). Therefore, equipping individuals with skills to critically assess online information, engage positively with digital tools, and manage their digital footprints is essential (Majnoor & Vinayagam, 2023). Educational initiatives that promote digital literacy not only foster resilience but also prepare individuals to navigate complexities safely and effectively (Reyes et al., 2021). This foundation empowers users to harness technology productively while guarding against the risks associated with the digital landscape (Urs et al., 2024).

Fostering Community and Connection: Building strong communities is a vital foundation for enhancing well-being (Valionienė & Zuperkienė, 2024). Social support is crucial during challenging times characterized by volatility and uncertainty (Alzoubi et al., 2024). In the digital age, online forums, social media groups, and virtual support networks can significantly contribute to a sense of belonging and shared experience (Baron et al., 2018). Efforts must be made to cultivate sincerity, empathy, and authenticity in these online interactions (Lisdiono et al., 2022). Facilitating meaningful connections can counteract feelings of isolation and help individuals feel grounded amid rapid technological shifts and societal pressures (Melián-Alzola et al., 2020).

Prioritizing Mental Health: The foundation of well-being must include a robust emphasis on mental health (Rani & Widyowati, 2021). In VUCA and BANI conditions, the prevalence of anxiety, depression, and stress can hinder individuals from thriving (Bellis et al., 2024). Organizations, policy-makers, and educators must prioritize mental health awareness and resources, providing easy access to support services and promoting a culture that destigmatizes mental health challenges (Finn et al., 2023). This effort includes fostering environments that allow open discussions about mental health and championing self-care practices as essential components of daily life (Livingstone & Anthony, 2023).

THE ROLE OF AI AND ML IN ENHANCING WELL-BEING IN THE DIGITAL AGE

Artificial Intelligence (AI) and Machine Learning (ML) are transforming the landscape of well-being, offering tools that can greatly enhance mental health (Raney, 2014), emotional resilience, and overall quality of life, especially in the context of VUCA and BANI environments (Wang et al., 2022). These complex and dynamically changing scenarios require adaptive solutions (Carter & Cordero, 2022); AI and

ML can provide insights and interventions that are tailored to individual needs, thus fostering well-being in unprecedented ways (Helmrich & Chester, 2022).

One of the most immediate applications of AI and ML in enhancing well-being is in mental health support (Hong et al., 2020). Algorithms can analyze data from various sources, such as user interactions with apps or wearables, to identify patterns indicative of stress, anxiety, or mood swings (Kachniewska & Para, 2023). This allows for personalized mental health interventions, such as prompts for mindfulness exercises or connections to live mental health professionals (Zarif et al., 2024). In a VUCA world where stressors are constant and changeable, such timely and personalized interventions can significantly mitigate feelings of helplessness and anxiety (Bodenheimer, 2018).

Another critical role of AI and ML tools lies in the prevention of misinformation, which can adversely affect mental health (de Graaff et al., 2020). AI-driven content moderation systems can help identify and flag harmful or misleading information on social media platforms, fostering healthier online environments (Debellis et al., 2021). This is particularly important in VUCA and BANI scenarios, where misinformation can exacerbate confusion and anxiety (Lane et al., 2021). By using technology to create safer information spaces, AI can contribute to enhancing overall societal well-being (Zhang-Zhang et al., 2022).

IMPACT OF ENHANCING WELL-BEING IN THE DIGITAL AGE IN THE VUCA & BANI WORLD

In the contemporary landscape characterized by VUCA (Volatility, Uncertainty, Complexity, Ambiguity) and BANI (Brittle, Anxious, Nonlinear, Incomprehensible), enhancing well-being has transcended a mere personal benefit and developed into a societal necessity (Alzoubi et al., 2024). The digital age, marked by rapid technological advancements and ever-evolving information streams, profoundly influences the way individuals relate to their own well-being (Rani & Widyowati, 2021).

One of the most significant impacts of enhancing well-being in this context is the ability to foster resilience (Bellis et al., 2024). Individuals who prioritize their mental health and emotional well-being tend to cope more effectively with the fluctuations inherent in a VUCA environment (Finn et al., 2023). The digital tools available today facilitate support networks, access to mental health resources, and community engagement (Carter & Cordero, 2022). Social media platforms and online forums create spaces for individuals to share experiences, seek advice, and find like-minded communities (Kachniewska & Para, 2023). This social connectivity can buffer against feelings of isolation and anxiety fueled by uncertainty and complexity (Bodenheimer, 2018).

Moreover, the digital age offers a wealth of information and resources that can equip individuals with the knowledge needed to pursue their well-being actively (Zhang-Zhang et al., 2022). Online courses, webinars, and applications designed to promote mindfulness, exercise, and nutrition empower people to take control of their health and well-being (Hegde et al., 2023). In a BANI world, where anxiety and incomprehensibility can dominate personal experience, these resources provide clarity and strategies for coping, helping individuals navigate their challenges more effectively (Niemeyer-Rens et al., 2022).

On a broader scale, enhancing well-being can lead to improved productivity and innovation within organizations (Debellis et al., 2021). Companies that prioritize employee mental health create a culture of support, which, in turn, can lead to enhanced job satisfaction and increased retention rates (Yeomans & Bowman, 2021). As employees feel valued and understood, their commitment to the organization often strengthens, fostering an environment conducive to creative problem-solving amidst volatility and complexity (Chingwena & Scheepers, 2022).

However, the digital age also poses challenges that must be navigated to enhance well-being effectively (Grobler & Flotman, 2021). The constant influx of information can lead to information overload, contributing to the very anxiety and stress that well-being initiatives aim to alleviate (Hegde et al., 2023). A critical component of enhancing well-being in a VUCA and BANI world is therefore educating individuals about effective information management and promoting digital literacy (Raney, 2014) helping them discern credible sources, set boundaries around tech use, and engage in self-care that mitigates the adverse effects of excessive screen time (Millings et al., 2019).

CHALLENGES

In today's increasingly complex and fast-paced environment, characterized by the acronyms VUCA (Volatile, Uncertain, Complex, Ambiguous) and BANI (Brittle, Anxious, Non-linear, Incomprehensible), enhancing well-being presents a myriad of challenges (Slater et al., 2014). These frameworks underscore a reality where change is rapid and unpredictable, necessitating adaptable strategies for well-being initiatives (Yu et al., 2022).

First and foremost, information overload is a significant challenge in the digital age (Zarif et al., 2024). With the proliferation of digital devices and platforms, individuals are inundated with information from countless sources (Chingwena & Scheepers, 2022). This can lead to cognitive overload, where the ability to make decisions and maintain mental health is compromised (Del Sisto et al., 2024). For instance, while social media offers avenues for connection and engagement (Meyerowitz et al., 2018), it can also propagate anxiety through the constant comparison with curated online personas and experiences (Fonsén et al., 2024). Consequently, individuals may find it difficult to discern beneficial information that promotes well-being versus content that exacerbates feelings of inadequacy (Charteris et al., 2022).

Additionally, the inherent volatility in social structures and economies, as highlighted by the VUCA model, affects mental health and well-being (Fernandes et al., 2023). The unpredictability of job markets and the rise of gig economies force individuals into precarious employment situations, leading to anxiety and stress (Kligyte & Barrie, 2014). This economic uncertainty, coupled with rapid technological advancements, can make many feel obsolete, as they struggle to keep pace with the skills required in a swiftly changing job landscape (Whyte et al., 2022).

Moreover, the complex nature of digital ecosystems complicates efforts to enhance well-being (Mitchell, 2016). Many digital interventions aimed at promoting mental health, such as apps and online therapy, while innovative, often lack empirical backing (Candrasa et al., 2024). This lack of standardization can lead to varied efficacy among users (Indrianti et al., 2024). Additionally, the complexity of human emotions and interactions cannot easily be distilled into algorithms or data-driven approaches (Lisdiono et al., 2022), potentially oversimplifying the genuine challenges individuals face in improving their well-being (Mustofa & Tjahjono, 2024).

There is also the issue of access and equity (Silalahi et al., 2023). Not everyone has equal access to digital tools and mental health resources, which creates disparities in well-being outcomes (Eng et al., 2023). Particularly vulnerable populations, including those in low socioeconomic statuses, may not have the means to utilize technologies that could enhance their well-being (Mahajan et al., 2023). This creates an ethical dilemma as efforts to promote digital interventions may inadvertently widen the gap between those who can benefit from such tools and those who cannot (Zahari et al., 2023).

In a BANI world, where anxiety and uncertainty are prevalent, the psychological impacts of technology consumption must not be overlooked (Dhandapani & Kaur, 2024). The constant connectivity facilitated by digital technology can lead to what is often referred to as "digital fatigue" (Mahajan et al., 2023). This syndrome, characterized by weariness from being perpetually plugged into digital environments, can hinder an individual's ability to engage in meaningful offline relationships and experiences (Majnoor & Vinayagam, 2023).

Lastly, the rapid pace of change leads to an often reactive approach to mental health and well-being, rather than a proactive one (Rowe et al., 2024). Organizations, the media, and communities may respond to crises in mental health stimulated by the digital age but are less effective in implementing long-term strategies that foster resilience and well-being before crises arise (Chaudhri et al., 2024).

ETHICAL CONSIDERATIONS

As society seeks to enhance well-being amid the complexities of the digital age, particularly within the frameworks of VUCA and BANI, a range of ethical considerations emerge (de Graaff et al., 2020). These concerns require careful navigation to ensure that well-being initiatives are effective, equitable, and respectful of individual rights (Faling & Biesbroek, 2019).

Privacy and Data Security is one of the foremost ethical considerations (Weerts et al., 2018). Many digital mental health solutions collect sensitive personal data, often without transparent consent processes (de Diego Ruiz et al., 2024). This raises ethical dilemmas regarding user privacy and the potential misuse of data (Desrues, 2020). Users may not be fully aware of how their information is being used or stored, leading to breaches of trust (Melián-Alzola et al., 2020). Ethical engagement requires organizations to adopt robust data security measures and promote transparent practices, sharing exactly how data will enhance user experiences while safeguarding personal information (Zhang-Zhang et al., 2022).

Furthermore, the use of AI and algorithms in well-being solutions introduces additional ethical dimensions (Baron et al., 2018). These technologies can perpetuate biases inherent in their programming and training data, potentially marginalizing specific groups (Hoff et al., 2023). An algorithm designed to assess mental health could, for instance, overlook cultural nuances or specific demographic factors, leading to misdiagnoses or inappropriate treatments (Rodina et al., 2024). Thus, it is imperative for creators of these digital wellness tools to ensure inclusive design practices, rigorous bias testing, and ongoing evaluation to mitigate the risks of algorithmic discrimination (Mustofa & Tjahjono, 2024).

Commercialization of Well-Being also poses ethical challenges (Silalahi et al., 2023). The rise of the wellness industry, often heavily marketed through digital platforms, can commodify mental health, reducing it to a transactional experience rather than a fundamental human right (Long et al., 2022). This commercialization might prioritize profit over genuine care, causing well-being interventions to become inaccessible, ineffective, or overly prescriptive (Zahari et al., 2023). Ethical considerations necessitate a focus on building products that genuinely support users rather than exploit vulnerabilities for profit (Al-Hawamleh, 2024).

Moreover, there exists a risk of over-reliance on technology, leading to a detriment in the nuanced understanding of mental health care (Cheruiyot & Venter, 2024). While digital tools can provide valuable resources, a heavy reliance on them can overshadow the importance of human connection and professional guidance in mental health care (Hamsal et al., 2022). Ethical considerations should ensure that digital tools complement, rather than replace, traditional methods of care, promoting a balanced approach to mental well-being (Lisdiono et al., 2022).

Another critical area of ethical concern is equity and access to well-being resources (Ledesma, 2014). The digital divide means that not everyone benefits equally from technological advancements (Montasser et al., 2023). For example, marginalized communities with limited access to the internet may miss out on digital mental health services (Ramadan et al., 2023). It is vital for policymakers and tech developers to prioritize inclusivity, ensuring that all communities have equal access to tools that promote mental well-being (Schulze & Pinkow, 2020). This includes recognizing systemic inequities that may impact an individual's ability to utilize digital resources effectively (De Sisto et al., 2024).

Lastly, the impact of misinformation presented through digital channels compels ethical scrutiny (Hube et al., 2022). In a VUCA and BANI world where emotional distress is common, easily accessible yet erroneous information about mental health can spread rapidly, causing harm (Alzoubi et al., 2024). Ethical engagement necessitates robust fact-checking, transparency, and collaboration with mental health professionals to curate accurate resources that promote well-being (Scheepers et al., 2023).

IMPLEMENTATION OF REGULATORY FRAMEWORKS

The complexities of modern society, characterized by volatility, uncertainty, complexity, and ambiguity (VUCA), along with the fragility, anxiety, nonlinearity (Hong et al., 2020), and incomprehensibility (BANI) dimensions, necessitate robust regulatory frameworks aimed at enhancing well-being in the digital age (Viano et al., 2024). The pervasive influence of technology on everyday life has created significant challenges while also presenting opportunities for societal growth (Bakar & Dorasamy, 2023). It is essential to establish regulations that not only protect individuals but also promote mental health, encourage digital literacy, and safeguard privacy (Cheruiyot & Venter, 2024).

To implement effective regulatory frameworks, it is vital to first identify key areas that require oversight (Hamsal et al., 2022). This involves addressing issues such as data protection, misinformation, mental health impact, and the digital divide (Lisdiono et al., 2022). Regulatory frameworks must be adaptive, recognizing the rapid pace of technological advancement (Safitra et al., 2023). For instance, policies that address data privacy, such as the General Data Protection Regulation (GDPR) in the European Union, serve as a model (Yamin, 2021). These regulations establish basic rights related to personal data, promoting individual agency and well-being (Ledesma, 2014).

In addition to data protection, regulations addressing misinformation are increasingly essential (Montasser et al., 2023). The spread of false information can have serious implications for public health and safety, as well as mental well-being (Zahari et al., 2023). Governments and organizations must collaborate with technology providers to create transparent mechanisms for combating misinformation while allowing for freedom of expression (Ramadan et al., 2023). Frameworks for content moderation should prioritize accuracy and prevent harm, thereby fostering a healthier digital environment (Schulze & Pinkow, 2020).

Mental health considerations must be woven into regulatory mechanisms as well (De Sisto et al., 2024). The digital landscape can exacerbate mental health issues such as anxiety and depression, particularly among young people (Hube et al., 2022). Practitioners and regulators should advocate for the inclusion of mental health resources in online platforms and promote mental well-being initiatives (Alzoubi et al., 2024). Regulations that require technology companies to assess and mitigate their platforms' impact on user mental health remain critical in this context (Silalahi et al., 2023).

The digital divide also reflects a significant area where regulatory frameworks must exert influence (Magwegwe, 2021). Disparities in access to technology and digital literacy can exacerbate inequalities and limit opportunities for youth and marginalized communities (Scheepers et al., 2023). Governments can establish regulations and incentives to foster equitable access to the internet and digital education, ensuring that all individuals can participate in and benefit from the digital age (Hong et al., 2020). Successful initiatives might include subsidized internet service, community technology programs, and partnerships with educational institutions (Viano et al., 2024).

Moreover, implementing any regulatory framework requires collaboration among multiple stakeholders, including governments, industry leaders, civil society, and academia (Lane et al., 2021). Public consultations play a significant role in developing regulations, as they allow for a diverse range of perspectives to be taken into account (Thude et al., 2019). Continuous policy review and adaptation should also be integral components, ensuring that regulations remain relevant as technology and societal norms evolve (Mariani et al., 2023).

CONCLUSION

The digital age presents an intricate tapestry of challenges and opportunities, particularly shaped by the VUCA and BANI environments (Bernatovic et al., 2022). As society grapples with the transformative influence of technology, enhancing well-being emerges as a crucial objective, guiding the development of coherent strategies and policies (Song et al., 2023). Achieving this objective necessitates a multi-faceted approach that integrates the unique characteristics of these volatile, uncertain, complex, ambiguous (VUCA) and fragile, anxious, non-linear, and incomprehensible (BANI) landscapes (Zupic & Čater, 2015).

Ultimately, the intersection of technology, well-being, and regulatory frameworks in the digital age calls for holistic and adaptive strategies (Schoemaker & Heaton, 2018). By fostering collaboration among governments, industry leaders, and civil society, we can develop innovative solutions that address the multifaceted challenges presented by the VUCA and BANI environments (Gu & Day, 2013).

SUGGESTIONS

In the rapidly evolving landscape characterized by Volatility, Uncertainty, Complexity, Ambiguity (VUCA), and the fragility, anxiety, non-linearity, and incomprehensibility (BANI), enhancing well-being has become an urgent need (Pasmore et al., 2019). Individuals and organizations alike are grappling with the psychological and emotional toll imposed by the digital age (Ledesma, 2014).

Digital Detox and Mindfulness Practices: One of the foremost strategies is to promote digital detox periods that allow individuals to step away from screens and engage in face-to-face interactions, nature walks, or mindfulness activities (Debellis et al., 2021). Encouraging mindfulness practices such as meditation, yoga, and journaling can help individuals become more aware of their mental states, reduce anxiety, and cultivate a sense of presence and calm (Safitra et al., 2023).

Supportive Digital Communities: Leverage the digital environment to create supportive online communities (Ramli et al., 2023). While the internet can sometimes foster isolation, it can also connect people with shared experiences (Khan & Lew, 2018). These virtual spaces can serve as platforms for sharing stories, providing emotional support, and exchanging coping strategies (Mäkinen, 2018). Facilitated forums, discussion groups, or mental well-being apps could engage users in meaningful interactions (Fernandes et al., 2023).

Workplace Well-Being Initiatives: Employers need to take initiative in enhancing employee well-being (Yu et al., 2022). By implementing flexible work arrangements, promoting work-life balance, and providing mental health resources, workplaces can significantly improve employee satisfaction and productivity (Whyte et al., 2022). Regular workshops focusing on resilience training and stress management can equip employees with tools to navigate a VUCA/BANI world more effectively (Meyerowitz et al., 2018).

Media Literacy and Digital Citizenship: As misinformation proliferates in the digital space, fostering media literacy becomes critical (Baron et al., 2018). Educational institutions can implement curriculums focused on critical thinking, digital literacy, and healthy consumption of digital media (Ferreira et al., 2016). This will enable individuals to differentiate between credible and unreliable information, reducing anxiety and confusion caused by digital overload (Trujillo & Long, 2018).

Technology Design Considerations: Tech companies must be increasingly aware of the psychological impacts of their products (Yang et al., 2019). Designing software and applications with user well-being in mind such as incorporating features that promote breaks, limit notifications, and present curated content can help mitigate negative impacts on mental health (Grames et al., 2019; Lozano et al., 2019).

Holistic Health Integration: Collaboration between healthcare providers and digital platforms can lead to the development of integrated solutions to support mental and physical health (Mustak et al., 2021). Apps that combine fitness tracking with mental health resources, nutritional advice, and community support can create a more holistic approach to well-being (Indrianti et al., 2024).

FUTURE RESEARCH DIRECTIONS

As we look to the future, it is vital to conduct research that informs our understanding of well-being in a digital landscape marked by VUCA and BANI characteristics (Wowk et al., 2023). By identifying gaps in current knowledge and exploring new areas of inquiry, researchers can contribute to the development of strategies that enhance mental and emotional health in these dynamic contexts (Florek-Paszkowska et al., 2021).

Longitudinal Studies on Digital Well-Being: There is a need for comprehensive longitudinal studies exploring the long-term effects of digital consumption on mental health (Helmrich & Chester, 2022). Understanding how prolonged engagement with digital technology influences stress, anxiety, and interpersonal relationships over time can provide crucial insights into developing effective interventions (Nayak et al., 2023).

Impact of AI and Automation on Mental Health: As artificial intelligence and automation continue to redefine work, their psychological implications are of paramount importance (Walch, 2019). Future research should explore how these technologies affect job satisfaction, worker participation, emotional well-being, and the potential for job displacement and anxiety about the future (Zabaniotou, 2020).

Cross-Cultural Studies of Digital Well-Being: Different cultures respond uniquely to digital influences, shaping the perception of well-being (Rejeb et al., 2023). Research comparing well-being practices and outcomes across diverse cultural contexts can illuminate effective strategies that honor and integrate cultural values into digital well-being initiatives (Mervyn et al., 2019).

Understanding the Role of Virtual Reality (VR) in Mental Health: Emerging technologies like VR offer new ways to support mental health and well-being (Finn et al., 2023). Investigating the effectiveness of VR interventions for issues such as anxiety, PTSD, and social isolation can provide novel avenues for therapeutic development (Nkurunziza et al., 2019).

Interdisciplinary Approaches to Mental Health: Future research should embrace interdisciplinary collaboration, drawing from psychology, sociology, technology, and public health (Govers & van Amelsvoort, 2023). Collaborative studies can produce holistic frameworks that consider various societal factors influencing well-being in a VUCA/BANI world (Cogan et al., 2022).

REFERENCES

Al-Hawamleh, A. (2024). Cyber resilience framework: Strengthening defenses and enhancing continuity in business security. *International Journal of Computing and Digital Systems*, *15*(1), 1315–1331. DOI: 10.12785/ijcds/150193

Alzoubi, R. H., Shnaikat, B., Abo Murad, M., Al Kloub, R., Alhmesat, M., & Alkharabsheh, O. H. M. (2024). Rethinking growth and survival of hospitality industry in Jordan: The role of responsible and resilient leadership styles in managing crisis. *Cogent Business & Management*, *11*(1), 2402080. Advance online publication. DOI: 10.1080/23311975.2024.2402080

Amiri, Z., Heidari, A., Zavvar, M., Navimipour, N. J., & Esmaeilpour, M. (2024). The applications of nature-inspired algorithms in Internet of Things-based healthcare service: A systematic literature review. *Transactions on Emerging Telecommunications Technologies*, *35*(6), e4969. DOI: 10.1002/ett.4969

Awah, L. S., Belle, J. A., Nyam, Y. S., & Orimoloye, I. R. (2024). A systematic analysis of systems approach and flood risk management research: Trends, gaps, and opportunities. *International Journal of Disaster Risk Science*, *15*(1), 45–57. DOI: 10.1007/s13753-024-00544-y

Bakar, S. A., & Dorasamy, M. (2023). From adoption to sustainability: A journey of large-scale agile implementation. *International Journal of Technology*, *14*(6), 1367–1379. DOI: 10.14716/ijtech.v14i6.6645

Baron, L., Rouleau, V., Grégoire, S., & Baron, C. (2018). Mindfulness and leadership flexibility. *Journal of Management Development*, *37*(2), 165–177. DOI: 10.1108/JMD-06-2017-0213

Bellis, P., Cunial, M., & Trabucchi, D. (2024). Mastering hybrid worlds through digital leadership: The role of agility in fostering innovation. *Business Horizons*, *67*(4), 369–380. DOI: 10.1016/j.bushor.2024.04.002

Bernatovic, I., Slavec Gomezel, A., & Černe, M. (2022). Mapping the knowledge-hiding field and its future prospects: A bibliometric co-citation, co-word and coupling analysis. *Knowledge Management Research and Practice*, *20*(3), 394–409. DOI: 10.1080/14778238.2021.1945963

Bodenheimer, L. (2018). Leadership reflections: Resiliency in library organizations. *Journal of Library Administration*, *58*(4), 364–374. DOI: 10.1080/01930826.2018.1448651

Bronk, K. C., Reichard, R. J., & Qi Li, J. (2023). A co-citation analysis of purpose: Trends and (potential) troubles in the foundation of purpose scholarship. *The Journal of Positive Psychology, 18*(6), 1012–1026. DOI: 10.1080/17439760.2023.2168563

Candrasa, L., Cahyadi, L., Cahyadi, W., & Cen, C. C. (2024). Change management strategies: Building organizational resilience in the digital era. *Journal of Ecohumanism, 3*(7), 4125–4135. DOI: 10.62754/joe.v3i7.4534

Carter, E. D., & Cordero, M. L. (2022). Salir adelante: Social capital and resilience during the Covid-19 pandemic in Argentina. *Health & Place, 77*, 102870. DOI: 10.1016/j.healthplace.2022.102870

Charteris, J., Smardon, D., & Kemmis, S. (2022). Leadership in the built spaces of innovative learning environments: Leading change in people and practices in the perfectly self-managing society. *Studies in Continuing Education, 44*(2), 212–231. DOI: 10.1080/0158037X.2021.1928051

Chaudhri, V., Armbrust, N., & Nguyen, S. (2024). Permanently on standby: Practitioner perspectives on the complexities of crisis planning. *Journal of Contingencies and Crisis Management, 32*(3), e12600. Advance online publication. DOI: 10.1111/1468-5973.12600

Chen, C., & Song, M. (2019). Visualizing a field of research: A methodology of systematic scientometric reviews. *PLoS One, 14*(10), e0223994. DOI: 10.1371/journal.pone.0223994

Cheruiyot, R., & Venter, R. (2024). Complex systems and sustainable leadership: Enhancing resilience and sustainability of community-based social enterprises in Soweto, South Africa. *Sustainability (Basel), 16*(19), 8555. DOI: 10.3390/su16198555

Chingwena, T., & Scheepers, C. B. (2022). Dramatic social change (COVID-19) moderating complexity leadership and organisational adaptability in Zimbabwean SMEs. *European Business Review, 34*(6), 749–775. DOI: 10.1108/EBR-01-2022-0015

Cogan, N., Kennedy, C., Beck, Z., McInnes, L., MacIntyre, G., Morton, L., Tanner, G., & Kolacz, J. (2022). ENACT study: What has helped health and social care workers maintain their mental well-being during the COVID-19 pandemic? *Health & Social Care in the Community, 30*(6), e6656–e6673. DOI: 10.1111/hsc.13992

de Diego Ruiz, E., Almodóvar, P., & Birkinshaw, J. (2024). The effects of a firm's internationalization, age, and environmental turbulence on the capabilities that comprise strategic agility. *The International Entrepreneurship and Management Journal, 20*(3), 1935–1961. DOI: 10.1007/s11365-024-00971-9 de Graaff, N., ten Brink, T., & Parmar, I. (2020). China's rise in a liberal world order in transition–introduction to the FORUM. *Review of International Political Economy*, *27*(2), 191–207. DOI: 10.1080/09692290.2019.1709880

De Sisto, M., Shearing, C., Heffernan, T., & Sanderson, D. (2024). Reshaping disaster management: An integrated community-led approach. *Australian Journal of Public Administration*, •••, 1–28. DOI: 10.1111/1467-8500.12668

Debellis, F., De Massis, A., Messeni Petruzzelli, A., Frattini, F., & Del Giudice, M. (2021). Strategic agility and international joint ventures: The willingness-ability paradox of family firms. *Journal of International Management*, *27*(1), 100739. DOI: 10.1016/j.intman.2020.100739

Desrues, T. (2020). Authoritarian resilience and democratic representation in Morocco: Royal interference and political parties' leaderships since the 2016 elections. *Mediterranean Politics*, *25*(2), 254–262. DOI: 10.1080/13629395.2018.1543038

Dhandapani, S., & Kaur, K. (2024). Instructional leadership: The key to addressing psychological issues post-covid-19. *Journal of Interdisciplinary Studies in Education*, *13*(S1), 16–33. DOI: 10.32674/p547wx36

Döner, E., & EfeoGlu, İ. E. (2023). Being affected by VUCA factors? Developing the 'Perceived VUCA Exposure' Scale. *GAB Akademi,* 3(2), 28–53.

Dwivedi, R., Nerur, S., & Balijepally, V. (2023). Exploring artificial intelligence and big data scholarship in information systems: A citation, bibliographic coupling, and, co-word analysis. *International Journal of Information Management Data Insights*, *3*(2), 100185. DOI: 10.1016/j.jjimei.2023.100185

Eberz, S., Lang, S., Breitenmoser, P., & Niebert, K. (2023). Taking the lead into sustainability: Decision makers' competencies for a greener future. *Sustainability (Basel)*, *15*(6), 4986. DOI: 10.3390/su15064986

Eng, T.-Y., Mohsen, K., & Wu, L.-C. (2023). Wireless information technology competency and transformational leadership in supply chain management: Implications for innovative capability. *Information Technology & People*, *36*(3), 969–995. DOI: 10.1108/ITP-06-2021-0489

Evans, S., & Bahrami, H. (2020). Super-flexibility in practice: Insights from a crisis. *Global Journal of Flexible Systems Managment*, *21*(3), 207–214. DOI: 10.1007/s40171-020-00246-6

Evseeva, S., Evseeva, O., & Rawat, P. (2022). Employee development and digitalization in BANI world. *Communications in Computer and Information Science,* 1619(CCIS), 253–264. DOI: 10.1007/978-3-031-14985-6_18

Faling, M., & Biesbroek, R. (2019). Cross-boundary policy entrepreneurship for climate-smart agriculture in Kenya. *Policy Sciences*, *52*(4), 525–547. DOI: 10.1007/s11077-019-09355-1

Fernandes, V., Wong, W., & Noonan, M. (2023). Developing adaptability and agility in leadership amidst the COVID-19 crisis: Experiences of early-career school principals. *International Journal of Educational Management*, *37*(2), 483–506. DOI: 10.1108/IJEM-02-2022-0076

Fernández-Pérez de la Lastra, S., & Sánchez-Gardey, G. (2024). Organizational ambidexterity: A reconceptualization and research agenda for the VUCA international context. *Journal of Contingencies and Crisis Management*, *32*(2), e12565. Advance online publication. DOI: 10.1111/1468-5973.12565

Ferreira, J. J. M., Fernandes, C. I., & Ratten, V. (2016). A co-citation bibliometric analysis of strategic management research. *Scientometrics*, *109*(1), 1–32. DOI: 10.1007/s11192-016-2008-0

Finn, D., Chandrasekhar, D., & Xiao, Y. (2023). A region recovers: Planning for resilience after superstorm Sandy. *Journal of Planning Education and Research*, *43*(1), 136–149. DOI: 10.1177/0739456X19864145

Florek-Paszkowska, A., Ujwary-Gil, A., & Godlewska-Dzioboń, B. (2021). Business innovation and critical success factors in the era of digital transformation and turbulent times. *Journal of Entrepreneurship. Management and Innovation*, *17*(4), 7–28. DOI: 10.7341/20211741

Gantman, E. R., & Dabós, M. P. (2018). Research output and impact of the fields of management, economics, and Sociology in spain and france: An analysis using google scholar and scopus. *Journal of the Association for Information Science and Technology*, *69*(8), 1054–1066. DOI: 10.1002/asi.24020

Gao, S., Meng, F., Gu, Z., Liu, Z., & Farrukh, M. (2021). Mapping and clustering analysis on environmental, social and Governance field a bibliometric analysis using scopus. *Sustainability (Basel)*, *13*(13), 7304. DOI: 10.3390/su13137304

Govers, M., & van Amelsvoort, P. (2023). A theoretical essay on socio-technical systems design thinking in the era of Digital transformation. *Gruppe. Interaktion. Organisation. Zeitschrift Für Angewandte Organisationspsychologie*, *54*(1), 27–40. DOI: 10.1007/s11612-023-00675-8

Grames, E. M., Stillman, A. N., Tingley, M. W., & Elphick, C. S. (2019). An automated approach to identifying search Terms for systematic reviews using keyword co-occurrence networks. *Methods in Ecology and Evolution*, *10*(10), 1645–1654. DOI: 10.1111/2041-210X.13268

Grobler, A., & Flotman, A.-P. (2021). Servant leadership, team-based learning and hope and optimism: A sectoral Comparative study. *South African Journal of Business Management*, *52*(1), a2444. DOI: 10.4102/sajbm.v52i1.2444

Gu, Q., & Day, C. (2013). Challenges to teacher resilience: Conditions count. *British Educational Research Journal*, *39*(1), 22–44. DOI: 10.1080/01411926.2011.623152

Haddaway, N. R., Page, M. J., Pritchard, C. C., & McGuinness, L. A. (2022). PRISMA2020: An R package and Shiny app For producing PRISMA 2020-compliant flow diagrams, with interactivity for optimised digital transparency and Open Synthesis. *Campbell Systematic Reviews*, *18*(2), e1230. DOI: 10.1002/cl2.1230

Hamsal, M., Dwidienawati, D., Ichsan, M., Syamil, A., & Trigunarsyah, B. (2022). Multi-perspective approach to building Team resilience in project management—A case study in Indonesia. *Sustainability (Basel)*, *14*(20), 13137. DOI: 10.3390/su142013137

Harrison, R. T. (2017). Leadership leadership development and all that jazz. *Leadership*, *13*(1), 81–99. DOI: 10.1177/1742715016681120

Hegde, S., Larsen, E., Torbett, O., Ponnala, S., Pohl, E., Sze, R., & Schaeubinger, M. M. (2023). A proactive learning Approach toward building adaptive capacity during COVID-19: A radiology case study. *Applied Ergonomics*, *110*, 104009. DOI: 10.1016/j.apergo.2023.104009

Helmrich, A., & Chester, M. (2022). Navigating exploitative and explorative leadership in support of infrastructure resilience. *Frontiers in Sustainable Cities*, *4*, 791474. DOI: 10.3389/frsc.2022.791474

Hickman, X. (2023). A challenge for higher education leadership: The multi-faceted and irreversible impact of the COVID-19 pandemic. *Journal of Higher Education Policy and Leadership Studies*, *4*(4), 147–158. DOI: 10.61186/johepal.4.4.147

Hoff, R., Helmrich, A., Dirks, A., Kim, Y., Li, R., & Chester, M. (2023). Dynamic criticality for infrastructure prioritization In complex environments. *Environmental Research: Infrastructure and Sustainability*, *3*(1), 015011. DOI: 10.1088/2634-4505/acbe15

Hong, J., Cross Francis, D., Wang, Q., Lewis, L., Parsons, A., Neill, C., & Meek, D. (2020). The role of trust: Teacher capacity during school leadership transition. *Frontiers in Education*, *5*(108), 1–9. DOI: 10.3389/feduc.2020.00108

Hube, B., Stockport, G., & Soutar, G. (2022). A cogwheel model of dynamic capabilities: Evidence from an Australian university. *Australian Journal of Public Administration*, *81*(4), 569–588. DOI: 10.1111/1467-8500.12554

Indrianti, Y., Sasmoko, S., Abdinagoro, S. B., & Rahim, R. K. (2024). Building bridges to entrepreneurial resilience: Exploring the mediating role of business model innovation capacity in ambidextrous leadership and entrepreneurial mindfulness. *Uncertain Supply Chain Management*, *12*(4), 2245–2252. DOI: 10.5267/j.uscm.2024.6.007

Jun, L., & Weilin, W. (2024). A bibliometric analysis of the evolving mechanisms of shadow education research. *European Journal of Education*, *59*(3), 1–18. DOI: 10.1111/ejed.12646

Juncos, A. E. (2018). Resilience in peacebuilding: Contesting uncertainty, ambiguity, and complexity. *Contemporary Security Policy*, *39*(4), 559–574. DOI: 10.1080/13523260.2018.1491742

Kachniewska, M., & Para, A. (2023). Feminine vs. masculine: Expectations of leadership styles in hotels during the COVID-19 pandemic. *Sustainability (Basel)*, *15*(13), 10602. DOI: 10.3390/su151310602

Kaya, Y. (2023). Agile leadership from the perspective of dynamic capabilities and creating value. *Sustainability (Basel)*, *15*(21), 15253. DOI: 10.3390/su152115253

Kelly, S. E., Moher, D., & Clifford, T. J. (2016). Quality of conduct and reporting in rapid reviews: An exploration of Compliance with PRISMA and AMSTAR guidelines. *Systematic Reviews*, *5*(1), 79. DOI: 10.1186/s13643-016-0258-9

Khan, Z., & Lew, Y. K. (2018). Post-entry survival of developing economy international new ventures: A dynamic capability perspective. *International Business Review*, *27*(1), 149–160. DOI: 10.1016/j.ibusrev.2017.06.001

Kligyte, G., & Barrie, S. (2014). Collegiality: Leading us into fantasy-the paradoxical resilience of collegiality in academic leadership. *Higher Education Research & Development*, *33*(1), 157–169. DOI: 10.1080/07294360.2013.864613

Krupinski, E. A. (2019). Writing systematic reviews of the literature—It really is a systematic process. *Journal of Digital Imaging*, *32*(2), 199–200. DOI: 10.1007/s10278-018-00176-x

Lane, P. J., Clay-Williams, R., Johnson, A., Garde, V., & Barrett-Beck, L. (2021). Creating a healthcare variant CYNEFIN Framework to improve leadership and urgent decision-making in times of crisis. *Leadership in Health Services, ahead-of-print(ahead-of-print),* 454–461. DOI: 10.1108/LHS-03-2021-0013

Ledesma, J. (2014). Conceptual frameworks and research models on resilience in leadership. *SAGE Open*, *4*(3), 1–8. DOI: 10.1177/2158244014545464

Lisdiono, P., Said, J., Yusoff, H., & Hermawan, A. A. (2022). examining leadership capabilities, risk management practices, and organizational resilience: The case of state-owned enterprises in Indonesia. *Sustainability (Basel)*, *14*(10), 6268. DOI: 10.3390/su14106268

Livingstone, O., & Anthony, B. P. (2023). Documenting fisheries co-management in action: Case of the Grand Cape Mount County co-management Association, Liberia. *Marine Policy*, *157*, 105844. DOI: 10.1016/j.marpol.2023.105844

Long, N. D. B., Ooi, P. T., Le, T. V., Thiet, L. T., Ai, T. V., An, L. Q., Hudson, A., Tan, K. S., & Van, N. T. L. (2022). Leading in the age of the fourth industrial revolution–a perspective from Vietnam. *International Journal of Technology*, *13*(5), 949–957. DOI: 10.14716/ijtech.v13i5.5839

Lozano, S., Calzada-Infante, L., Adenso-Díaz, B., & García, S. (2019). Complex network analysis of keywords co-occurrence in the recent efficiency analysis literature. *Scientometrics*, *120*(2), 609–629. DOI: 10.1007/s11192-019-03132-w

Lyu, P., Liu, X., & Yao, T. (2023). A bibliometric analysis of literature on bibliometrics in recent half-century. *Journal of Information Science*, 01655515231191233. Advance online publication. DOI: 10.1177/01655515231191233

Magwegwe, F. (2021). Ayanda Mbatha: Resilience in action during the COVID-19 pandemic (A). *Emerald Emerging Markets Case Studies*, *11*(4), 1–12. DOI: 10.1108/EEMCS-06-2021-0201

Mahajan, R., Lim, W. M., Kumar, S., & Sareen, M. (2023). COVID-19 and management education: From pandemic to endemic. *International Journal of Management Education*, *21*(2), 100801. DOI: 10.1016/j.ijme.2023.100801

Majnoor, N., & Vinayagam, K. (2023). The ascendency of the paradigm shift from organizational change management to change agility. *International Journal of Professional Business Review*, *8*(4), e01151. DOI: 10.26668/businessreview/2023.v8i4.1151

Mäkinen, E. I. (2018). Complexity leadership theory and the leaders of transdisciplinary science. *Informing Science*, *21*, 133–155. DOI: 10.28945/4009

Mariani, M. M., Machado, I., Magrelli, V., & Dwivedi, Y. K. (2023). Artificial intelligence in innovation research: A systematic review, conceptual framework and future research directions. *Technovation*, *122*, 102623. DOI: 10.1016/j.technovation.2022.102623

Maupin, C. K., Mohan, G., Choudhury, A., Deepak, P., & Jin, F. (2024). Network-based approaches to leadership: An organizing framework review, and recommendations. *The Leadership Quarterly*, *35*(1), 101753. DOI: 10.1016/j.leaqua.2023.101753

Melián-Alzola, L., Domínguez-Falcón, C., & Martín-Santana, J. D. (2020). The role of the human dimension in organizational agility: An empirical study in intensive care units. *Personnel Review*, *49*(9), 1945–1964. DOI: 10.1108/PR-08-2019-0456

Mervyn, K., Amoo, N., & Malby, R. (2019). Challenges and insights in inter-organizational collaborative healthcare networks: An empirical case study of a place-based network. *The International Journal of Organizational Analysis*, *27*(4), 875–902. DOI: 10.1108/IJOA-05-2018-1415

Meyerowitz, D., Lew, C., & Svensson, G. (2018). Scenario-planning in strategic decision-making: Requirements, benefits and inhibitors. *foresight,* 20(6), 602–621. DOI: 10.1108/FS-04-2018-0036

Millings, M., Burke, L., & Robinson, G. (2019). Lost in transition? The personal and professional challenges for probation leaders engaged in delivering public sector reform. *Probation Journal*, *66*(1), 60–76. DOI: 10.1177/0264550518820120

Mitchell, M. (2016). Authentic leadership in illness blogs: What we can learn from jaw surgery bloggers. *Cogent Social Sciences*, *2*(1), 1271270. DOI: 10.1080/23311886.2016.1271270

Mohamed Shaffril, H. A., Samsuddin, S. F., & Abu Samah, A. (2021). The ABC of systematic literature review: The basic Methodological guidance for beginners. *Quality & Quantity*, *55*(4), 1319–1346. DOI: 10.1007/s11135-020-01059-6

Montasser, D., Prijadi, R., & Balqiah, T. E. (2023). The mediating effect of IT-enabled dynamic capabilities and organizational readiness on the relationship between transformational leadership and digital business model innovation: Evidence from Indonesia incumbent firms. *SAGE Open*, *13*(2), 21582440231181588. DOI: 10.1177/21582440231181588

Morales, S. N., Martínez, L. R., Gómez, J. A. H., López, R. R., & Torres-Argüelles, V. (2019). Predictors of organizational resilience by factorial analysis. *International Journal of Engineering Business Management*, *11*, 1–13. DOI: 10.1177/1847979019837046

Mustak, M., Salminen, J., Plé, L., & Wirtz, J. (2021). Artificial intelligence in marketing: Topic modeling, scientometric analysis, and research agenda. *Journal of Business Research, 124*, 389–404. DOI: 10.1016/j.jbusres.2020.10.044

Mustofa, A., & Tjahjono, H. K. (2024). Improving organizational agility through ethical leadership, employee job satisfaction and Islamic OCB. *Asian Journal of Business Research, 14*(2), 129–150. DOI: 10.14707/ajbr.240176

Nayak, B., Bhattacharyya, S. S., & Krishnamoorthy, B. (2023). Integrating the dialectic perspectives of resource-based view and industrial organization theory for competitive advantage–a review and research agenda. *Journal of Business and Industrial Marketing, 38*(3), 656–679. DOI: 10.1108/JBIM-06-2021-0306

Nguyen, T., Le, C. V., Nguyen, M., Nguyen, G., Lien, T.T.H., & Nguyen, O. (2024). The organisational impact of agility: A systematic literature review. *Management Review Quarterly,* 1-49. DOI: 10.1007/s11301-024-00446-9

Niemeyer-Rens, R. F., Sullivan, G., & Self, S. W. (2022). Leadership adaptability within higher education. *Journal of Higher Education Theory and Practice, 22*(9), 1–15. DOI: 10.33423/jhetp.v22i9.5357

Nkurunziza, G., Munene, J., Ntayi, J., & Kaberuka, W. (2019). Business process re-engineering in developing economies: Lessons from microfinance institutions (MFIs) in Uganda. *Innovation & Management Review, 16*(2), 118–142. DOI: 10.1108/INMR-03-2018-0010

Øby, E. (2024). Assessing transparency and methodological precision in variable measurement within organizational research: Implications for validity. *Quality & Quantity*, •••, 1–18. DOI: 10.1007/s11135-024-01991-x

Pasmore, W., Winby, S., Mohrman, S. A., & Vanasse, R. (2019). Reflections: Sociotechnical systems design and organization change. *Journal of Change Management, 19*(2), 67–85. DOI: 10.1080/14697017.2018.1553761

Ramadan, M., Bou Zakhem, N., Baydoun, H., Daouk, A., Youssef, S., El Fawal, A., Elia, J., & Ashaal, A. (2023). Toward digital transformation and business model innovation: The Nexus between leadership, organizational agility, and knowledge transfer. *Administrative Sciences, 13*(8), 185. DOI: 10.3390/admsci13080185

Ramli, K., Spigel, B., Williams, N., Mawson, S., & Jack, S. (2023). Managing through a crisis: Emotional leadership strategies of high-growth entrepreneurs during the COVID-19 pandemic. *Entrepreneurship and Regional Development, 35*(1-2), 24–48. DOI: 10.1080/08985626.2022.2143905

Raney, A. F. (2014). Agility in adversity: Integrating mindfulness and principles of adaptive leadership in the administration of a community mental health center. *Clinical Social Work Journal, 42*(3), 312–320. DOI: 10.1007/s10615-014-0487-0

Rani, I. H., & Widyowati, L. A. (2021). The role of visionary leader competency in bridging company's levers of controls and dynamic capabilities. *Journal of Logistics. Informatics and Service Science, 8*(2), 1–18. DOI: 10.33168/LISS.2021.0201

Rejeb, A., Rejeb, K., Simske, S. J., & Treiblmaier, H. (2023). Drones for supply chain management and logistics: A review and research agenda. *International Journal of Logistics, 26*(6), 708–731. DOI: 10.1080/13675567.2021.1981273

Rethlefsen, M. L., Kirtley, S., Waffenschmidt, S., Ayala, A. P., Moher, D., Page, M. J., Koffel, J. B., Blunt, H., Brigham, T., Chang, S., Clark, J., Conway, A., Couban, R., de Kock, S., Farrah, K., Fehrmann, P., Foster, M., Fowler, S. A., Glanville, J., & Young, S. (2021). PRISMA-S: An extension to the PRISMA Statement for Reporting Literature Searches in Systematic Reviews. *Systematic Reviews, 10*(1), 1–19. DOI: 10.1186/s13643-020-01542-z

Reyes, D. L., Bisbey, T., Day, D., & Salas, E. (2021). Translating 6key insights from research on leadership and management in times of crisis. *BMJ Leader, 5*(4), 291–294. DOI: 10.1136/leader-2020-000349

Roblek, V., Dimovski, V., Jovanov Oblak, K., Meško, M., & Peterlin, J. (2024). Leadership and managerial challenges to ensure agile management as a method to enable business success: A Delphi study of the Slovenian health organisations. *Measuring Business Excellence, 28*(1), 39–51. DOI: 10.1108/MBE-09-2023-0122

Rodina, L., Harris, L., Ziervogel, G., & Wilson, J. (2024). Resilience countercurrents: Water infrastructures, informality, and inequities in Cape Town, South Africa. *World Development, 180*, 106619. DOI: 10.1016/j.worlddev.2024.106619

Rowe, S. W., Arghode, V., & Bhattacharyya, S. S. (2024). A study on adaptive performance, work-related psychological health and demographics in Episcopal Church bishops. *Journal of Work-Applied Management, 16*(1), 31–47. DOI: 10.1108/JWAM-02-2023-0015

Safitra, M. F., Lubis, M., & Fakhrurroja, H. (2023). Counterattacking cyber-threats: A framework for the future of cyber-security. *Sustainability (Basel), 15*(18), 13369. DOI: 10.3390/su151813369

Sampson, M., McGowan, J., Cogo, E., & Horsley, T. (2006). Managing database overlap in systematic reviews using batch citation matcher: Case studies using Scopus. *Journal of the Medical Library Association,* 94(4), 461–463, e219. +E.

Saputra, N., Putera, R. E., Zetra, A., Valentina, T. R., & Mulia, R. A. (2024). Capacity building for organizational performance: A systematic review, conceptual framework, and future research directions. *Cogent Business & Management*, *11*(1), 2434966. Advance online publication. DOI: 10.1080/23311975.2024.2434966

Scheepers, C. B., Mathibe, M. S., & Verachia, A. (2023). Localisation of PPE manufacturing during COVID-19 (local manufacturing partnership and Kingsgate Clothing South Africa). *Emerald Emerging Markets Case Studies*, *13*(1), 1–13. DOI: 10.1108/EEMCS-10-2022-0352

Schell, J. A. (2023). Designing for academic resilience in hands-on courses in times of crisis: Two models for supporting hands-on online learning drawn from the COVID-19 Pandemic. *The American Behavioral Scientist*, *67*(13), 1632–1654. DOI: 10.1177/00027642221118292

Schoemaker, P. J. H., Heaton, S., & Teece, D. (2018). Innovation, dynamic capabilities, and leadership. *California Management Review*, *61*(1), 1–28. DOI: 10.1177/0008125618790246

Schulze, J. H., & Pinkow, F. (2020). Leadership for organisational adaptability: How enabling leaders create adaptive space. *Administrative Sciences*, *10*(3), 37. DOI: 10.3390/admsci10030037

Silalahi, S. A. F., Junaidi, A., Nasution, L. Z., Sutoto, A., Hutomo, A. S., Sutrisno, J., Lukas, A., Ginting, J., Suripto, S., & Lestari, R. (2023). Grit and business survivability among small business during the COVID-19 pandemic: The moderating role of servant leadership. *Cogent Business & Management*, *10*(3), 2284442. Advance online publication. DOI: 10.1080/23311975.2023.2284442

Siregar, A. A., Afiff, A. Z., & Halim, R. E. (2023). Linking agile leadership and business sustainability through the mediation of political and social capabilities. *Journal of Open Innovation*, *9*(4), 100153. DOI: 10.1016/j.joitmc.2023.100153

Slater, S. F., Mohr, J. J., & Sengupta, S. (2014). Radical product innovation capability: Literature review, synthesis, and illustrative research propositions. *Journal of Product Innovation Management*, *31*(3), 552–566. DOI: 10.1111/jpim.12113

Song, Y., Lei, L., Wu, L., & Chen, S. (2023). Studying domain structure: A comparative analysis of bibliographic coupling analysis and co-citation analysis considering all authors. *Online Information Review*, *47*(1), 123–137. DOI: 10.1108/OIR-12-2020-0540

Song, Y., Sun, X., & Duan, W. (2023). Co-word analysis of first-authored publications written by top cited scholars of social work. *Research on Social Work Practice*, *33*(8), 929–943. DOI: 10.1177/10497315221148661

Thude, B. R., Juhl, A. G., Stenager, E., von Plessen, C., & Hollnagel, E. (2019). Staff acting resiliently at two hospital wards. *Leadership in Health Services (Bradford, England)*, *32*(3), 445–457. DOI: 10.1108/LHS-09-2018-0042

Trujillo, C. M., & Long, T. M. (2018). Document co-citation analysis to enhance transdisciplinary research. *Science Advances*, *4*(1), e1701130. DOI: 10.1126/sciadv.1701130

Urs, N., Roja, A., & Nisioi, I. (2024). Digital transformation in organizational management: A bibliometric analysis. *NISPAcee Journal of Public Administration and Policy*, *17*(1), 198–227. DOI: 10.2478/nispa-2024-0009

Utomo, F. C., & Latukismo, T. H. (2022). Trends and patterns in workforce agility literature: A scopus-based bibliometric analysis. *Journal of Wireless Mobile Networks, Ubiquitous Computing and Dependable Applications*, *13*(4), 211–224. DOI: 10.58346/JOWUA.2022.I4.014

Viano, S., Shahrokhi, F., & Hunter, S. B. (2024). Improvement science and school leadership: The precarious path to dynamic school improvement. *Frontiers in Education*, *9*, 1371664. Advance online publication. DOI: 10.3389/feduc.2024.1371664

Walch, C. (2019). Adaptive governance in the developing world: Disaster risk reduction in the State of Odisha, India. *Climate and Development*, *11*(3), 238–252. DOI: 10.1080/17565529.2018.1442794

Wang, D., Yang, Y., Yang, L., & Yang, H. (2023). Bibliometric analysis and visualization of endocrine therapy for breast cancer research in the last two decade. *Frontiers in Endocrinology (Lausanne)*, *14*, 1287101. DOI: 10.3389/fendo.2023.1287101

Wang, X., Reardon, L., & To, L. S. (2022). Visualizing Nepal's electricity supply resilience from a whole-systems perspective: A participatory approach. *Energy Research & Social Science*, *85*, 102409. DOI: 10.1016/j.erss.2021.102409

Weerts, K., Vermeulen, W., & Witjes, S. (2018). On corporate sustainability integration research: Analysing corporate leaders' experiences and academic learnings from an organisational culture perspective. *Journal of Cleaner Production*, *203*, 1201–1215. DOI: 10.1016/j.jclepro.2018.07.173

Weng, H., & Feng, Y. (2022). Discussion of 'cocitation and coauthorship networks of statisticians. *Journal of Business & Economic Statistics*, *40*(2), 486–490. DOI: 10.1080/07350015.2022.2037432

Whyte, J., Naderpajouh, N., Clegg, S., Matous, P., Pollack, J., & Crawford, L. (2022). Project leadership: A research agenda for a changing world. *Project Leadership and Society*, *3*, 100044. DOI: 10.1016/j.plas.2022.100044

Winkler, P., Kretschmer, J., & Wamprechtsamer, P. (2024). Navigating through digitalization challenges in strategic communication: Introducing the VUCA radar. *Journal of Communication Management (London)*. Advance online publication. DOI: 10.1108/JCOM-11-2023-0119

Wowk, K., Adams, M., & Martinez, E. (2023). Translating the complexity of disaster resilience with local leaders. *Frontiers in Communication*, *8*, 1100265. DOI: 10.3389/fcomm.2023.1100265

Yamin, M. A. (2021). Investigating the drivers of supply chain resilience in the wake of the COVID-19 pandemic: Empirical evidence from an emerging economy. *Sustainability (Basel)*, *13*(21), 11939. DOI: 10.3390/su132111939

Yang, L., Han, L., & Liu, N. (2019). A new approach to journal co-citation matrix construction based on the number of co-cited articles in journals. *Scientometrics*, *120*(2), 507–517. DOI: 10.1007/s11192-019-03141-9

Yeomans, L., & Bowman, S. (2021). Internal crisis communication and the social construction of emotion: University leaders' sensegiving discourse during the COVID-19 pandemic. *Journal of Communication Management (London)*, *25*(3), 196–213. DOI: 10.1108/JCOM-11-2020-0130

Yu, M., Wen, J., Smith, S. M., & Stokes, P. (2022). Building-upHresilience and being effective leaders in the workplace: A systematic review and synthesis model. *Leadership and Organization Development Journal*, *43*(7), 1098–1117. DOI: 10.1108/LODJ-09-2021-0437

Zabaniotou, A. (2020). A systemic approach to resilience and ecological sustainability during the COVID-19 pandemic: Human, societal, and ecological health as a system-wide emergent property in the Anthropocene. *Global Transitions*, *2*, 116–126. DOI: 10.1016/j.glt.2020.06.002

Zahari, A. I., Abdul Manan, D. I., Mohamed, N., & Said, J. (2023). Impact of dynamic leadership and marketing planning on organizational resilience during Covid-19: Higher learning institutions. *SAGE Open*, *13*(4), 1–12. DOI: 10.1177/21582440231212042

Zarif, A., Bandyopadhyay, S., Miller, G., & Malawana, J. (2024). Delivering medical leadership training through the Healthcare Leadership Academy: A four year analysis. *BMC Medical Education*, *24*(1), 194. DOI: 10.1186/s12909-024-05031-y

Zhang, J., Liu, Y., Gu, X., & Chai, J. (2024). A bibliometrics and visualization analysis of ropivacaine research from 2000 to 2023. *Frontiers in Medicine*, *11*, 1465308. DOI: 10.3389/fmed.2024.1465308

Zhang-Zhang, Y., Rohlfer, S., & Varma, A. (2022). Strategic people management in contemporary highly dynamic VUCA contexts: A knowledge worker perspective. *Journal of Business Research*, *144*, 587–598. DOI: 10.1016/j.jbusres.2021.12.069

Zupic, I., & Čater, T. (2015). Bibliometric methods in management and organization. *Organizational Research Methods*, *18*(3), 429–472. DOI: 10.1177/1094428114562629

Chapter 8
Digitalization of Metropolis:
Resources, Risks, Capitalization

Ekaterina Vladimirovna Kargapolova
Plekhanov Russian University of Economics, Russia

Arina Sergeevna Dikanova
https://orcid.org/0009-0006-6205-9018
Plekhanov Russian University of Economics, Russia

Anna Jurievna Aryasova
https://orcid.org/0009-0004-0702-2656
Astrakhan State University of Architecture and Civil Engineering, Russia

Julia Victorovna Georgievskaya
Astrakhan State University of Architecture and Civil Engineering, Russia

ABSTRACT

Digitalization of metropolis transforms social connections and human capital structures. Relevance of the study on the resource base, content and negative consequences of using Internet resources by residents of the metropolis is obvious. This survey presents results of a sociological study of Moscow residents, conducted according to the Standard Methodology of the All-Russian Scientific Research Program "Sociocultural Evolution of Russia and Its Regions" in December 2023. The assessments of the Internet resources used by respondents for entertainment, work, interaction with local authorities, opportunities for saving money and time are analyzed, risks and negative consequences are considered. Respondents use capabilities of online stores, various platforms for online communication, with a predominance of young and adult respondents, as well as respondents from the

DOI: 10.4018/979-8-3373-2240-7.ch008

wealthy category. Low level of benefit from using Internet resources is observed among elderly and low-income respondents. These categories also need adapted digital services and accessible opportunities for their needs.

INTRODUCTION

Digitalization of metropolis represents a significant aspect of modern urban development, especially in the conditions of uncertainty and instability characteristic of the VUCA and BANI era (Bounfour, 2016). In contemporary urban theory, metropolis is a space where digital systems have a highly real impact on inhabitants' experience, move, communication and social interaction. (Liu, Ju, Feng, Hu, 2019; Bolivar, Muñoz, 2019). Internet technologies are taking over all dimensions of urban life, transforming the metropolis into a "computer for living" (Ratti, Claudell, 2017, p. 37; Townsend, 2013; Song, Srinivasan, Sookoor, Jeschke, 2017).

Modern technologies such as 5G, Internet of Things (IoT) (Familiar, Barnes, 2017) and cloud computing form the basis for digitalization. These technologies provide real-time data collection and analysis, which is critical for the effective management of city systems (Song, Srinivasan, Sookoor, Jeschke, 2017). There is a growing need to integrate these technologies to improve the resilience of cities in conditions of uncertainty and for the self-organization of civil society. The ambiguity and complexity of the impact of digitalization processes on the development of cities and regions in the first quarter of the 21st century are emphasized by many experts (Kabanov, Chugunov, 2018; Sim, Hwang, Choi, 2019). Thus, among the positive aspects of the influence of digitalization, it can be highlighted that it contributes to the economic growth of megacities, development of start-ups and innovative enterprises, creating new jobs and improving the quality of life (Evans, 2017).

However, there are also serious problems. Thus, in the VUCA and BANI conditions it is difficult to predict how technologies and needs of the population of megacities will develop. In such realities no one is insured against ineffective investments. The problem of growing consumption of all types of material resources by metropolis is increasing year by year. The solution to this problem is seen in the development and widespread use of new resource-intensive, intelligent urban development technologies, taking into account both global trends, threats, challenges, and the specifics of a particular urban area.

Large business is interested in investing in promising industries, particularly, in the development of information systems and technologies. Digitalization requires significant financial investments, including both public funding and private investment, based on mutually beneficial partnerships between the public and private sectors for the successful implementation of digital initiatives (Ashmarina, Mesquita, Vochozka, 2020; Kotarba, 2018).

Big data analytics is becoming a key asset in the digital economy. The increasing level of digitalization leads to an increase in cyber threats, making the protection of city systems and private users' data a cybersecurity priority for city authorities. Effective data management, responding to the challenges associated with data management and protection in the context of digitalization are becoming critical to ensuring the security and privacy of both enterprises and ordinary users (Maheshwari, 2019; Nicoletti,2018).

Digitalization can become a factor in increasing social inequality if access to technologies is limited for certain groups of the population – issues of inclusiveness and accessibility of technology are among the most pressing (Van Dijk, 2020;Nesti, 2019).The contradictions and volatility of the digital age in which we live arise from the need to overcome the contradiction between the potential for new forms of engagement and self-organization of citizens and the tendency to be marginalized by new forms of technocratic control, which they often produce themselves (Pappas, Mikalef, Giannakos, Krogstie, Lekakos, 2018).

The increasing use of Internet technologies can lead to negative consequences for the environment. In such conditions, there is a priority of the need for friendly environment and sustainable development of the metropolis in the context of digitalization (Manjon-Antolin, Aouni,Crutzen, 2019).

The issues of using Internet resources by residents of megacities metropolises are becoming increasingly relevant in the light of digitalization and changes in the social structure. Metropolis residents actively use various Internet platforms to meet their needs for information, entertainment and social interactions through streaming services and online games, which highlights the influence of these platforms on leisure and cultural preferences (de Lange, Synnes, Leindecker, 2019; Johnson, Lee, 2020). Different age groups in metropolitan areas use social media to communicate and express themselves, which affects their social connections and identity (Smith et al., 2021; Ryazanova, 2020).

METHODS

In December 2023, a research group from the Department of Political Analysis and Socio-Psychological Processes of the Plekhanov Russian University of Economics, under the leadership of Doctor of Sociological Sciences, Professor E.V. Kargapolova, conducted an initiative sociological study among Moscow residents. This study was based on the Standard Methodology of the All-Russian Scientific Research Program “Sociocultural Evolution of Russia and its Regions”, which has been tested since 2006 in 30 regions of the country. The methodology was developed by the Center for the Study of Sociocultural Changes (CSSC) of the Institute of Philosophy of the Russian Academy of Sciences, under the guidance of such scientists as Corresponding Member of the Russian Academy of Sciences, Doctor of Philosophy N.I. Lapin and Doctor of Sociology, Leading Researcher, Acting Director of CSSC IF RAS L.A. Belyaeva.

Vologda Scientific Center of the Russian Academy of Sciences, headed by Doctor of Economics A.A. Shabunova, also made its own additions to the methodology, which made it possible to compare results of the study in Moscow with the data from other regions of Russia, where similar empirical studies were conducted. The research method was an online questionnaire, in which 1070 respondents took part. The sample was targeted, and for several years the research group led by E.V. Kargapolova developed an algorithm for implementing an online survey taking into account the specified quota characteristics (details can be found in the work of Kargapolova and Davydova, 2022). The age and gender structure of respondents corresponds to the general structure of the Moscow population: 43.2% men and 56.8% women; 27.7% youth (aged 14 to 35), 46.7% adults (aged 36 to 59) and 25.6% elderly people (over 60).

The questionnaire contained questions allowing to assess the frequency and possibilities of using Internet resources by residents of the metropolis, in particular, concerning the use of social networks, entertainment, educational and news resources, e-mail, online banking services, online stores and the State Services resource. Answers were also received to the question about the possibilities of earning or saving resources (material resources and time) with the help of the Internet. The question of whether respondents had encountered negative consequences of using the Internet made it possible to identify the existing challenges and dangers of using the "world wide web". The data analysis included the study of linear distributions by gender and age. In accordance with the recommendations for calculating the index proposed in the Standard Methodology, the scale for assessment included positive values.

RESULTS

When answering the question "What Internet resources and how often do you use almost daily?" 82.9% of respondents noted social networks, for example, VKontakte, Odnoklassniki, Telegram (Table 1). This option is the most popular among young people (96%), and the least popular among older people (63.5%). Women use social networks more often than men (84.4% versus 80.8%). A study of the financial situation of respondents demonstrates that the share of those who reports almost daily use of social networks is the highest among rich people (88.5%), while poor people chose this Internet resource the least often (75.5%).

Table 1. Distribution of respondents' answers to the question "What Internet resources do you use and how often?" (% of respondents which chose the answer option "almost daily")

Answer options	Gender		Age			Total
	men	women	youth	adult	elderly	
Social networks, for example, VKontakte, Odnoklassniki, Telegram	80,8	**84,4**	**96,0**	85,9	63,5	82,9
Online banking services	65,3	64,9	**80,8**	68,9	40,9	65,1
News resources	**66,4**	62,8	57,6	**68,5**	63,9	64,3
E-mail	54,5	55,4	59,9	**64,9**	31,8	55,0
Entertainment, games, movies	**41,2**	32,8	**55,2**	31,5	25,2	36,4
Educational resources	33,0	32,3	**46,1**	31,7	19,7	32,6
Online stores	26,3	**34,3**	**39,4**	31,5	20,1	30,8
Government online services	14,9	18,0	16,5	16,9	16,1	16,7

The second most popular Internet resource is online banking services, used by 65.1% of respondents. Young people also dominate over other age groups in using these services, while the share of elderly people using this resource is almost half as much (80.8% versus 40.9%). The difference in the share of men and women is statistically insignificant. As for the financial status of respondents, here we can see a clear prevalence of rich people over other groups conditionally identified by us on this basis - in this group, 83.2% of respondents use online banking services. For comparison: among the people we define as poor, only 51% of respondents choose this Internet resource. News resources are used the least by young people (57.6%), and the most by adults (68.5%). The share of men using Internet news resources is slightly higher than that of women (66.4% versus 62.8%). As for the financial status of respondents, news resources are the most often consumed by people belonging to the poor and wealthy strata of the population (68.4 and 68.5%, respectively). This resource is chosen the least by the poor (52.8%).

More than half of the Muscovites surveyed (55%) mention e-mail as an Internet resource. The majority of respondents who choose this option are adults (68.5%), while the share of elderly people is more than twice less than the share of adults and amounts to 31.8%. E-mail as an Internet resource is the most often used by the rich (71.7%), and the least often by the poor (48%).Internet resources as a source of entertainment, games and movies are noted by 36.4% of respondents. The greatest interest in this method of using the Internet is demonstrated by men compared to women (41.2% versus 32.8% of women) and young people compared to adults and the elderly (55.2% versus 31.5% and 25.2% respectively). Interestingly, polar groups in terms of financial status – the rich and the poor – prefer this type of Internet use equally (43.4% and 42.9% respectively). At the same time, poor people choose Internet resources as a source of entertainment less often than other groups distinguished by financial status (28.6%). Educational resources are used by 32.6% of respondents. This option is chosen by almost every second young person (46.1%) and every fifth elderly person (19.7%). An analysis of the financial situation of respondents who noted this option demonstrated the same situation as in the case of the option "Internet resources as a source of entertainment, games and films", namely, that educational resources are most popular among two diametrically opposed groups - the rich and the poor (43.4% and 38.8%, respectively). Less than a third (30.8%) of respondents said they use online stores. There are almost 10% more women than men who have noted this Internet resource (34.3% versus 26.3%). The share of young people using online stores exceeds the share of older people by almost two times (39.4 and 20.1%, respectively). It is important to note that online stores are the most often used by the rich and the poor (50.4% and 38.8%, respectively). Despite the fact that the difference between the indicators of these groups is more than 10%, these strata still stand out significantly from other groups identified by financial status. The last place in the list of resources used by Muscovites surveyed is occupied by government services websites, which are used by 16.7% of respondents. Interestingly, this option is equally noted by both the poor and the rich (26.5% each).

In response to the question "Have you managed to earn or save money and time using the Internet?", the largest number of respondents (49.1%) stated that they managed to save time using the Internet by receiving state and municipal services online, ordering home delivery, etc. (Table 2). Most Muscovites (55.2%) who noted this option belong to the adult age group. As for financial status, there is a clear difference between those who are more and less well-off in material terms. Thus, almost every second person among the well-off, prosperous and rich noted this option (52.4, 54.9 and 52.2% respectively), while among the poor, the share of respondents who managed to save money and time was 16.3%, among the poor 38.8% and among the disadvantaged - 41.7%.

Table 2. Distribution of respondents' answers to the question "Have you been able to earn or save money and time using the Internet?" (% of respondents)

Answer options	Gender		Age			Total
	men	women	youth	adult	elderly	
Yes, with the help of the Internet I have saved my time (received state and municipal services online, ordered home delivery, etc.)	48,9	49,2	49,2	**55,2**	37,6	49,1
Yes, I have made great purchases online.	48,5	48,5	55,6	54,8	29,2	48,5
Yes, I have worked/work using the Internet	**37,3**	29,3	**42,8**	37,6	13,1	32,8
Yes, I have successfully sold/exchanged my items using online resources.	**30,8**	25,2	**41,1**	28,9	10,9	27,7
No, I have not been able to earn and/or save money and time using the Internet	18,3	**23,4**	12,8	16,9	**38,3**	21,2

About half (48.5%) of respondents made profitable purchases on the Internet. At the same time, only 29.2% of the elderly noted this option. An analysis of the financial situation demonstrated a significant difference between the poor and the rich when choosing this answer option. Only every fourth poor noted that he managed to make profitable purchases on the Internet (26.5%), while among the rich, every second (54.9%) chose this option. Every third person has worked or works using the Internet and thus earns or saves money. Among them, one can observe the predominance of men over women (37.3 and 29.3%, respectively). Interestingly, the share of young people exceeds the number of elderly people by almost four times. In groups by financial status, the rich and well-off most often choose this option (39.8% and 39.2%, respectively). Less than a third of respondents answered that they successfully sold or exchanged their things using Internet resources. It is important to note that more than 40% of young people noted this method of saving, while among the elderly this option appears in the answers of only every tenth. As for the division of groups based on the financial situation of the respondent's family, there is a problem with the implementation of this mechanism among the poor and the needy (18.4% and 16.3%, respectively). These results are almost half as many as in more financially prosperous groups.

Among respondents who answered that they were unable to earn or save money and time when using the Internet, there are every fifth. The largest number of such people are among the elderly - 38.3%, while the share of young people giving such an answer is only 12.8%. An analysis of the financial situation of respondents showed that the poor more often choose this option (46.9%). Their share is more than four times higher than the share of the rich who gave the same answer (11.5%). Muscovites' answers to the question "Have you and your immediate circle encountered the following negative consequences of using the Internet?" were distributed as follows (Table 3). In the first place is the option "Deterioration of physical health (in terms of vision, spine, blood vessels, etc.)", which was chosen by about a third of respondents. More often, this option is chosen by women and young people. Among the groups identified by material characteristics, the poor and the rich are the least likely to encounter this consequence (24.5 and 22.1%, respectively).

Table 3. Distribution of respondents' answers to the question "Have you and your immediate environment encountered the following negative consequences of using the Internet" (% of respondents)

Answer options	Gender		Age			Total
	men	women	youth	adult	elderly	
Deterioration of physical health (visual organs, spine, blood vessels, etc.)	24,4	**30,3**	29,0	27,7	26,6	27,7
The emergence of dependence on the use of Internet resources	25,0	27,4	**35,7**	26,7	15,7	26,4
Deterioration of psychological well-being (apathy, fear, melancholy, resentment)	21,3	**25,4**	**33,3**	23,1	14,2	23,6
Conflicts, quarrels in the family	15,7	15,6	18,2	16,5	11,3	15,6
Difficulties, problems at work/study	10,6	9,5	**15,8**	8,2	6,9	10,0
My environment and me have not encountered such consequences	28,9	28,9	29,0	27,5	31,4	28,9
I find it difficult to answer	**23,7**	19,7	15,5	21,7	**27,0**	21,4

The second most popular answer option is "development of dependence on the use of Internet resources", which was indicated by every fourth respondent, and mostly by young people. The share of the poor who note this consequence is less than the share of more prosperous groups, almost 10% and is 16.3%. In the third place there is an option "deterioration of psychological well-being (apathy, fear, melancholy, resentment)", which was chosen by every fourth or fifth respondent. This consequence is more typical for young people. Among the elderly, the share of those who indicated this answer option is more than two times less. We also see that among the rich, only 15% of respondents choose this option, which is the lowest figure among groups of respondents differentiated by financial status.

Every sixth or seventh respondent chose the answer option "conflicts, quarrels in the family". Young people face this consequence more frequently, and the elderly more rarely. When analyzing the respondents' answers, taking into account their financial situation, we saw that the poor choose this option more often, and the rich choose it less (21.4 versus 7.1%). Every tenth respondent faces consequences in the form of difficulties, problems at work/study. Moreover, this option most often appears in the answers of young people. Least often this option is noted by people we characterize as rich (6.2%). It is also necessary to consider the answer option "I and my environment have not encountered such consequences", which appears in the answers of almost the third of respondents. This option is the most popular among older people, and the least popular among adults. An analysis of the financial situation of respondents demonstrated a clear prevalence of those who chose this option among the rich compared to other groups (34.5%). About the fifth of respondents found it difficult to answer this question and there are more men and older people among them.

CONCLUSION

In conclusions, we can draw the following statements.

1. The results of our study confirm the popularity of using various Internet resources as the main source of information. At the same time, young people dominate over other age groups. The study of the financial situation of respondents demonstrated the prevalence of rich people over other groups conditionally identified by us on this basis. The difference between men and women in the consumption of various Internet content is in many cases not significant.
2. The main reason that motivates Russians to receive online services, make purchases, and work using the Internet is saving money and time. The most active online consumers are also concentrated in the youth group. Among respondents who indicate experience working using the Internet, men prevail over women. Half of those surveyed have already assessed the advantages and benefits of using online services and online stores. In this case, only among the elderly category is there a slight lag.
3. Among the negative consequences mentioned by ordinary users in the Internet environment, we can mention a dependence on the Internet and gadgets, as well as deterioration of emotional and psychological well-being stand out. It is confirmed by other studies (Vorobyeva, 2022), that these are the greatest risks felt by young respondents, and the lowest by the elderly. It is noteworthy that every third or fourth respondent in all age groups notes a deterioration in well-being due to the use of Internet resources.

REFERENCES

Bolivar, M. P. R., & Muñoz, L. A. (2019). *E-Participation in Smart Cities: Technologies and Models of Governance for Citizen Engagement*. Springer. DOI: 10.1007/978-3-319-89474-4

Bounfour, A. (2016). *Digital Futures, Digital Transformation*. Springer Cham Heidelberg., DOI: 10.1007/978-3-319-23279-9

de Lange, M., Synnes, K., & Leindecker, G. (2019). Smart Citizens in the Hackable City: On the Datafication, Playfulness, and Making of Urban Public Spaces Through Digital Art. In. Lecture Notes in Computer Science: Vol. 11380. *Smaniotto Costa, C., et al. CyberParks – The Interface Between People, Places and Technology* (pp. 157–166). Springer., DOI: 10.1007/978-3-030-13417-4_13

Evans, N. D. (2017). *Mastering digital business: How powerful combinations of disruptive technologies are enabling the next wave of digital transformation*. BCS - The Chartered Institute for IT.

Familiar, B., & Barnes, J. (2017). *Business in Real-Time Using Azure IoT and Cortana Intelligence Suite: Driving Your Digital Transformation*. Apress., DOI: 10.1007/978-1-4842-2650-6

Johnson, Lee 2020 News Consumption across Media Platforms and Content https://www.researchgate.net/publication/344619753

Kabanov, Yu. A., & Chugunov, A. V. (2018). Conceptualization of concepts used in "smart city" and "e-governance" studies: An experience of scientometric analysis. *International Journal of Open Information Technologies*, *6*(11), 54–58.

. Kargapolova E.V., DavydovaYu.A. (2022). Well-being of the population of a Russian region in the context of socio-economic restrictions (on the example of the Moscow agglomeration) // CITISE. No. 3. P. 301-316.

Kotarba, M. (2018). Digital Transformation of Business Models. *Foundations of Management.*, *10*(1), 123–142. DOI: 10.2478/fman-2018-0011

Liu, L., Ju, J., Feng, Y., & Hu, Q. 2019. Impact of Governance Structure Characteristics of Public-private Partnerships on Smart City Project Success: Evidence from a Multi-case Study in China // Proceedings of the 52nd Hawaii International Conference on System Sciences. DOI: DOI: 10.24251/HICSS.2019.397

Maheshwari, A. (2019). *Digital Transformation: Building Intelligent Enterprises*. Wiley.

. Manjon-Antolin, M., Aouni, Z., Crutzen, N. (2019). Green and Digital Entrepreneurship in Smart Cities // In Annals of Regional Science. Volume 68. Pp. 429-462 DOI: .DOI: 10.1007/s00168-021-01080-z

Nesti, G. (2019). Mainstreaming gender equality in smart cities: Theoretical, methodological and empirical challenges. *Information Polity*, *24*(3), 289–304. DOI: 10.3233/IP-190134

Nicoletti, B. (2018). Agile Procurement: Volume II: Designing and Implementing a Digital Transformation. Cham, Switzerland. *Palgrave Macmillan., 281p*. Advance online publication. DOI: 10.1007/978-3-319-61085-6

Pappas, I. O., Mikalef, P., Giannakos, M. N., Krogstie, J., & Lekakos, G. (2018). Big data and business analytics ecosystems: Paving the way towards digital transformation and sustainable societies. *Information Systems and e-Business Management*, *16*(3), 479–491. DOI: 10.1007/s10257-018-0377-z

Ratti, K., & Claudel, M. (2017). *Cities of Tomorrow: Sensors, Networks, Hackers and the Future of Urban Life*. Gaidar Institute Publishing House.

Ryazanova, L. V. (2020). *2020 Digital transformation of state and society: collection of articles*. Rusains.

Sim, K., Hwang, W.-S., & Choi, M.-R. (2019). Building a New Smart City: Integrating Local Culture and Technology. *Journal of Digital Convergence.*, *17*(9), 193–198. DOI: 10.14400/JDC.2019.17.9.193

Song, H., Srinivasan, R., Sookoor, T., & Jeschke, S. (2017). *Smart Cities: Foundations, Principles, and Applications*. Wiley., DOI: 10.1002/9781119226444

Townsend, A. M. (2013). *Smart Cities: Big Data, Civic Hackers, and the Quest for a New Utopia*. W. W. Norton & Company, Inc.

van Dijk, J. (2020). *The Digital Divide*. Polity Press., DOI: 10.1002/asi.24355

Vorobyova, I. V., & Zhdanova, N. E. Personal predictors of students in the perception of stress factors of the real and virtual urban environment // Pedagogical education in Russia. 2022. No. 1. P. 129-136. DOI: DOI: 10.26170/2079-8717_2022_01_15

Chapter 9

Research, Development, and Comparative Characterization of Methods for Invisible Embedding Digital Watermarks in Electronic Text Documents

Maxim Martemyanov
https://orcid.org/0009-0004-1906-1150
Faculty of Mathematics and Computer Sciences Named After Prof. Nikolay Chervyakov, Russia

Maria Lapina
https://orcid.org/0000-0001-8117-9142
Faculty of Mathematics and Computer Sciences Named After Prof. Nikolay Chervyakov, Russia

Mary Anita E. A.
Independent Researcher, India

ABSTRACT

This article discusses the concept of digital watermark (DWC), its classification, the main file formats for storing textual information electronically, types of attacks on DWC, and methods of embedding digital watermarks in electronic text documents

DOI: 10.4018/979-8-3373-2240-7.ch009

based on formatting changes, using metadata, structural and linguistic changes. The practical development of each method is carried out, their advantages, disadvantages and resistance to various attacks are identified, and their comparative characterization is presented. On the basis of the analysis, we experimentally developed our own methods of introducing DWC into text documents, including the approach using zero-width spaces (ZWSP) (U+200B), which provides high resistance to various attacks and imperceptibility for the user. The results of the work demonstrate the effectiveness of each implementation and provide recommendations for their application depending on the requirements for copyright protection and authentication of text files.

INTRODUCTION

This is an era in which the information technology industry is placing great importance on electronic document processing. The ever-increasing mass of digital data and the ease of access via the Internet create new risks of unauthorised copying and distribution. However, at this stage, digital watermarks (DWC) act as a defense against threats and are very effective in such cases, thus ensuring that copyright is not infringed by unwanted interference from third parties.

Research results highlight the importance of digital watermarks as a security measure for electronic information. The inclusion of digital watermarks in text documents is an invisible marking that can only be identified using specialised technologies, thus reducing piracy (Ochnev & Chirkin, n.d.). In this way, DWC increase the level of information security by embedding themselves in the text in such a way that it is difficult to remove or forge it without compromising the integrity of the document. DWC can be used to track changes and identify those who have edited or distributed a document.

The role of these technologies is growing, and with it the development of digital multimedia. The introduction of digital watermarks in text documents serves not only to combat piracy, but also to identify the owner, including his rights, to the document; thus, it helps to avoid plagiarism and misappropriation of others' work.

DWC has two main directions: steganography, aimed at hiding the facts of transmission or storage of confidential information – digital watermarks can be used for hidden transmission, and cryptography, the purpose of which is to encrypt the data itself. Currently, the embedding of invisible digital watermarks in digital images and electronic documents, which carry information about the copyright holder, is being actively used. In connection with the above, the need to create and implement methods of embedding DWC in electronic documents, resistant to various attacks, to increase the effectiveness of protection against threats of theft is an urgent scientific task.

A review of existing methods will help to determine which of them are most suitable for different applications, and to propose recommendations for their use, depending on the specific requirements for protecting textual information.

Digital watermark is a technology created to protect the copyrights of multimedia files. It is information in the form of text or image that is embedded in digital content (for example, in text documents) in such a way that it becomes invisible or barely noticeable during normal viewing, but can be detected and restored using special methods. Watermarks are widely used for copyrights protection, authentication and to prevent unauthorised copying, misuse and distribution of digital content (Popov, 2024).

Thus, DWC is some information (label) that is added to digital content and can be detected or extracted to claim copyright on this content. The main differences between digital watermarks and conventional watermarks (paper watermarks) are that DWC are usually invisible and the task of attacker is not to imitate the watermark as accurately as possible, but, on the contrary, to destroy it completely. So, invisible DWC are embedded in digital data in such a way that it is difficult for the user to detect the added mark if he is not familiar with its format (Borisenko, 2007).

Let's consider the basic classification of digital watermarks – digital marking methods (figure 1). DWC are categorised by perception (visible and invisible: fragile – for authentication, persistent – for confirmation of authorship, semi-fragile – for integrity control), reversibility (reversible – with possibility of restoring the original, irreversible – with minimal distortion), embedding method (linear – hidden spaces and symbols, non-linear – complex algorithms, fractal coding – self-similarity of the image) and embedding area (spatial – visual changes, with container transformation: frequency-based – through Fourier transform, moment-based – change of statistical characteristics) (Oreshkina & Favorskaya, n.d.).

Figure 1. Classification of digital watermarks (digital labelling methods)

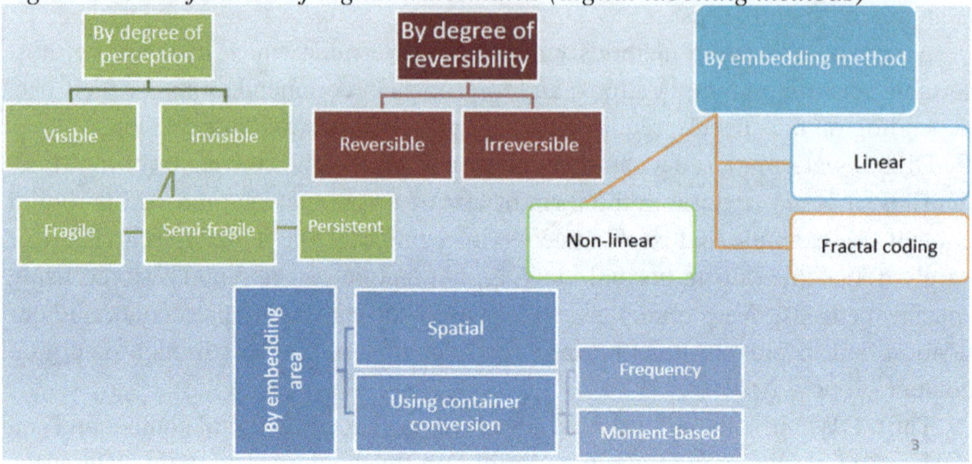

Nowadays, electronic text information can be stored in a variety of file formats, each of which has its own unique features and purposes. This has a significant impact on the choice of methods for implementing digital watermarks. Let us consider the most common and popular file formats for storing text information in electronic form:

1) TXT – the most basic text format that does not support formatting and metadata. The famous application «Notepad» in Windows works with this format. Its simplicity makes it easy to embed digital watermarks, but the lack of formatting limits the methods of hiding them, the main methods of embedding DWC in TXT include using invisible characters and changing letter spacing, methods can also be based on changing the structure of the text, for example, embedding through manipulation of spaces or lines. Text files in TXT format contain only characters and spaces, this is what makes them easy and compatible with any text editors and operating systems, due to the lack of complex formatting, the capacity for watermark embedding in TXT is limited (Istishev, 2021).
2) DOCX – a Microsoft Word document format that supports complex formatting, metadata, and document structure. It has advanced capabilities for embedding DWC, allows to use both formatting changes and metadata to be used to hide information, making DOCX one of the most optimal formats for protecting electronic documents. This format also allows embedding watermarks through style changes and in more complex elements such as tables and images, but this requires more complex algorithms and approaches to embedding DWC (Blinova & Suschenia, n.d.).
3) PDF – a widely used format for documents developed by Adobe that supports text, images, interactive elements, video and complex formatting, and includes

built-in security features such as encryption and digital signatures. It also supports layers and metadata, making it effective for embedding invisible watermarks, providing high security and resistance to deletion attempts that can be integrated at various levels of the document. However, the complexity of the format can complicate both embedding and detection of digital watermarks (Martsinkevich, Larionova, Tereshchenko, Sitnikova, & Gorlushkina, n.d.).

Each of these formats provides unique opportunities and challenges for implementation of digital watermarks, which requires careful analysis and selection of appropriate methods depending on the purpose of information protection. Depending on the specific requirements for information security, the choice of format for implementing DWC should be based on a balance between the complexity of implementation and the required level of protection against attacks.

The resistance of digital watermarks to attacks is the ability of a watermark to retain its properties and remain detectable and retrievable after exposure to various types of attacks. The main types of attacks include (Ilyushechkin, n.d.):

- Formatting changes. Formatting modification attacks include adding or removing spaces, changing line and letter spacing, changing font and other text formatting options.
- Compression. File compression, especially lossy compression, can remove or change watermarks embedded using simple formatting modification techniques.
- Converting formats. Converting documents from one format to another (e.g., from DOCX to PDF) may cause the embedded watermarks to be lost or changed.
- Text editing. Making changes to the text (adding, deleting or replacing words) can affect embedded watermarks, especially if they are based on formatting changes or statistical characteristics of the text.
- Metadata attacks. Deleting or modifying document metadata can remove watermarks embedded using this data.

Let's consider the main methods of embedding DWC in text files.

Method 1. «Formatting changes». Watermark embedding methods based on text formatting changes involve minor changes in the visual representation of an electronic document that remain invisible to the ordinary person, but can be detected by specialised software. These changes may include changes in line and letter spacing, font, text size and other formatting parameters (Blinova, n.d.).

The embedding algorithm is as follows. First, the text document in which the digital watermark will be embedded is defined. Next, the formatting parameters that will be modified to embed the watermark (e.g., letter spacing) are selected. After the DWC is generated based on a predefined algorithm and key and its embedding is done by slightly modifying the selected formatting parameters. Let's take a closer look at examples of the changes themselves – watermark embedding can be done through minor changes in letter spacing. For this purpose, certain letters are selected in the text and the spacing between them is increased or decreased. These changes encode the watermark information in a binary notation, where the letter just above is «1» and just below is «0», which can be decoded by analysing the letter spacing. A similar approach can be similarly applied to line spacing, where the spacing between lines of text is varied. In some cases, minor changes in font or text size can be applied to encode the watermark, although this approach is less commonly used due to the increased visibility of the changes. After all the manipulations, the final step of the algorithm is to save the document with the changes made, i.e., with the DWC protection.

The «Text formatting changes» method has several advantages. It can be easily implemented and applied in most text editors. Also, the changes made are visually invisible to an ordinary user and the method can be applied to various text document formats (DOCX, PDF).

However, here are disadvantages to this method. It is vulnerable to formatting changes, and, accordingly, there is limited (low) resistance to formatting attacks is also observed: the watermarks may be lost when copying text without formatting; changing fonts and formatting may destroy watermarks; when converting a document to another format, DWC may also be lost. Also, digital watermarks carry a limited amount of information, as it will depend on the amount of text in the original file.

The method of changing text formatting gives invisible, fragile and irreversible DWC. It is used in general documents where minor changes are acceptable.

Let's take a practical look at an example implementation of «Method 1».

For this approach and subsequent ones, we will use the author's first name and surname as a watermark, for example, «Иван Иванов», and for greater protection we will encrypt it in the code using Caesar's algorithm (figure 2) – «Легр Легрсе». The full name is sufficient to determine the authorship, we will also use the Russian alphabet and texts. Pseudocodes are implemented in the Python language using the python-docx library, translated using the online editor «Overleaf».

Figure 2. Algorithm «Encryption of DWC by Caesar's Algorithm»

```
Algorithm 0: Text encryption by Caesar's algorithm for Russian
alphabet
---------------------------------------------------------------
Input: text, shift
Output: encrypted_text

Function caesar_cipher_rus(text, shift);
begin
    encrypted_text ← ;          // Initial value of the ciphertext
    foreach char in text do
        if 'а' ≤ char ≤ 'я' then
            start ← ord('а');
            alpha_size ← 32 ; // Number of letters in the Russian
             alphabet
            encrypted_text +← chr((ord(char) - start + shift) %
             alpha_size + start);
        else
            if 'А' ≤ char ≤ 'Я' then
                start ← ord('А');
                alpha_size ← 32;
                encrypted_text +← chr((ord(char) - start + shift) %
                 alpha_size + start);
            else
                encrypted_text +← char ;     // Other symbols remain
                 unchanged

    return encrypted_text ;            // Return the encrypted text

author_name ← "Иван Иванов";
shift ← 3;
encrypted_name ← caesar_cipher_rus(author_name, shift);
Print "Encrypted name: encrypted_name;
watermark_length ← len(encrypted_name);
Print "Watermark length: watermark_length;
```

Next, let's implement one of the examples of the algorithm for embedding DWC using the «Formatting changes» method in a DOCX file (figure 3).

Figure 3. Algorithm for implementing DWC in text using the «Formatting changes» method

```
Algorithm 1: Embedding a watermark in text by changing the line
spacing
Input: input_filename, watermark, output_filename
Output: File with embedded watermark
Function text_to_bits(text);
begin
    return ''.join(format(ord(char), '08b') for char in text) ;
     // Converts text to a sequence of bits
Function embed_watermark(input_filename, watermark,
 output_filename);
begin
    doc ← Document(input_filename) ;            // Open the source
     document
    watermark_bits ← text_to_bits(watermark);
    bit_index ← 0 ;        // Initial index of the watermark bit
    foreach para in doc.paragraphs do
        if bit_index < len(watermark_bits) then
            bit ← watermark_bits[bit_index];
            p ← para._element ;      // Get the XML-element of the
             paragraph
            pPr ← p.get_or_add_pPr();
            spacing ← OxmlElement('w:spacing') ;  // Create element
             'spacing'
            if bit == '1' then
                spacing.set(qn('w:line'), '242') ;        // Increase the
                 interval for encoding '1'
            else
                spacing.set(qn('w:line'), '240') ;  // Leave the default
                 interval for '0'
            pPr.append(spacing);
            bit_index += 1 ;  // Moving on to the next bit of the
             watermark
    doc.save(output_filename) ;  // Save the watermarked document
    Print "Embedded watermark in letters: watermark;
    return The watermarked file has been successfully created
```

First, we convert the encrypted watermark into a binary sequence of bits «1» and «0» (figure 4).

Figure 4. Encryption of DWC into binary code

Зашифрованное имя: Легр Легрсе 2. Implementation

Водяной знак в двоичной записи: 100000110111000011010110000110011100010000000001000001000001101110000110101100001100111000100000 010001000000110000110101

Next, going through each paragraph of the source document, e.g., DOCX-format (figure 3), we embed DWC (set of zeros and ones): get the current watermark bit – if it is equal to «1», then we increase the paragraph line spacing by 0.1, and if it is «0», we leave the default line spacing. Finally, save the document with the already embedded digital watermark (figure 5).

Figure 5. Embedding encrypted DWC in paragraphs of a text document

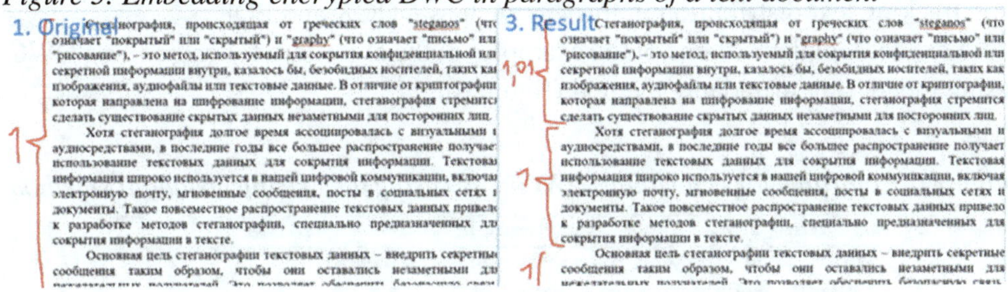

As a result, the changes to the original text are hardly noticeable to humans.

Method 2. «Using metadata». The method of embedding digital watermarks through metadata is based on adding information to the metadata of a text document. Metadata contains various attributes of a file such as author, creation date, software used to create the document, and others. Watermarks can be embedded in these attributes, making them invisible to the average user (Ganzhur, Dzyuba, & Panchenko, n.d.).

DWC is embedded through the document metadata according to some algorithm. Initially, the text document (text file format) in which the digital watermark will be embedded is defined. Next, the metadata to be used for watermark embedding is defined (e.g., document properties in DOCX or PDF format). After the DWC is generated based on a predefined algorithm and key, and it is embedded in the file by adding or modifying metadata attributes. The digital watermark can be embedded as hidden information in metadata fields such as «Author», «Keywords» or «Comments» or by making minor changes to the data in existing-complete metadata fields to encode the watermark in a specific way. The end result is that the document is saved with the changes made to the metadata – the DWC.

The advantages of the «Using metadata» method are: digital watermarks do not affect the visual representation of the text; the method can be used with different document formats (for example, DOCX, PDF); DWC are preserved when the text is changed if the metadata remains unchanged; medium resilience to copy and paste – metadata can be lost when copying a document without preserving the metadata; high resistance to changes in text formatting, as metadata is not directly linked to the text.

The disadvantages of this method 2 include: metadata may be deleted, when copying or transformation documents; not all text editors and programs display or save metadata; low resistance to format conversion – metadata may be lost.

The method of using text metadata applies invisible, semi-fragile and reversible DWC. It is used in general documents where minor changes are acceptable. It is widely used in business and official documents, where it is important to preserve the visual representation of the text. The method is used for copyright protection and tracking text distribution in formats such as DOCX and PDF.

Next, let's implement one of the examples of the algorithm for embedding DWC using the «Using metadata» method (figure 6).

Figure 6. Algorithm for embedding DWC in text using the «Using metadata» method

```
Algorithm 2: Embedding a watermark in document metadata
Input: input_filename, watermark, output_filename
Output: Watermarked file in metadata
Function embed_watermark_in_metadata(input_filename,
 watermark, output_filename);
begin
    doc ← Document(input_filename) ;              // Open the source
     document
    core_props ← doc.core_properties ;            // Getting document
     metadata
    core_props.comments ← watermark ;             // Add watermark to
     metadata comment field
    doc.save(output_filename) ;          // Saving a document with a
     watermark in metadata
    Print "Embedded watermark in metadata: watermark;
    return The watermarked file has been successfully created
```

First, we open the source document (DOCX format) and use the built-in functions of the required library to add a digital watermark to the document metadata in the «Comments» field and save the document with the DWC (figure 7).

Figure 7. Embedding encrypted DWC in text document metadata

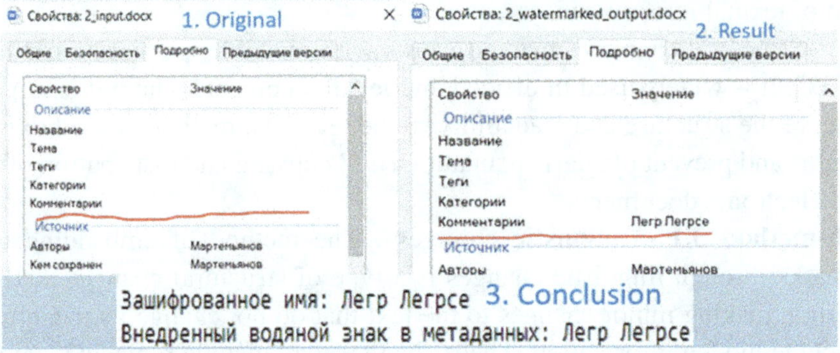

As a result, the original text of the document was not affected, so the changes are visible only in the properties, metadata of the document.

Method 3. «Structural changes». The structural change method involves embedding watermarks by modifying the structure of the document, which may include changing the order of paragraphs and adding invisible characters. These changes are usually invisible to the user, but can be detected by specialised analysis of the document structure (Kumar & Ranga, 2015).

The algorithm for embedding DWC in a text file is as follows. The text document (and document format) in which the digital watermark will be embedded is defined. Next, the structure of the document is analysed to identify possible places for embedding changes, after which DWC itself is generated based on a predefined algorithm and key. Then the document is modified, such as: changing the order of paragraphs in the document to encode the watermark (shifting one position down – «1», up – «0»); or inserting invisible characters (e.g., spaces) in certain places of the text. Finally, the document is saved with the embedded DWC and the corresponding structural changes.

Some advantages of this method can be noted: imperceptibility of changes for the user; watermarks are preserved when the document structure changes; and the method does not require complex mathematical transformations and can be implemented using simple algorithms. The method provides high robustness to copy and paste, i.e. DWC is preserved under these manipulations. Also, the use of structural changes provides medium resistance to formatting changes, which means that DWC can be partially destroyed in case of significant formatting changes. The resistance to format conversion is also medium, as some structural changes may be lost with this technique.

There are also some disadvantages of this approach. A detailed analysis of the text structure may reveal changes made and the amount of information that can be embedded is limited, as there is a dependence of the length of the DWC on the amount of textual information.

The «Structural changes» method works with invisible, persistent and reversible DWC. It is widely used in official and legal documents where it is important to preserve the structure and readability of the text. The method is used to protect copyrights and prevent plagiarism, unauthorised copying and distribution of copyrighted electronic documents.

Sub-method 3.1 «Linguistic changes». The method of embedding digital watermarks through linguistic changes is a type of structural changes method, it is based on making minor changes to the text that do not change its meaning, but also include hidden information. These changes may include replacing synonyms across the text, rearranging words, using alternative grammatical constructions according to special algorithms that remain invisible to an ordinary reader, but can be recognised during analysis (Gerling & Ahrameeva, n.d.).

The algorithm of embedding the DWC using linguistic changes is carried out as follows. Firstly, the text document in which the watermark will be embedded is determined. Next, the text is analysed visually or by a special program to identify possible places for making changes. After that, a watermark is generated based on a predefined algorithm and key, and changes are made to the text, such as replacing words with synonyms, rearranging words in sentences, etc., according to the watermark and the analysed file. If we consider how synonyms are replaced, then we can say that certain words in the text are selected and replaced, so that the changes correspond to the watermark (for example, «the weather was great» is replaced by «the weather was excellent», where the first letter of the replaced word corresponds to the first letter of the DWC). Changing the order of words in sentences according to certain rules to encode the watermark is also often used. At the end of all changes, the document is saved with the linguistic changes made to encode the digital watermark.

The advantages of the «Linguistic changes» method, first of all, lie in their imperceptibility for an average user, as these changes do not change the overall meaning of the text. Also, embedded watermarks can be preserved even with minor text changes, such as formatting or copying, which indicates the high robustness of the DWC to formatting changes and to copy and paste. And the method can be applied to many types of texts, but some linguistic changes may be lost when formats are converted (medium robustness).

The disadvantages of linguistic changes are that they can be detected by comparison with the original text. In addition, the amount of information that can be embedded is limited compared to the size of the text data.

The method of using linguistic changes in the text creates invisible, persistent and reversible DWC. It is often used in scientific and literary works, news articles and educational materials where it is important to preserve the meaning, integrity, readability and originality of the text. It helps to protect copyright, prevent plagiarism, identify the source and resist unauthorised distribution of the text.

Next, let's create an effective algorithm for embedding DWC in TXT file using «Structural changes» method. To embed the digital watermark, we will use the structural change method using zero-width spaces (ZWSP) (U+200B). The algorithm itself for embedding DWC using ZWSP is shown in figure 8.

Figure 8. Algorithm of DWC implementation in text with ZWSP (U+200B), «Structural changes» method

```
Algorithm 3: Embed Watermark into a Text File
Input  : input_filename, watermark, output_filename
Output: A file with the embedded watermark
file = Open(input_filename, mode='r');
lines = ReadLines(file);
watermarkedLines = [];
watermarkIndex = 0;
watermarkLength = len(watermark);
foreach line in lines do
    words = line.split();
    watermarkedWords = [];
    foreach word in words do
        Append(watermarkedWords, word);
        numOfSeparators = ord(watermark[watermarkIndex %
          watermarkLength]) - 1000;
        watermarkedWords[-1] += \u200B * numOfSeparators;
        watermarkIndex += 1;
    Append(watermarkedLines, ' '.join(watermarkedWords));
watermarkedText = '\n'.join(watermarkedLines);
outputFile = Open(output_filename, mode='w');
Write(outputFile, watermarkedText);
```

This method allows to covertly integrate additional information into the text without changing its visual representation. The main application of this approach is in copyright protection and document authentication (Ivanenko & Ushakov, n.d.). In particular, the chosen approach, which includes the insertion of zero-width spaces after each word, where the number of these spaces determines the letter of the digital watermark (author's full name), provides additional security and resistance to various attacks. This method remains invisible to the user and allows efficient embedding of DWC in text documents.

A zero-width space, U+200B – it is an invisible character in Unicode that does not create a visible space between characters. It is used to indicate places where it is safe to break a line when transferring text, as well as to separate words in continuous scripts such as Thai or Khmer. In texts and programming, it can be used to embed hidden information, as in the case of watermarks. This method is effective because most text formats support the Unicode character table.

The embedding algorithm is as follows. As in the previous algorithms, we take the author's full name as DWC, in this case we will limit ourselves to the surname and first name. This is enough to confirm the copyright. Create a watermark «Иван Иванов».

We have the original TXT file (figure 9) for simplicity and clarity (the method works for both DOCX and PDF), a file for saving the result and a digital watermark – the encrypted author's name. The watermark is encrypted using the Caesar cipher, which shifts the letters by a set number of positions. This provides a second layer of protection, if the DWC is discovered, then it cannot be extracted and forged without the original key. Let's add a letter-by-letter encrypted watermark after each word to understand how this will work (figure 10).

Figure 9. Source text document

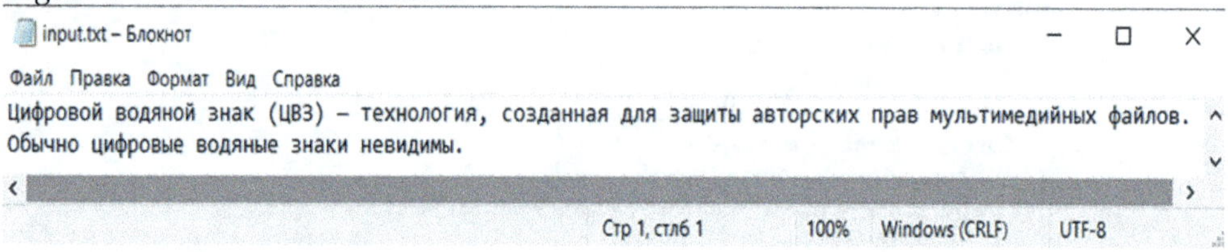
input.txt – Блокнот

Файл Правка Формат Вид Справка

Цифровой водяной знак (ЦВЗ) – технология, созданная для защиты авторских прав мультимедийных файлов.
Обычно цифровые водяные знаки невидимы.

Стр 1, стлб 1 100% Windows (CRLF) UTF-8

Figure 10. Text document with embedded visible DWC

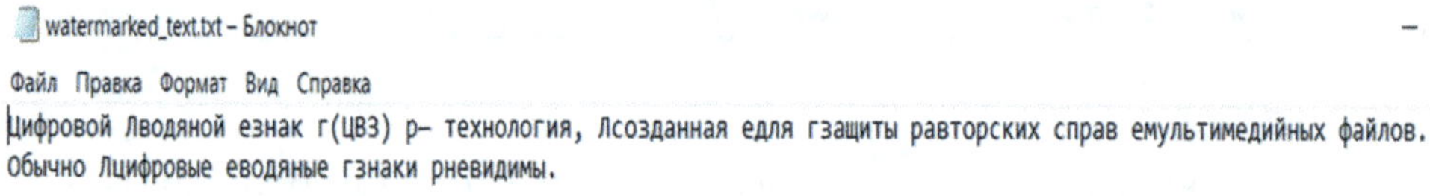
watermarked_text.txt – Блокнот

Файл Правка Формат Вид Справка

Цифровой Лводяной езнак г(ЦВЗ) р– технология, Лсозданная едля гзащиты равторских справ емультимедийных файлов.
Обычно Лцифровые еводяные гзнаки рневидимы.

Then open the file and read its contents line by line. Each row is saved in a list. Go through each line from the source file. Split the string into words. For each word, we add the appropriate number of invisible characters (ZWSP) calculated based on the current watermark character. This number is defined as the character code minus 1000. The watermark is embedded cyclically, i.e., after reaching the end of the watermark, it starts again from the first character (figure 11). This protects the whole document, even if there will be changes in the document, if at least a part of the document with the DWC is saved, it can be extracted to confirm the authorship and other people's interference.

Figure 11. Text document with embedded invisible DWC

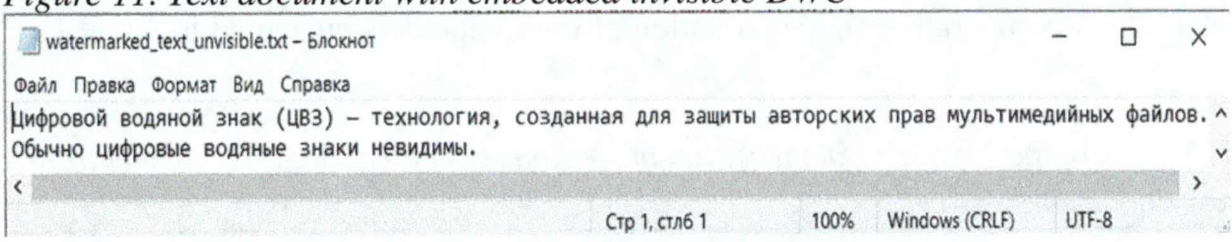

Figure 12 shows the number of embedded zero-width spaces. Now, knowing the original key and the watermark length, we can extract it (figure 13). It can be seen that the initial length of the watermark is 11 characters, and the watermark is cyclically embedded with 18 characters. After that, the required length of the watermark is extracted.

Figure 12. Number of embedded zero-width spaces

```
Количество невидимых символов U+200B в файле: 1167
```

Figure 13. Code algorithm data

```
Зашифрованное имя: Легр Легрсе
Длина водяного знака: 11
Количество невидимых символов U+200B в файле: 1167
Водяной знак: Иван Иванов
Зашифрованный водяной знак: Легр Легрсе
Внедренные и извлеченные символы водяного знака: Легр ЛегрсеЛегр Ле
Расшифрованный водяной знак: Иван Иванов
11 11 18 11
```

Thus, based on the analysis performed, we have developed our own method of introducing DWC using U+200B. This method provides high resistance to attacks aimed at changing the formatting (editing) and structure of text, remains invisible to the user and does not require complex mathematical transformations. The method is also suitable for working with different text document formats, providing flexibility and versatility of its application. So, the proposed method is an effective and practical solution for copyright protection of electronic text documents.

Comparing the methods of embedding DWC in text, the following conclusion can be formed: 1. Text formatting changes and using metadata are easy to implement and unnoticeable to the user, but they are vulnerable to formatting changes and metadata cleaning. 2. Linguistic and structural changes provide high resistance to copying and editing, but require more complex algorithms and can be detected by careful analysis. A comparative characterisation of the methods is presented in Table 1.

Table 1. Comparative characterisation of methods of embedding DWC in electronic text documents

Criteria/methods	Formatting changes	Using metadata	Structural changes	Linguistic changes
Method realisation	Changing letter spacing and line spacing	Embedding information in metadata	Rearranging paragraphs, invisible characters	Replacing synonyms, rearranging words
Advantages	Simple, discreet, compatible	High compatibility, does not affect visual presentation	Imperceptible, copy-proof	Resistance to formatting
Limitations	Vulnerability to formatting changes	Metadata can be cleansed	Can be detected in the analysis	Limited capacity, can be detected
Resistance to copying and pasting	Low	Medium	High	High
Resistance to formatting changes	Low	High	High	High
Resistant to format conversion	Low	Low	Medium	High
Invisibility	✓	✓	✓	✓
Type of resistance DWC	Fragile	Semi-fragile	Persistent	Persistent
Reversibility		✓	✓	✓

From the comparative analysis of different methods of embedding digital watermarks in text documents, it was found that all the methods provide invisible watermarks. However, their robustness and reversibility differ. The methods of formatting changes and using metadata showed the least resistance to copying and formatting, and low reversibility. Linguistic and structural changes showed high resistance to copying and formatting changes, while maintaining reversibility. Thus, for text documents, the structural modification method is the most effective and practical, providing a balance between attack resistance and watermark reversibility.

Future plans are to create more resilient methods against attacks by combining several techniques simultaneously, including the study and application of cryptographic signatures to fully secure electronic text documents.

REFERENCES

Blinova, E. A. (n.d.). Steganographic method based on the line-shift coding method on non-displayed symbols of the electronic text document. *CyberLeninka.* https://cyberleninka.ru/article/n/steganograficheskiy-metod-na-osnove-izmeneniya-mezhdustrochnogo-rasstoyaniya-neotobrazhaemyh-simvolov-strok-elektronnogo-tekstovogo/viewer

Blinova, E. A., & Suschenia, A. A. (n.d.). Several steganographic methods using for embedding of hidden data in electronic text documents. *CyberLeninka.* https://cyberleninka.ru/article/n/primenenie-neskolkih-steganograficheskih-metodov-dlya-osazhdeniya-skrytyh-dannyh-v-elektronnyh-tekstovyh-dokumentah/viewer

Borisenko, B. (2007). Digital watermarks in images. *Information Security/ Информационная безопасность, (4).* https://lib.itsec.ru/articles2/Oborandteh/cifrov_vodyan_znaki_v_izobrazheniyah#:~:text=Основные%20отличия%20цифровых%20водяных%20знаков,наоборот%2C%20в%20полном%20его%20уничтожении

Ganzhur, M. A., Dzyuba, Y. V., & Panchenko, V. A. (n.d.). Features of digital steganography as a method of ensuring data concealment. *CyberLeninka.* https://cyberleninka.ru/article/n/osobennosti-tsifrovoy-steganografii-kak-metoda-obespecheniya-sokrytiya-dannyh/viewer

Gerling, E. Yu., & Ahrameeva, K. A. (n.d.). Method of linguistic steganography based on keyword word. *CyberLeninka.* https://cyberleninka.ru/article/n/metod-lingvisticheskoy-steganografii-osnovannyy-na-opornom-slove/viewer

Ilyushechkin, E. A. (n.d.). Research of robustness of digital watermarking methods to distortions. *CyberLeninka.* https://cyberleninka.ru/article/n/issledovanie-ustoychivosti-metodov-vstraivaniya-tsifrovyh-vodyanyh-znakov-k-iskazhayuschim-vozdeystviyam/viewer

Istishev, V. (2021). Data protection with digital watermarks. *Habr.* https://habr.com/ru/companies/droider/articles/568776/

Ivanenko, V. G., & Ushakov, N. V. (n.d.). Digital watermarks in electronic document management. *ResearchGate.* https://www.researchgate.net/publication/318831123_Digital_watermarks_in_electronic_document_circulation

Kumar, S., & Ranga, V. (2015). A novel approach of zero watermarking for text documents. *International Journal of Advanced Research in Computer Science.* https://scholar.archive.org/work/cvvyhadj4fadrohwpc3ycavyha/access/wayback/http://ijarf.com/wp-content/uploads/2015/07/IJARF-18-21.pdf

Martsinkevich, V. I., Larionova, G. S., Tereshchenko, V. V., Sitnikova, K. A., & Gorlushkina, N. N. (n.d.). Analysis of the possibilities of parsing electronic text documents for the automation of standard control. *CyberLeninka.* https://cyberleninka.ru/article/n/analiz-vozmozhnostey-parsinga-elektronnyh-tekstovyh-dokumentov-dlya-avtomatizatsii-normokontrolya/viewer

Ochnev, D. V., & Chirkin, E. S. (n.d.). Digital watermarks as a method of protection of text printed documents. *CyberLeninka.* https://cyberleninka.ru/article/n/tsifrovye-vodyanye-znaki-kak-metod-zaschity-tekstovyh-pechatnyh-dokumentov/viewer

Oreshkina, E. I., & Favorskaya, M. N. (n.d.). Classification of embedding digital watermarking methods. *CyberLeninka.* https://cyberleninka.ru/article/n/klassifikatsiya-metodov-naneseniya-tsifrovyh-vodyanyh-znakov/viewer

Popov, A. E. (2024). Methods of steganographic information hiding in images. *Science and Technology, (5).* https://na-journal.ru/5-2024-informacionnye-tekhnologii/11724-metody-steganograficheskogo-skrytiya-informacii-v-izobrajeniyah

Chapter 10
Megalopolis as a Space of Challenges in the Context of Digital Transformation:
Real vs. Virtual

Arina Dikanova
https://orcid.org/0009-0006-6205-9018
Plekhanov Russian University of Economics, Moscow, Russia

Ekaterina Kargapolova
Plekhanov Russian University of Economics, Moscow, Russia

Victor Mikhailovich Nemchinov
https://orcid.org/0000-0002-1242-8325
Institute of Oriental Studies, Russian Academy of Sciences, Russia

Irina Gavrilova
Moscow State University, Moscow, Russia

ABSTRACT

In modern conditions of very high change rates socio-cultural space of megacities is often characterized by VUCA-risks. Hence research relevance of citizens' adaptation to everyday life challenges. This study presents results of risks investigation of Moscow residents, conducted in December, 2023. Respondents are most protected from harassment by gender, age, nationality and religion, and are least protected from crime, poverty, job loss, arbitrariness of officials and of the law enforcement agencies. The study has revealed that respondents feel insecurity from the Internet

DOI: 10.4018/979-8-3373-2240-7.ch010

risks and online scammers being much higher than dangers existing in real life. This conclusion requires further verification with identical questions and relevant answers. But the obtained results make us think about paradoxes of transitive society and culture, in which real world with all its problems is more understandable and closer to residents of a megalopolis than the challenges of online services and the cyberspace attractions.

INTRODUCTION

Since the beginning of the 21st century national educational systems around the world have encountered a range of unpredictable nonlinear digital electronic challenges where simple and apparent solutions frequently fail to provide the expected reliable or prolonged pedagogical results; and yet the educators need to cope with existing split duality of coexisting knowledge formats. Time to straighten out what has happened with the way e-transmuted culture tends to regulate human civilization and condition. Prior to the explosive digitalization era, we all had lived in the world of analog artifacts. They permeated every bit of human relations with past or present material world, made them sturdy, predicable and reliable. This time proven mode of communicative interaction with Nature and humankind ensured soundness and veracity of perception. Any distortion of information on an analog carrier (be it on film, stone or paper) had left palpable, noticeable and detectable traces of fraud. Universal duplication of the world cultural heritage in the electronic screen pixel coding has substituted solid analog reality with its malleable digital shadows. Analog and virtual cultural facets are currently coexisting. But communicative situation has universally become viciously inconceivable, brittle and full of ambiguousness.

With the e-knowledge instantly available to any student, human memory qualities and personal lived experiences concern educators no longer. Edutainment replaced the essence of cramming. Ease of fakes, targeted functionality of bots, predominant spread of cloud data networks and frequently dubious student neural cheat sheets have become the legitimate fad of our times. Convenience of working with e-generated ready-made information superseded reliability and authority of traditional knowledge and media sources. Laptops, tablets and interactive AI multifunctional gadgets are so easy to operate at the will of our finger tips that mobile phones (similarly to underwear) cling to human bodies round the clock. And every personal query, wish or whim expressed in any corner of the world enriches neural networks with incremental machine data beyond individual or collective control. Hence, particularly in our entrenched and robust educational systems we seem to have recognized the urge of charting VUCA, BANI and SHIVA vulnerability zones in order to lay out appropriate safe road maps into the future.

National (primarily European) educational systems based on ideals of Enlightenment opted for kindergarten, class, school and college instruction that made a rational linear formative road of educational progress tailored to the requirements of the third and fourth technological orders. Industrialization, urbanization, mass production and consumption had been nurtured by the universal educational model that still remains in place today. It is not going to leave even though its time has already passed. Entrenched institutional educational infrastructure, the industry of teaching with long traditions of skills and knowledge transfer simultaneously represents a corner stone and the stumbling block for national and international adaptation to digital changes in the ways human consciousness is evolving. That is the main challenge and the predicament of pedagogues worldwide.

Do we, the educators, have courage and stamina to ask ourselves, our students and our political leadership what kinds of education are relevant now and will be acclaimed in the next decade? Are we indeed so enlightened as to dare set up the goal of nurturing individual talent in our children? Do we sincerely want to have a new really talented and versatile full-fledged young generation among us? Can we afford this urge for human divinity that promotes planetary values and not the small-minded selfish interests? Or shall capable kids forever have to fight their way through and yet frequently perish in scores nipped in the bud as creators. Shall we indifferently watch how the young downgrade as decentered subjects losing their precious subjective "*in*" individuality of the "*individuus*" that turns into malleable "*dividuus*"? What educational strategy to select Oriental of Western: the one that guides the disciple in gradual personal approximations to the essence of dignified living, or the one that trains the student to accrue a new authentic bit of scholarship elongating the hitherto accumulated bulk of advanced knowledge? Creativity pursues the road map to an anticipated result that can be replicated. Talented endeavor bears no promises and when attained will not be repeated. AI mechanically reflects and aggregates all these evolutions in scores of neural responses. And besides that, today the educators need to make their mind and get to know what kinds of labor force their society will need: the working hands, excellent professionals or bright students for the next wave of brain drain?

Whatever a balance of answers will be established there exists a growing demand for personalized educational practices with edifier, mentor and guru type dialogical exchange between teachers and students nourishing curiosity, ingenuity and dedication among the new generation and offering life-long learning opportunities for the adults. Naturally, conditions differ in distant mountain and isolated areas, in landlocked territories, in rural and urban settings, in cultural historical centers. Thus this research work rests within the power of international teams of dedicated professionals. Keeping such ideas in mind the first step for the project brought together a university student, a Master's degree, an experienced educator and a seasoned academic engaged in comparative cultural studies. The authors decided to launch their research in "the Moscow megalopolis, that includes Moscow and the Moscow Region, ranking among the ten largest mega-cities in the world and being the largest in Europe".

Modern city is a multitier communicative urban landscape that has complex nonlinear structure. This is a volatile socio-cultural organism in which the system of convoluted cause-and-effect urban relationships becomes much more ambiguous when these accelerating relationships are activated simultaneously (Dulina and Tokarev, 2005, p. 43). This is especially true of megacities with their historically dense population concentration and vicious antithetical dynamism.

On the one hand, the social space of megacities is characterized by significant opportunities associated with the noticeable increase of labor productivity, open access to education, and impressive growth of human capital quality (Shapiro, 2008). Researchers note fundamental importance of megacities in any country as engines of human development, centers of innovation and investment, drivers of economic growth, focused on resolving problems of unemployment, and on dissemination of modern lifestyle (Akubue, 2019; Bettencourt, 2014; Hussaini, Hussaini, 2023).

On the other hand, the megalopolis is characterized by an unprecedented scaling of livelihood problems. This "dramatizes the trend towards the ever growing megacities size and their preeminence around the world" (Ford, 1965). Such cities become the space of extreme challenges, global risks, and dangers, which distinguishes them from areas of smaller settlements (Barnes, 2002; Sergeev and Preobrazhenskaya, 2023). Among such challenges are the problems of life quality (traffic, energy supply), environment condition (air and water pollution, exhaustion of natural resources, subsidence of the soil), which lead to "depletion of spaces left for further development" (Akubue, 2019).

There are problems in the field of social safeguarding, which lead to severe socio-economic split and fragmentation, uncontrolled spread of informal employment, marginalization, poverty, increased social conflicts, crime and violence (Akubue, 2019; Hussaini, Hussaini, 2023; Keck, Etzold, 2013; Li, Ma, Cheng, 2018; Rodwin, 1981). Unrestrained expansive growth leads to the loss of manageability, control and planning capabilities for megacity development (Ford, 1965; Hussaini, Hussaini, 2023; Keck, Etzold, 2013). As Bettencourt (2014) noted, "integrated urban planning is computationally intractable (i.e., almost inconceivable) in large cities, regardless of the amount of data available". This dilemma between the need for planning and coordination and its impossibility to be done in detail is solved by recognizing the fact that cities are primarily self-organizing social networks embedded in certain space and are supported by their urban infrastructure and services.

Rapid growth of megacities blurs the boundaries between the city and the countryside, leads to personal identity instability (Sergeev, Preobrazhenskaya, 2023;

Urso, 2020), and hampers self-presentation and self-affirmation chances (Kutsepal, Zinchenko, Perebyynis, 2020). Subsequent dejection may lead to stigmatization (Romero-Lankao, Hughes, Qin, Hardoy, Rosas-Huerta, Borquez, Lampis, 2014), that causes health problems (Hussaini, Hussaini, 2023), enhances psychological, psychological, informational, environmental, and occupational stress (Aksenova, Kamynina, Turzin, 2023). Consequently this leads to an increase of "anxious feelings of isolation and alienation" (Sanusi, 2019). S. Kutsepal, N. Zinchenko, and Y. Perebyynis (2020) conclude that in the long run intrinsic "values of megacity's chronotope, become transformed or ignored, and only the value of consumption becomes the predominant one."

The above listed problems of modern megacities are further exacerbated by an ambiguous impact of unpredicted digitalization of public services. Computer illiteracy of elderly people, general unpreparedness to encounter and deal with robotic voice assistants and with online responses of distant call centers' operators employed from the far away locations without knowledge of the city's real space – all this causes additional anxiety and frustration. Disorientation is further increased by the lack of appropriate education facilities that should have been provided on the part of municipalities, city educators and civil servants. Scammers alert from banks and law enforcing agencies are not very effective against constantly improving attacks of electronic imposters who leave multiple victims scammed out of their money and property.

Similar grievances relate to the lack of clear step by step instructions and the on-demand assistance to multiple digital innovations and to hybrid interfaces between analog original documents and their electronic substitutes, particularly when complicated password requirements and restrictions have to be passed in order to get the work started. Rules of applicability are constantly improved, they are frequently incomprehensible, not easy to follow and these alterations make social relations rather fragile. Last but not the least rampant electronic fraud increasingly spreads VUCA fears among citizens. This is a humanitarian problem that needs to be resolved.

Some social researchers view the prospects for digitalization of megacities in a very positive way as a factor for growth opportunities, increased management efficiency, coordination, integration, and trust (Batty, 1995; Bettencourt, 2019; Kirby, 2012; Kraus, 2023; Li, Ma, Cheng, 2018). They recon it is creating a comfortable environment for "smart, intelligent cities", computer cities and for direct public participation in this process (Batty, 1995; Bokolo, 2020; Fomina, 2023). This may be true on a very limited scale of a neighborhood domestic chats, but turns to become formal of a larger social scale. There is an opinion that pretty soon the main criterion for defining the concept of a "city" will not be based on density of population and traffic flows, but on density of cell phone users' geolocation data (Dong, 2024).

At the same time direct correlation between the prevalence of Facebook usage and the level of violent crimes has been revealed by researchers (Asongu, Nwachukwu, Orim, Pyke, 2019). In addition, the results of empirical studies demonstrate that the willingness of residents of megacities to participate in megaprojects in the context of digitalization of the economy and the use of artificial intelligence is ambiguous and depends on many contradictory factors (Vinichenko, 2021).

It should be also noted that the townspeople "demonstrate a higher degree of adaptability to change than the villagers, primarily due to the higher level of technology development" (Dulina, 2005, p. 36). It seems relevant to investigate the level of adaptation of megalopolitan residents to the challenges of the high-tech social space of a large city, taking into account the awareness of "widening social and spatial gaps that may arise as a result of uneven adaptation" (Alfasi, Margalit, 2021).

METHOD

In December 2023, a research group led by Professor E. V. Kargapolova, Doctor of sociology in the Department of political analysis and socio-psychological processes at Plekhanov Russian University of Economics conducted an initiative sociological study among Moscow residents using standard methodology of the All-Russian research program "Sociocultural evolution of Russia and its Regions", developed at the Center for the Study of Sociocultural changes in the Institute of physics and technology, RAS. This research methodology was developed by Professor N.I. Lapin, Doctor of philosophy, Corresponding member of the Russian Academy of Sciences and by L.A. Belyaeva, Doctor of sociology, leading research fellow at the Center for the Study of Sociocultural Changes at the Institute of Philosophy, RAS. Since 2006 their standard methodology was tested at the federal level in 30 regions of Russia. The research methodology has been further adapted for comparative studies conducted by an interregional team of scholars who continue to apply Standard methodology in the Vologda scientific center, RAS headed by A.A. Shabunova, Doctor of economics. Their studies made it possible for authors to compare the data they obtained in Moscow with the results from other regions of Russia where empirical studies using similar methodology have already been conducted.

The research method is represented by an online questionnaire (N=1070). The type of sample is a target one. Our research group for several years has been developing the algorithm for implementing online questionnaires based on specified quota criteria (for more details, see Kargapolova, Davydova 2022). Selected gender and age structure of the respondents corresponds to the general population structure of Moscow (43.2 per cent of men and 56.8 per cent of women were surveyed; 27.7

per cent of young people (14 to 35 years old), 46.7 per cent of adults (aged 36 to 59), 25.6 per cent of the elderly respondents (over 60 years old).

The questionnaire included questions that made it possible to analyze the level of adaptation of residents of the metropolitan area to the challenges of the big city, namely, the assessment of Muscovites' protection from social dangers and from risks in real life and in virtual space. The level of citizens' protection was evaluated basing on responses to the following two questions: "How much do you personally feel protected from various dangers?" (Scale of answers included the following options: "not protected at all", "perhaps not protected", "it's hard to say", "perhaps protected", "protected") and "What do you think, to what extent are modern people protected from the risks of using the Internet?" (Answer options were: "absolutely not protected", "rather not protected", "I find it difficult to answer", "rather protected", "absolutely protected").

The survey data analysis included the study of linear distributions and subsequent scaling. In accordance with the recommendations for calculating the index of social hazards developed in the Standard methodology the scale proposed for evaluation of the question "How do you personally feel protected from various dangers?" could include only positive figures. Thus, the answer option "not protected at all" received the value "1", "perhaps not protected" - "2", "it's hard to say" - "3", "perhaps protected" - "4", protected" – "5". The obtained responses were summed up and the final index value was calculated using the formula K= the total amount divided by 5 divided by 100 (Regions of Russia 2009: p. 793). Similarly, we calculated the value of risk protection indices in the virtual space (the question was: "How do you think modern people are protected from the following risks of using the Internet?").

RESULTS

According to respondents answers they feel most protected from harassment due to age and religious beliefs. Level of security from this type of harassment ranks first and second in the structure of self-assessments of the level of protection from real-life threats (index values are 0.83) (Table 1). Analysis of the gender and age aspects of this safety indicator reveals that men feel more protected from harassment due to their age than women, and the elderly and adults feel more protected than young people. There is no statistically significant difference between men and women in their assessments of protection from harassment due to religious beliefs, while the feeling of security also increases with the increase of respondents' age.

Next follows feeling of protection from harassment due to gender and nationality (index values 0.8). Men and young people feel more protected from harassment because of their gender. Men and the elderly people feel to be more protected from harassment because of their nationality.

By the wide margin of larger insecurity compared to the above listed dangers come environmental threats (index value 0.67). This result correlates with the conclusions obtained from the long-term surveys conducted under the aegis of the All-Russian research project "Sociocultural evolution of Russian regions". Pollution and environmental threats rank among the five most acute social dangers (see, for example, Kogai, Dremova 2021; Regions in Russia 2009; Romashkina, Khudyakova 2020). Our sample shows that men have a higher self-assessment of being protected from this type of threat than women. Young people feel themselves much less scared by environmental threats than adults and the elderly respondents.

The next in the increasing level of insecurity come self–assessments of being protected from persecution for political beliefs (0.66). The younger the respondents are, the less protected they feel from persecution for their political beliefs. Same results relate to the fear of loneliness. According to self–assessments of respondents men feel somewhat more protected from loneliness than women, and young people think to be more protected from loneliness than adults.

Even less secure feel the respondents to be protected from the threat of crime and arbitrariness of law enforcement agencies (index value 0.61). Within these indicator men, youth, and the elderly feel to be more protected from crime than women, while the elderly feel themselves more protected from the arbitrariness of law enforcement agencies.

Muscovites feel even less protected from poverty and have fear of job loss (index values 0.6). Again, women feel themselves more vulnerable on these indicators than men. Adults regard themselves to be less protected from poverty and job loss than respondents from other age groups. This finding indicates that the age group of adults considers probable this risk of poverty and job loss as a likely outcome of their future life events. The research results confirm the conclusion that poverty in urban areas is accompanied by negative phenomena: such as high unemployment rate, social pathologies, crime increase and "the high level of spatial deprivation, including poor condition of housing facilities and of municipal infrastructure" (Kisiała, Rącka, 2021).

The threat to real life, from which respondents feel themselves to be the least protected relates to the arbitrariness of officials (0.58). Women, adults, and the young people feel to be least protected from this threat.

Table 1. Self-assessment of Moscow residents' protection from the dangers of the real world (indeces)

Answer options	Gender		Age			Total
	Male	Female	Youth	Adult	Elderly	
Harassment due to your gender	0,88	0,79	0,79	0,83	0,86	0,83
Harassment due to your religious beliefs	0,83	0,83	0,80	0,82	0,87	0,83
Harassment due to your age	0,84	0,77	0,82	0,79	0,80	0,80
Infringement due to your nationality	0,82	0,80	0,80	0,79	0,85	0,80
Fear of an environmental threat	0,70	0,64	0,72	0,64	0,66	0,67
Fear of persecution for political beliefs	0,66	0,66	0,62	0,65	0,71	0,66
Fear of loneliness	0,68	0,65	0,68	0,64	0,68	0,66
Fear of crime	0,63	0,59	0,64	0,59	0,61	0,61
Fear of law enforcement agencies' arbitrariness	0,61	0,62	0,59	0,60	0,66	0,61
Fear of poverty	0,61	0,59	0,62	0,57	0,62	0,60
Loss of work fear	0,62	0,58	0,61	0,56	0,64	0,60
Fear of the arbitrariness of officials	0,58	0,57	0,57	0,56	0,62	0,58
Column average	*0,70*	*0,67*	*0,69*	*0,67*	*0,71*	*0,69*

As for the answers to the question "How much do you think modern people are protected from the following risks of using the Internet?" the structure of risk assessments ranges in descending order in the following way. The top level in terms of safety is occupied by the level of protection from coercion to commit illegal acts (including sexual ones) with an index value of 0.48 (Table 2). A significant difference was registered in the assessment of this risk by men and women (0.51 versus 0.45). As the age of respondents increases, their assessment of the level of protection against this risk decreases.

On a lower level of protection are consumer risks (abuse of consumer rights, purchase of low-quality goods, forgeries and falsifications). They got the index value of 0.41. Women feel less protected from this type of threat than men. In the responses of the young people risk indicator reaches its highest value (0.44), while the elderly respondents show the value of 0.4 and the adults cling to the mark of 0.39.

On the third and fourth places down the scale of security assessments from the dangers existing in the virtual space come the risks of negative communication (humiliation, intolerance, bullying, discussion of one's personal characteristics, harassment, including that on gender, appearance, on ethnic and nationality grounds), scamming and fraud (theft of personal information, financial resources, blackmail) (index value 0.38.) However, even under such circumstances men feel themselves more protected from these risks than women. As for the age-related risk characteristics of negative communication, researchers detect equal values of security index among young and adult respondents, with the higher index values among the elderly people. As for the risks of fraud, the assessment of the level of protection among young people exceeds the assessment of the same indicator among adults and elderly respondents.

The last place in the structure of assessments of security level of on the Internet is occupied by protection from receiving negative information (violence, aggression, obscene language, propaganda of racial hatred, drug addiction, alcohol, unhealthy behavior, etc.) (Index value 0.36). Analysis of the obtained results demonstrates higher level of protection from this threat in the men ratings. Differentiation of responses by age groups makes it possible to observe the decrease in the level of security as the age of the respondents increases.

Table 2. Respondents' perceptions of people's security from the dangers of virtual space (indices)

Answer options	Gender		Age			Total
	Male	Female	Youth	Adult	Elderly	
Coercion to commit illegal acts (including sexual ones)	0,51	0,45	0,49	0,48	0,47	0,48
Consumer risks (abuse of consumer rights, purchase of low-quality goods, counterfeiting, falsification)	0,43	0,39	0,44	0,39	0,40	0,41
Negative communication. (humiliation of dignity, intolerance, discussion of personal characteristics, harassment, including on gender and national grounds)	0,40	0,37	0,38	0,38	0,39	0,38
Fraud (theft of personal information, financial resources, blackmail)	0,41	0,36	0,41	0,37	0,37	0,38
Receiving negative information (violence, aggression, obscene language, propaganda of racial hatred, drug addiction, alcohol, unhealthy behavior, etc.)	0,38	0,34	0,37	0,36	0,35	0,36
Column average	*0,43*	*0,38*	*0,42*	*0,40*	*0,40*	*0,40*

CONCLUSION

Survey of challenges perceived by respondents of megalopolis in the context of evolving digital transformation can be summarized in the following conclusions.

1. Population of megalopolis on the whole is gradually getting adapted to the digital transformation of urban life with its increasing communicative and informational risks. Comparison of the presented study results with the conclusions of long-term studies conducted in other regions of Russia using the Standard methodology confirms (with some variants) similarity of respondents' attitude on the national scale to the new digital dangers and social risks. The structure of adaptation to social dangers and challenges of the real world appear to be

similar in various surveys. Their respondents are feeling most protected from harassment by gender, age, nationality and religion and least protected - from crime, poverty, job loss, arbitrariness of officials and law enforcement agencies (see, for example, Regions in Russia... 2009; Romashkina, Khudyakova 2020). As had been noted by Battistelli and Galantino (2019), the hazards that we had surveyed belong to a group of risks that "lead to social discord at the subnational level."

2. Men are somewhat more adapted to the electronic challenges of the social space in the city, as they assess their level of protection from various threats to be slightly higher than women, both in the case of real-life dangers and in the case of the Internet risks (their average self–assessment of real-life dangers is 0.7 versus 0.67 by women, correspondingly threat assessments differences in virtual space are 0.43 against 0.38). The only exception is the danger of arbitrariness on the part of law enforcement agencies, where women rate their rate of protection slightly higher than men.
3. When dangers of real life are reflected by the respondents, elderly people in nine out of twelve cases demonstrate the highest value of security index compared to the youth and adults. Paradoxically, in this way, it is the elderly who recon that they are somewhat more adapted to the challenges of megalopolis (perhaps due to less exposure to mundane encounters with urban dangers) while the active adults feel less protected than other age groups that the authors have tested.
4. Young people rate their security against risks on the Internet slightly higher than the adults and the elderly people. At the same time, among young people who are much more involved in online activities, the difference in the index values compared to that of other age groups representatives is not so significant (0.42 versus 0.4). These figures do not indicate that a qualitatively different level of adaptation to the Internet space exists between generations.
5. The authors observe a clear lack of confidence among respondents regarding the safety of online activities. The average value of the indices of protection from dangers in real life and from the risks of the Internet space are 0.79 versus 0.4. It means that in general respondents feel insecurity from risks on the Internet to be twice as higher than they perceive dangers in the real life. This conclusion is tentative as the authors could only compare relevant answers conditionally, since the questions in various surveys sounded a bit different and the response scales were also not identical. This conclusion needs to be further tested on identical questions with identical response structure.

The following concluding remarks can be drawn from this stage of research. It turns out that real world with all its cumbersome problems is more understandable and closer to the perception of the respondents than the online digital world with its newly emerging virtual culture, with the challenges of anonymity, nonlinear discrepancies and with distancing from habitual ways of real life contact communication. At the same time the cyberspace is both dangerous and lucrative with its broad variety of evadable rules that provoke "expanding and violating the limits of permissible and not permissible actions" (Martynova and Bachkova, 2023, p. 57). The time of rapid changes discussed in the beginning of this paper makes the authors think with great concern about the dichotomies of the present evolving polycentric world, in which online virtual space is often positioned as a wonderful world of opportunities as opposed to the risky globalized offline life. But in the third decade of the 21st century adult residents of many megalopolises still feel to be more comfortable and familiar with the way of life that had been set by the offline era of mass consumption. Even the younger generation seems to clearly separate the real and the virtual which they regard as a space of significant risks but prefer to restrict their offline communicative practices preferring mobile chats. Results of the presented survey make the authors suggest that at this stage the coexistence of analog cultural activities with interactive cyberspace and its virtual reality imitating traditional human modes of communication via robots, avatars and neural assistants is fraught with unanticipated challenges and threats that humanity is not yet ready to accommodate. Overt and covert reality full of BANI characteristics provokes e-risks to a greater extent than was expected. These developments should get into the purview of policy makers and need to be encompassed in the educators' new worldview, mission and practices.

REFERENCES

Aksenova, E. I., Kamynina, N. N., & Turzin, P. S. (2023). Urban stress of a megalopolis resident: Stress factors and correction. *Probl Sotsialnoi Gig Zdravookhranenniia i Istor Med, 31*(Special Issue 2), 1087–1091. DOI: 10.32687/0869-866X-2023-31-s2-1087-1091

Akubue, J. (2019). Challenges of mega cities. *Modern Economy, 10*(9), 2031–2050. DOI: 10.4236/me.2019.109128

Alfasi, N., & Margalit, T. (2021). Toward the sustainable metropolis: The challenge of planning regulation. *Sustainability, 13*(15), 1–21. DOI: 10.3390/su13158189 PMID: 34804601

Asongu, S. A., Nwachukwu, J. C., Orim, S. M., & Pyke, C. (2019). Crime and social media. *Information Technology & People, 32*(5), 1215–1233. DOI: 10.1108/ITP-06-2018-0280

Barnes, P. (2002). Approaches to community safety: Risk perception and social meaning. *Australian Journal of Emergency Management, 17*(1), 15–23. https://search.informit.org/doi/10.3316/informit.370758269308829

Battistelli, F., & Galantino, M. G. (2019). Dangers, risks and threats: An alternative conceptualization to the catch-all concept of risk. *Current Sociology, 67*(1), 64–78. DOI: 10.1177/0011392118793675

Batty, M. (1995). The computable city. In *Fourth International Conference on Computers in Urban Planning and Urban Management* (July 11–14). www.acturban.org/biennial/doc_planners/computable_city.htm

Bettencourt, L. M. A. (2014). The uses of big data in cities. *Big Data, 2*(1), 12–22. DOI: 10.1089/big.2013.0042 PMID: 27447307

Bettencourt, L. M. A. (2019). The digital transformation of metropolises. *Metropolis Observatory, 8*, 1–16. https://www.metropolis.org/sites/default/files/resources/Observatory_Issue-Paper-08_Digital-transformation-metropolises.pdf

Bokolo, A. (2020). Managing digital transformation of smart cities through enterprise architecture: A review and research agenda. *Enterprise Information Systems, 15*(3), 299–331. DOI: 10.1080/17517575.2020.1812006

Dong, L., Duarte, F., Duranton, G., Santi, P., Barthelemy, M., Batty, M., Bettencourt, L., Goodchild, M., Hack, G., Liu, Y., Pumain, D., Shi, W., Verbavatz, V., West, G. B., Yeh, A. G. O., & Ratti, C. (2024). Defining a city — Delineating urban areas using cell-phone data. *Nature Cities*, *1*(2), 117–125. Advance online publication. DOI: 10.1038/s44284-023-00019-z

Dulina, N. V. (2005). The state of the modern Russian city and the need to develop a strategy for its development. *Regional Economics: Theory and Practice*, *6*, 36–37.

Dulina, N. V., & Tokarev, V. V. (2005). Monitoring as a means of information support for social management of a modern city. *Regional Economics: Theory and Practice*, *6*, 43.

Fomina, A. (2023). Comfortable urban environment of the megapolis and digitalization. *AIP Conference Proceedings*, *2560*(1), 040005. Advance online publication. DOI: 10.1063/5.0134578

Ford, J. B. (1965). The challenge of the megalopolis. *The International Migration Digest*, *2*(1), 41–51. DOI: 10.2307/3002979

Hussaini, W., & Hussaini, S. G. (2023). Risks of a megacity. *Journal of Survey in Fisheries Sciences*, *10*(1), 1576–1588. DOI: 10.53555/sfs.v10i1.2414

Kargapolova, E. V., & Davydova, Yu. A. (2022). The well-being of the population of the Russian region in conditions of socio-economic constraints (on the example of the Moscow agglomeration). *CITISE*, *3*, 301–316.

Keck, M., & Etzold, B. (2013). Risks and resilience in Asian megacities: Guest editorial. *Erdkunde*, *67*(1), 1–3. DOI: 10.3112/erdkunde.2013.01.01

Kirby, A. (2012). Current research on cities and its contribution to urban studies. *Cities (London, England)*, *29*(1), 3–8. DOI: 10.1016/j.cities.2011.12.004

Kisiała, W., & Rącka, I. (2021). Spatial and statistical analysis of urban poverty for sustainable city development. *Sustainability (Basel)*, *13*(2), 858. DOI: 10.3390/su13020858

Kogai, E. A., & Dremova, L. A. (2021). The state of civil society in the Kursk region in 2020. In *Effectiveness of national projects: Socio-economic processes, civil society institutions, education* (pp. 11–16). Kursk Institute of Management, Economics and Business.

Kraus, K. (2023). *The nature of digital transformation: Energy of the economy. How to feel, understand and use it.* Baltija Publishing., DOI: 10.30525/978-9934-26-287-6

Kutsepal, S., Zinchenko, N., & Perebyynis, Y. (2020). Specificity of social existence in megalopolis. *SHS Web of Conferences, 85*, 01003. https://doi.org/DOI: 10.1051/shsconf/20208501003

Lapin, N. I., & Belyaeva, L. A. (Eds.). (2009). *Regions in Russia: Socio-cultural portraits of regions in the all-Russian context*. Academia.

Li, D., Ma, J., Cheng, T., van Genderen, J. L., & Shao, Z. (2018). Challenges and opportunities for the development of megacities. *International Journal of Digital Earth, 12*(12), 1382–1395. DOI: 10.1080/17538947.2018.1512662

Martynova, M. D., & Bochkova, I. A. (2023). Transformation of moral values of youth in the field of digital communication. *Primo Aspectu, 3*(55), 55–64. DOI: 10.35211/2500-2635-2023-3-55-55-64

Rodwin, L. (1981). Problems of the metropolis. In *Cities and city planning. Environment, development, and public policy: Cities and development* (pp. 41–51). Springer., DOI: 10.1007/978-1-4684-1089-1_6

Romashkina, G. F., & Khudyakova, M. V. (2020). Sociological analysis of human capital factors and resources. *Economic and Social Changes: Facts, Trends. Forecast, 13*(6), 232–251.

Romero-Lankao, P., Hughes, S., Qin, H., Hardoy, J., Rosas-Huerta, A., Borquez, R., & Lampis, A. (2014). Scale, urban risk and adaptation capacity in neighborhoods of Latin American cities. *Habitat International, 42*, 224–235. DOI: 10.1016/j.habitatint.2013.12.008

Sanusi, Sh. (2019). Eyeing fear and anxiety: Postcolonial modernity and cultural identity in the urban space of The Eye. *Otherness: Essays and Studies, 7*(1), 70–99.

Sergeev, V. F., & Preobrazhenskaya, K. V. (2023). Megapolis as a risk space: Destructive factors of personal identity and methodologies for their description and research. *Bulletin of the Russian Christian Humanitarian Academy, 24*(1), 259–265.

Shapiro, J. M. (2008). Smart cities: Quality of life, productivity, and the growth effects of human capital. *The Review of Economics and Statistics, 88*(2), 324–335. DOI: 10.1162/rest.88.2.324

Urso, G. (2020). Metropolisation and the challenge of rural-urban dichotomies. *Urban Geography, 42*(1), 37–57. DOI: 10.1080/02723638.2020.1760536

Vinichenko, M. (2021). Megapolis residents' readiness to participate in megaprojects under the conditions of the digitalization of the economy and the use of artificial intelligence. *Revista Gestão Inovação e Tecnologias, 11*(4), 2926–2940. https://www.academia.edu/103501261/Megapolis_Residents_Readiness_to_Participate_in_Megaprojects_Under_the_Conditions_of_the_Digitalization_of_the_Economy_and_the_Use_of_Artificial_Intelligence

Chapter 11 Analysis of the Dynamics of the Multiplier Effect of Rural Tourism and Its Impact on the Integrated Rural Development

Natalia Sergeeva
https://orcid.org/0000-0001-7116-3526
Russian State Agrarian University - Moscow Timiryazev Agricultural Academy, Russia

Ludmila Evgrafova
https://orcid.org/0000-0003-2903-7782
Russian State Agrarian University - Moscow Timiryazev Agricultural Academy, Russia

ABSTRACT

Rural tourism, including agritourism, is a growing branch of the industry in rural areas. According to experts, in the near future the flow of rural tourists will increase and reach about seven million per year, while agricultural production enterprises will receive additional income from rural tourism in the amount of 5% of the total income of farms. Tools for the rural tourism development are also included in the state program "Tourism Development". With the growing demand, there is a challenge to ensure the supply of rural tourism services. Currently, the share of rural tourism in the total volume of tourism in Russia is estimated at about 2.0%. However, the situation will change as rural tourism is recognized as a priority trend of state policy in the near future. One of the priorities in promoting the rural tourism

DOI: 10.4018/979-8-3373-2240-7.ch011

development is to assess its multiplier effect on related industries. This will make it possible to determine the trends of the sustainable rural development and the main areas of the rural tourism development.

1. INTRODUCTION

Rural tourism, including agritourism, is a growing branch of the industry in rural areas. According to experts, in the near future the flow of rural tourists will increase and reach about seven million per year, while agricultural production enterprises will receive additional income from rural tourism in the amount of 5% of the total income of farms. For example, in 2023, 70% of users of the site svoe-zagorodom.ru searched for vacations in rural areas and 40% – for vacations in wineries. In addition, tools for the rural tourism development are also included in the state program "Tourism Development". With the growing demand, there is a challenge to ensure the supply of rural tourism services. Currently, the share of rural tourism in the total volume of tourism in Russia is estimated at about 2.0%. However, the situation will change as rural tourism is recognized as a priority trend of state policy in the near future. The development of this sector until 2030 is determined in the law adopted in 2021 (Federal Law No. 318-FZ dated 02.07.2021 "On Amendments to the Federal Law "On the Fundamentals of Tourism Activities in the Russian Federation" and Article 7 of the Federal Law "On the Development of Agriculture"). In our opinion, one of the priorities in promoting the rural tourism development is to assess its multiplier effect on related industries. This will make it possible to determine the trends of the sustainable rural development and the main areas of the rural tourism development. In the article, the authors present this assessment in the context of three blocks: infrastructural-economic block; socio-cultural block and ecological-climatic block. At the regional level, the strengths and weaknesses of the rural tourism development are identified, and the dynamics of the integrated indicator in the form of a "level" of the rural tourism development over two years is presented.

2. MATERIALS AND METHODS

The assessment of the dynamics of the multiplier effect of rural tourism is not related to the assessment of the ability of the constituent entities of the Russian Federation to meet their financial obligations. It does not provide a forecast of the probability of their financial difficulties in any given period. It represents an opinion on the factors influencing long-term sustainability and success in three key aspects: infrastructural-economic aspect, socio-cultural aspect and ecological-climatic aspect.

The methodology for assessing the dynamics of the multiplier effect of rural tourism in the constituent entities of the Russian Federation (hereinafter – methodology) is developed taking into account the current stage of the rural tourism development in Russia, as well as the level of involvement of regions in the rural tourism development agenda. The study focuses on the infrastructural-economic block of indicators, because first of all it is necessary to lay the foundations for further sustainable development of rural tourism: to form a regulatory legal framework; to include the concept of sustainable development as a priority in regional tourism development strategies; to develop effective measures and mechanisms of state support, stimulating the implementation of sustainable development projects; to provide specific measures for sustainable tourism development in state programs and to ensure their financial support.

That is why the methodology also includes a number of indicators "for growth", which have not yet been met by most regions. This will allow directing the focus of attention of regional administrations to the solution of a number of management tasks that will allow forming an institutional and regulatory environment, a system of strategic management of the rural tourism development.

As it has already been noted, the degree of dynamics of the multiplier effect of rural tourism involves depends on a balance between environmental, socio-cultural and economic objectives. In the conditions of global economic constraints, due to the growing geopolitical tensions at the territorial and corporate levels, the importance of economic sustainability is increasing. At the same time, as a result of the closure or restriction of many popular international tourist destinations, domestic rural tourism in Russia in the short and medium term receives additional incentives for development. In this regard, the presented study, takes into account the criteria of economic sustainability of the tourism industry in Russian regions. Relevant indicators are included in the methodology.

The set of the multiplier effect indicators used in this research was formed primarily on the basis of the possibility of their calculation on the basis of information from open official sources: Ministry of Agriculture of the Russian Federation, Federal State Statistics Service of the Russian Federation, regional official websites, Ministry of Culture of the Russian Federation, Ministry of Health of the Russian Federation, Ministry of Emergency Situations of the Russian Federation, Ministry of Natural Resources and Environment of the Russian Federation. Relevant indicators were used for each region. Based on the final assessment of the multiplier effect of rural tourism, the regions are divided into five levels of the rural tourism development according to the degree of integration of the indicators of rural tourism impact on the economic, social and environmental development of the region: beginning, developing, moderate, developed and advanced.

3. RESULTS

The dynamics of integration of the indicators from 2022 to 2023 was favorable. Three regions entered the advanced level: the Sverdlovsk, Tyumen and the Murmansk Regions (Table 1).

Table 1. Degree of integration of indicators of rural tourism impact on economic, social and environmental development of the region

Levels	Years		Dynamics of indicators
	2022	2023	
Advanced	5	8	+3
Developed	10	22	+12
Moderate	17	17	0
Developing	22	21	-1
Beginning	31	17	-14

In 2022, the achievements of five regions can be considered as the best practices in rural tourism, and in 2023 eight regions can be characterized by the "advanced" level (Table 2). It is important to note that the status of "advanced" is relative, and even regions with this level should continue to develop rural tourism and improve rural areas.

Table 2. Regions with an advanced level of integration of indicators of rural tourism impact on the economic, social and environmental development of the region, 2022-2023

Ranking place	Years	
	2022	2023
1	Krasnodar Krai	Leningrad Region
2	Leningrad Region	Krasnodar Krai
3	Moscow Region	Republic of Tatarstan
4	Republic of Tatarstan	Moscow Region
5	Kaliningrad Region	Tyumen Region

continued on following page

Table 2. Continued

Ranking place	Years	
	2022	2023
6		Sverdlovsk Region
7		Kaliningrad Region
8		Murmansk Region

The leading position in 2022 was taken by the Krasnodar Krai, where event tourism is most popular. Each district of the region can boast its own unique attractions and cultural events. The second place was occupied by the Leningrad Region, which created a favorable institutional environment for the development of both traditional and rural tourism. More than forty agritourism farms and over seventy guest houses operate in the region. Many of the support measures are unique in Russia. The third place was occupied by the Moscow Region, where the program for the development of agritourism is being implemented, which provides for the creation of sports and recreational routes, as well as hunting and fishing tours, master classes and gastronomic tours with tastings (Evgrafova & Ershova, 2020).

Among the leading regions in 2022 in terms of the level of integration of the indicators of rural tourism impact on economic, social and environmental development were the Republic of Tatarstan (4th place) and the Kaliningrad Region (5th place). In 2023, the first place was taken by the Leningrad Region, improving its position by one line. Representatives of the tourism and hotel industry, including guest houses and enterprises engaged in congress and exhibition activities, show active interest in obtaining environmental and social certification. The region also emphasizes the creation of a barrier-free environment for tourists (Evgrafova, 2015). Information on ecotourism is published on official online resources, and specially protected natural areas are available for physical and virtual visits.

The second place in 2023 was occupied by the Krasnodar Krai, which lost one position compared to 2022, due to the significant growth of the Leningrad Region. The third place was taken by the Republic of Tatarstan, which increased by one line. Tatarstan is implementing a number of initiatives to support the tourism industry, including rural tourism, which creates favorable conditions for launching projects in natural areas without harming the environment.

In 2023, the Moscow, Tyumen, Sverdlovsk, Kaliningrad and Murmansk Regions were included in the list of "advanced" regions. With the exception of the Moscow Region, all of these regions were among the leaders for the first time. They demonstrated a significant increase in the integration of the indicators of rural tourism impact on economic, social and environmental development. In addition, each of these regions has unique trends in tourism transformation. For example, the

Moscow Region actively applies the principles of "green" marketing, while for the Kaliningrad Region the most important aspect is the preservation of cultural and historical heritage (Trukhachev, Ivolga, & Varivoda, 2018).

The regions with the "developed" level for 2022 also show significant achievements in certain areas of rural tourism (Table 3). In order to join the group of leaders, these regions need to take several steps to improve the institutional structure and regulatory framework, as well as to increase the level of transparency of information on environmental, social and economic aspects of the rural tourism development, which will allow its objective assessment (Evgrafova, Sergeeva, Ismailova, Eremin, & Sherstyuk, 2022). It should be emphasized that among the "developed" regions one can find territories located in different parts of Russia and significantly different in size: from the traditionally popular destinations (in the south they are Crimea, the city of Sevastopol) to the increasingly popular destinations in the north (Arkhangelsk and Murmansk Regions) and in the Volga region (Chuvash Republic and Samara Region). This indicates that the transition to new forms of rural tourism is possible in regions with different "starting conditions" in terms of geographical location and climate.

Table 3. Regions with a developed level of integration of indicators of rural tourism impact on economic, social and environmental development of the region, 2022-2023

Ranking place	Years	
	2022	2023
1	Chuvash Republic	Ryazan Region
2	Tyumen Region	Republic of Kabardino-Balkaria
3	Pskov Region	Kaluga Region
4	Republic of Crimea	Perm Krai
5	Sverdlovsk Region	Kemerovo Region
6	Murmansk Region	Yamalo-Nenets Autonomous District
7	Arkhangelsk Region	Pskov Region
8	Republic of Kabardino-Balkaria	Republic of Khakassia
9	Samara Region	Kamchatka Krai
10	Sevastopol (city)	Republic of Crimea

continued on following page

Table 3. Continued

Ranking place	Years	
	2022	2023
11		Samara Region
12		Stavropol Krai
13		Rostov Region
14		Moscow Region
15		Ulyanovsk Region
16		Khabarovsk Krai
17		Nizhni Novgorod Region
18		Krasnoyarsk Krai
19		Arkhangelsk Region
20		Chuvash Republic
21		Vladimir Region
22		Novosibirsk Region

By the end of 2023, the twenty-two regions at the "developed" level had significant gains in some aspects of rural tourism. There was a high concentration of regions at the "developing" and "beginning" levels. At the same time, the number of regions at the "developing" level remained virtually unchanged (twenty-one regions), while the "beginning" level decreased significantly over the year from thirty-one to seventeen regions. Many regions that were previously at the "beginning" level were able to move up significantly, confirming that rapid progress is possible, especially through the creation of institutional conditions for tourism and the provision of up-to-date information (Agibalov, Zaporozhtseva, & Tkacheva, 2020). In some regions, the growing flow of tourists has led to social tensions and conflicts with the local population. In some places, such as the Chukotka and Nenets Autonomous Districts, the rural tourism development is hampered by harsh climatic conditions and limited transport accessibility and in the city of Sevastopol and the Republic of Ingushetia – by security threats. Significant growth was recorded in the Republic of Khakassia (+48 positions), which is associated with new information content in the field of ethno-tourism (Agibalov, Zaporozhtseva, & Tkacheva, 2019). At the same time, the Chukotka Autonomous District, the Republic of Karelia and the city of Sevastopol lost positions due to the lack of information and measures to develop rural tourism.

If we consider the changes over the past two years in the three main areas through which we assessed the multiplier effect of rural tourism on neighboring industries, we can see that in 2023, the decrease in the indicators was more often observed in the ecological-climatic block, while socio-cultural aspects deteriorated less often.

On the contrary, the positions of the regions in the infrastructural-economic block improved.

4. DISCUSSION

Let us now analyze the changes in the indicators of the multiplier effect of rural tourism by federal districts.

In the Central Federal District (Figure 1), almost all regions showed an increase in the indicators in comparison with 2022, with the exception of the Tver, Belgorod and Lipetsk Regions. The main obstacle to the development of rural green tourism in these regions is the mentality of the local population: the villagers are suspicious of tourists and are willing to accept them only if they see the active participation of their neighbors in this area.

Figure 1. Estimation of the multiplier effect of rural tourism in the Central Federal District, %

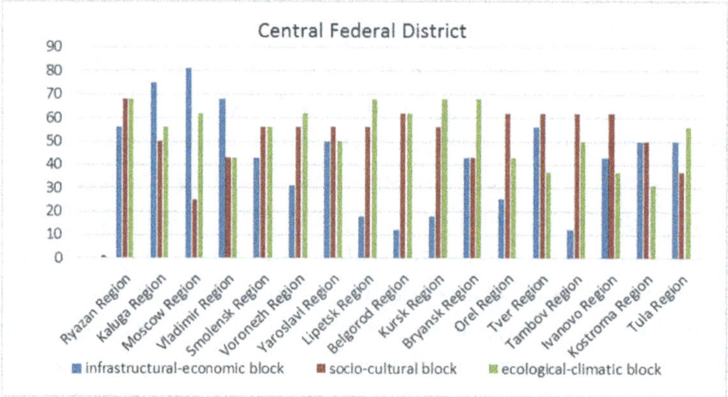

Of the seventeen constituent entities in the Central Federal District (excluding Moscow), ten were at the "developed" and "moderate" levels, while eight were at the "developing" level. The Ryazan, Kaluga, Smolensk, Yaroslavl, Tver, Kostroma and Tula Regions demonstrate the triune development. This aspect is fundamental for the development of rural areas. It is precisely the simultaneous development of all three blocks that is necessary to avoid slowdown in one of them. For example, if the economy is intensively developed, it can lead to excessive negative impact on rural areas and reduce the effectiveness of the ecological-climatic block. In the case of active growth of the socio-cultural block, which has a long payback period, there is a risk of worsening the results of infrastructural-economic development.

In the Northwestern Federal District (Figure 2) there is an increase in the indicators in all the constituent entities, except for the Arkhangelsk Regions and the Republic of Karelia. The key reasons slowing down the growth of rural tourism in these regions are the lack of quality paved roads (Besshaposhny et al., 2021. Despite the fact that rural tourism involves getting acquainted with the unique culture and lifestyle of the local population, it is often extremely difficult to get to many localities.

Figure 2. Estimation of the multiplier effect of rural tourism in the Northwestern Federal District, %

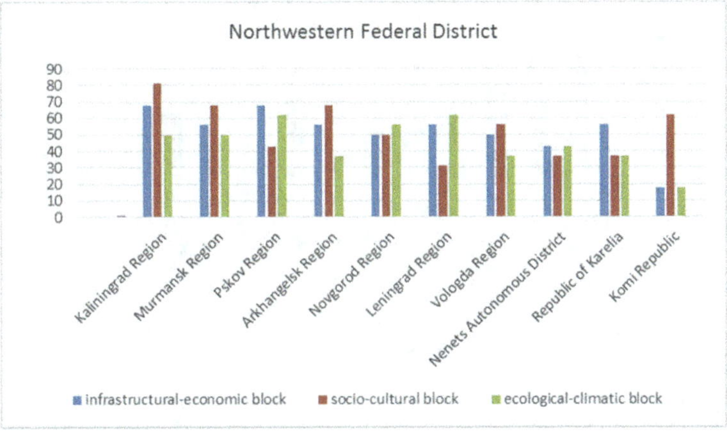

Of the ten constituent entities in the Northwestern Federal District (excluding St. Petersburg), four were at the "advanced" and "developed" levels, three were at the "moderate" level, and three were at the "beginning" level of the rural tourism development. The Nenets Autonomous District, the Komi Republic and the Republic of Karelia were at the "beginning" level. Their remoteness results in additional time and money spent on transportation for tourists. In addition, severe winters in the Nenets Autonomous District and the Komi Republic limit the year-round use of rural recreational areas. This creates problems with the profitability of this type of activity (Abrashkina, Agirbov, Andreev, et al., 2021). When organizing tours to remote historical and natural sites, the overall saturation of the program is reduced, which leads to a decrease in tourist interest, as well as to the creation of short routes that do not allow full immersion into the life of rural communities. At the same time, the Kaliningrad and Murmansk Regions were at the "advanced" level, where social, economic and environmental development is harmonious.

The Privolzhsky Federal District (Figure 3) also showed growth in the rural tourism development. In 2022, six constituent entities were at the "beginning" level, five were at the "moderate" and "developing" levels and only three were at the “developed” and “advanced” levels. In 2023, only the Saratov Region and the Republic of Mordovia were at the “beginning” level. In the published register of rural tourism sites in the Saratov Region, the majority are museums, including school and cultural museums. The rest of the routes are created by initiative citizens with farms or apiaries, but only a few receive tourists.

Figure 3. Estimation of the multiplier effect of rural tourism in the Privolzhsky Federal District, %

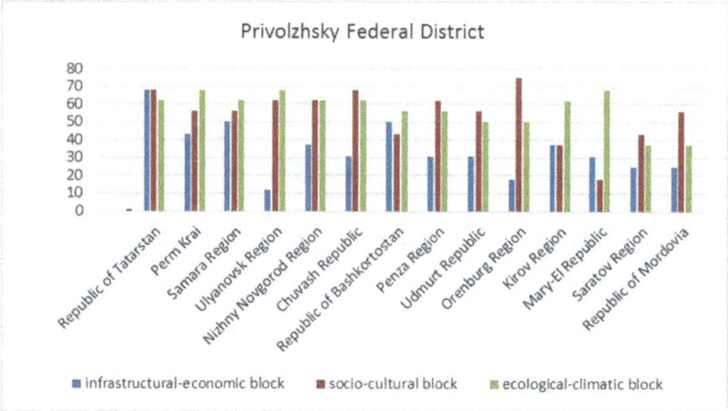

It should be noted that all the constituent entities in the Privolzhsky Federal District meet the minimum indicators in the infrastructural-economic block, with the exception of the Republic of Tatarstan. In Tatarstan, the rural tourism development shows a balanced and harmonious growth of all three blocks, which leads to a significant multiplier effect for related industries. In other regions, the share of socio-cultural development is much higher compared to the infrastructural-economic and ecological-climatic blocks, but in the Mary-El Republic this block has minimal values. It should be noted that all the regions of the Privolzhsky Federal District meet the minimum indicators in the infrastructural-economic block, with the exception of the Republic of Tatarstan.

In the Southern Federal District (Figure 4), the indicators of rural tourism remained practically unchanged compared to 2022, with the exception of the Rostov Region, where the level increased from “developing” to “developed”, and the city of Sevastopol, where the level decreased from “developed” to “beginning”.

Figure 4. Estimation of the multiplier effect of rural tourism in the Southern Federal District, %

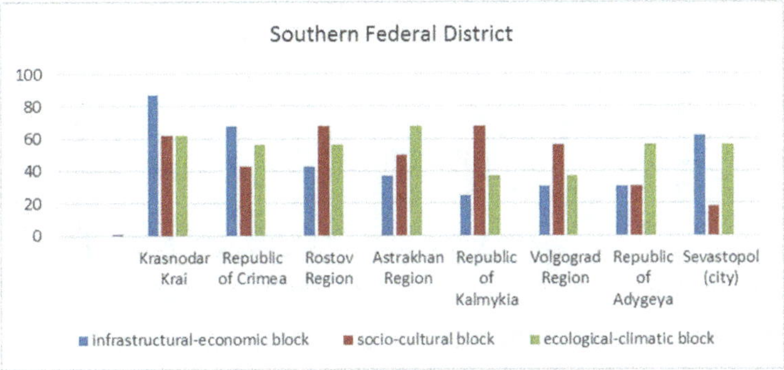

The optimal simultaneous development of all the components of the multiplier effect is observed in the Krasnodar Krai, the Rostov Region and the Republic of Crimea. In these regions, the agricultural sectors typical for rural areas are well represented. For example, in the Krasnodar Krai and the Rostov Region, sugar beet, grain and legumes, as well as beef cattle breeding, dairy production, vegetables and sunflowers dominate. Crimea's industries include grain farming, cattle breeding, viticulture, horticulture, vegetable growing and essential oil production. Agricultural enterprises and farms are actively engaged in developing tourist programs and receiving visitors.

In the North Caucasian Federal District (Figure 5) the level of the rural tourism development remained at the same level as in the Southern Federal District, with the exception of the Stavropol Krai (where the level increased from "moderate" to "developed") and the Republic of North Ossetia-Alania (where the level decreased from "moderate" to "developing"). The Stavropol Krai is considered one of the few "capitals" of rural tourism due to its developed agriculture, which creates significant tourism potential. In the Republic of North Ossetia-Alania, the decline in the rural tourism development is associated with socio-cultural security problems. The high level of the triune development was observed in the Republic of Kabardino-Balkaria, the Stavropol Krai and the Karachayevo-Cherkessian Republic, while only the Stavropol Krai and the Republic of Kabardino-Balkaria reached the "developed" level. The "beginning" level was observed in the Republics of Ingushetia and of Dagestan due to unfavorable ecological and climatic conditions.

Figure 5. Estimation of the multiplier effect of rural tourism in the North Caucasian Federal District, %

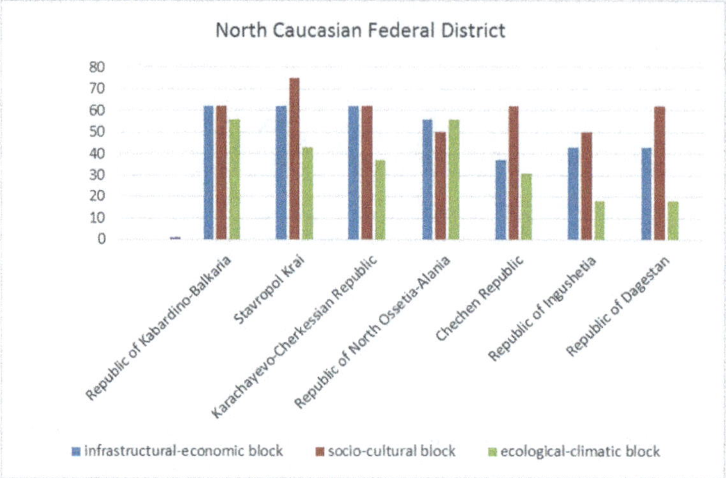

In the Urals Federal District (Figure 6), the Chelyabinsk Region lowered its level from "developing" to "initial". The federal subsidies for rural tourism planned for 2023 have been significantly reduced in this region. Due to the large number of industrial enterprises, the air in the region is polluted with heavy metals.

Figure 6. Estimation of the multiplier effect of rural tourism in the Urals Federal District, %

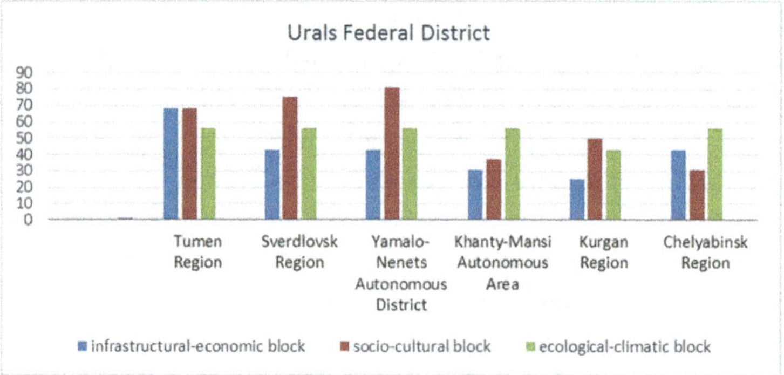

The Siberian Federal District (Figure 7) had no "advanced" regions, but four regions were recognized as "developed". They were the Kemerovo Region, the Republic of Khakassia, the Krasnoyarsk Krai and the Novosibirsk Region. In comparison, there were no "developed" regions in 2022. These regions show improvement in the socio-cultural indicators, while Khakassia shows improvement in all three blocks. All "developed" regions of Siberia have a high level of understanding of sustainable rural development issues, which is confirmed by their complete and detailed reports available in open sources. The lowest infrastructural-economic indicators are recorded in the Tomsk and Omsk Regions.

Figure 7. Estimation of the multiplier effect of rural tourism in the Siberian Federal District, %

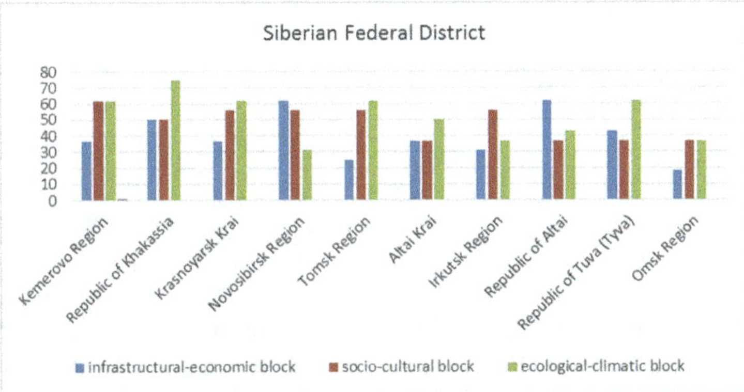

In the Far Eastern Federal District (Figure 8), two regions – the Kamchatka Krai and Khabarovsk Krai – were considered "developed". The Kamchatka Krai stands out as the most popular among the regions of the district and a preferred destination for ecotourism. There is a unique online resource that explains the basics of sustainable rural tourism in an accessible and visual way.

Figure 8. Estimation of the multiplier effect of rural tourism in the Far Eastern Federal District, %

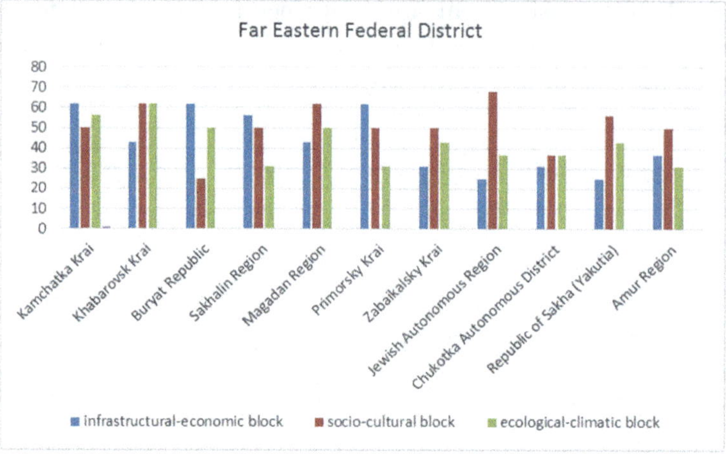

The Buryat Republic, the Sakhalin and Magadan Regions with a "moderate" level of development, and the Primorsky Krai and the Zabaikalsky Krai, which were categorized as "developing", were next. Four Far Eastern regions – the Republic of Sakha (Yakutia), the Jewish Autonomous Region, the Chukotka Autonomous District and the Amur Region – were classified as being at the "beginning" level.

The Kamchatka Krai significantly improved its positions in the sphere of social policy. At the same time, the Magadan and Amur Regions showed a decline in the ecologic-climatic indicators, while the Jewish Autonomous Region improved its results in the social sphere. The level of social efficiency slightly decreased in the Chukotka Autonomous District and the Republic of Sakha (Yakutia).

5. CONCLUSIONS

The results of the presented analysis of the integration of the indicators of the multiplier effect of rural tourism and its role in the sustainable development of rural areas confirm the positive trends in the implementation of this activity in the Russian regions. In comparison with the previous year, the average level of integration, calculated on the basis of 85 Russian regions, increased, and the number of regions with the improved indicators significantly exceeded the number of regions with the worsened indicators.

In 2023, the Leningrad Region took the first place in the ranking due to the creation of a favorable institutional environment for the sustainable development of rural tourism with unique practices. The Krasnodar Krai, which had been in the first place the previous year, fell back to the second place, but retained the status of an "advanced" region. This underlines the need for continuous work on improving the sustainable development and introducing new practices.

Eight regions could be singled out as the best examples of sustainable rural tourism, including new participants – the Tyumen, Sverdlovsk, Kaliningrad and Murmansk Regions. There were 22 regions with the "developed" indicators, and they have a high potential for the rural tourism development, but need better transparency of information.

CREDIT AUTHOR CONTRIBUTION STATEMENT

Ludmila Evgrafova: Conceptualization, Methodology, Data curation, Formal analysis, Investigation, Resources, Supervision, Validation, Writing - original draft preparation.

Natalya Sergeeva: Visualization, Writing - reviewing and editing.

DECLARATION OF COMPETING INTEREST

The authors declare no conflict of interest.

REFERENCES

Abrashkina, E. D., Agirbov, Y. I., Andreev, O. P., ... (2021). Agroindustrial Sector of Russia: Agriculture 4.0: Monograph in 2 volumes. Moscow: IPR Media, 379.

Agibalov, A. V., Zaporozhtseva, L. A., & Tkacheva, Yu. V. (2019). Scenario approach to setting strategy for rural sustainable development. *Vestnik of Voronezh State Agrarian University*, *12*(3(62)), 94–102. DOI: 10.17238/issn2071-2243.2019.3.94

Agibalov, A. V., Zaporozhtseva, L. A., & Tkacheva, Yu. V. (2020). Formation of a methodology for assessing the quality of sustainability of rural development. *International Agricultural Journal, 63*(1), 6. https://doi.org/DOI: 10.24411/2588-0209-2020-10130

Besshaposhny, M. N., Evgrafova, L. V., Lazar, V. V., Pichuzhkin, N. A., & Grachev, A. B. (2021). Innovative mechanism to increase the efficiency of indirect employment of the rural population. In *The Challenge of Sustainability in Agricultural Systems* (pp. 907–913). Springer International Publishing., DOI: 10.1007/978-3-030-73097-0_101

Evgrafova, L. V. (2015). Analysis of the efficiency of the dairy cattle breeding. *International Agricultural Journal*, *2*, 62–64.

Evgrafova, L. V., & Ershova, V. L. (2020). Methodical aspects of formation of agrotouristic cluster structure in the economy of the region. *International Scientific Journal*, *2*(2), 30–36. DOI: 10.34286/1995-4638-2020-71-2-30-36

Evgrafova, L. V., Sergeeva, N. A., Ismailova, I. Z., Eremin, V. I., & Sherstyuk, M. V. (2022). Assessing the functional efficiency of tourist and recreational clusters in the regions of Russia. *AIP Conference Proceedings*, *2467*, 070011. DOI: 10.1063/5.0092777

Trukhachev, A., Ivolga, A., & Varivoda, V. (2018). Analysis of agricultural tourism and study of its development potential. *Research Journal of Pharmaceutical, Biological and Chemical Sciences*, *9*(3), 792–801.

Chapter 12
Cultural Adaptation of Digital Payment Systems With a Focus on Community-Based Financial Practices

Nisha K. Kumar
https://orcid.org/0000-0001-5744-011X
Independent Researcher, India

N. Giribabu
https://orcid.org/0009-0004-8695-5349
Sri Venkateswara College of Engineering and Technology, India

Meera Sharma
https://orcid.org/0009-0002-6839-6817
USCS, Uttaranchal University, India

M. Siva Krishna Raju
https://orcid.org/0000-0002-5916-4937
SRKR Engineering College, India

V. Sri Krishna Chaturya
https://orcid.org/0009-0007-9489-5093
SRKR Engineering College, India

DOI: 10.4018/979-8-3373-2240-7.ch012

ABSTRACT

The emergence of digital payments is changing how financial transactions are carried out; however, cultural adoption of these practices within varying socio-economic frameworks is critical for success. This chapter analyzes the digital payments landscape using VUCA and BANI models and centers discussion on aspects like cultural pluralism, deficits in digital literacy, and trust gaps. M-Pesa, WeChat Pay, and Paytm serve as case studies for this investigation. The results suggest that culturally adaptive systems of digital payment enhance adoption, trust, and financial resilience. The framework argues for the necessity of truism-driven, inclusive FinTech. Further investigations are needed on AI, blockchain, and the regulatory components of these issues in order to build resilient ecosystems for digital finance.

INTRODUCTION

The global economy is now witnessing the upgrade and adaptation of traditional payment methods to digital payment systems, which have revolutionized the payments industry and the way businesses and individuals undertake transactions. These systems provide enhanced accessibility, convenience and economic opportunities for all, whether they be residents of metropolitan cities or those living in much more remote rural areas. Modern economies, especially amidst increasing smartphone penetration and wider internet availability, depend on digital payment systems for efficiency and cashless transactions (Huo et al., 2024).

The acceptance and effectiveness of digital payments systems, however, depend on the sociocultural context of a specific community. Every payment done is closely linked to the habits, customs, and norms of the societies members which means that a payment is much more than an exchange of goods and services. Payments done need to be contextualized with their culture for people to fully understand them. Ignoring these factors can lead to fragmentation amongst users, which tend to result in the reluctance or slow adoption of the system (Zhanga et al., 2011).

However, one such cultural activity that demonstrates the amalgamation of custom and modernity is Bohni, which is found in different parts of the market in South Asia. Bohni is the first sale and is made by the trader in the pursuit of making the highest turnover of sales. This act goes beyond just mere economic exchange, it is a symbolic act believed to forecast how successful one will be in the course of the day. In this transaction, merchants tend to prefer cash because with it, they feel more assured and operate with greater immediacy and sequent ritualistic significance.

Even though changing cash transactions to digital provides great benefits to people, it poses a challenge to traditional practices. Payment done with cash which is often said to lack humanity can sometimes interfere with the culture and feelings attached to Bohni. For cash reliant small traders, particularly those located in rural or semi-urban regions, these attitudes underline the necessity for care in designing financial technology that serves purpose but is also sensitive to culture.

This chapter examines the cultural and technological factors that certain traditions influence the financial behaviors and the impact that has on digital payments. By examining how such traditions could be incorporated into digital platforms, the chapter seeks to address the need for inclusivity at a greater scale. Bridging the divide between tradition and technology goes beyond improving user experience; doing so will enable trust, adoption, and economic empowerment in diverse cultures, which is absolutely essential.

Alongside the extensive cultural analysis, this chapter also advocates for the more important issue of digital transformations merging with local traditions for their cultures to be successful in the long run. It will analyze the opportunities and challenges of cultural adaptation to help fintech companies create systems that sustain traditions and modernity at the same time. In doing so, this chapter hopes to address the discourse on responsive financial technologies, and as a result, create a pathway toward a digital economy that is more inclusive and sustainable.

1.1 Background on Digital Payment Systems

There has been an integration of numerous financial technologies making digital payment systems more efficient in securing transactions for individuals, businesses, and governments and this has transformed the world's finances. Digital payment systems include mobile payment applications, online banking, digital wallets, point-of-sale systems, and other advanced technologies which have opposed the old-fashioned methods of cash and paper-based payments (Sunil & Nalwaya, 2023).

The growth of digital payment systems has been enhanced by various factors such as the adoption of smartphones, the improvement of technology, and advancement in security systems. Further, these systems have fostered payment conveniences in addition to enhancing financial inclusion especially in emerging economies (Goel & Nath, 2020). These systems have been evolving over time and their adoption rate by countries like India, Singapore, and Japan demonstrates the coincidental advance of Digital payments and the Fourth Industrial Revolution.

The integration of mobile devices and the advancement of technology has served as a catalyst to help promote novel idea of ease while making payments and purchases to consumers, especially during the global pandemic where social distancing and safety protocols were mandated. The need to protect public health during the COVID-19 pandemic helped to quickly normalize digital payments as the only feasible contactless method of payment (Sunil & Nalwaya, 2023).

The concept of trust and security has profound effects on the use and acceptance of digital payment systems. New and sophisticated payment methods such as tokenization and biometrics tend to enhance users' trust and security, which compels them to adopt more advanced payment systems. Security, reputation, and structural assurance of digital payments are said to increase the trust and continuance intention of the users (Goel & Nath, 2020).

This is further facilitated by the infrastructure and omnichannel nature of the digital payment systems — including card, e-payment, mobile payment systems and cryptocurrency (Khando et al., 2022). Each of these approaches has its own advantages and disadvantages. Non-banking mobile payments, for instance, have drastically changed the landscape by enabling financial inclusion. However, there are several social, technical, economical, awareness, and legal bound challenges that digital payment systems must overcome. These limitations will hinder progress on the adoption and acceptance of digital payment systems defining needs for more strategies.

1.1.1 Evolution and Adoption Trends

In the last twenty years, the use of digital payments has increased drastically as these systems have evolved from niche to widely utilized financial systems. The global digital payments market was valued at 96.07 billion USD in 2023 and is projected to grow at a compounding annual growth rate (CAGR) of 21.1% from 2024 to 2030 (Grand View Research, 2023). Developing regions, particularly Asian and African nations, have been leading this change due to the presence of revolutionary platforms like M-Pesa in Kenya and Paytm in India.

a) **Asia-Pacific Region:** Digital payments in Asia pacific region accounts for over 52% of global contribution (Euromonitor International, 2024). They are also the fastest growing in the world due to government policies aimed towards a cashless society, along with easy smartphone availability, and unrestricted access to the internet.
b) **Africa:** The sub-Saharan African region has seen a shift in financial inclusion due to mobile money platforms like M-Pesa, which enables users without bank accounts to access formal financial systems. The region is responsible for more than 70% of the volume of mobile money transfer, as stated by GSMA (2022).

This has broadened the scope of financial inclusivity along with driving economic growth by eliminating the costs associated with cash transactions.

1.1.2 Advantages of Digital Payment Systems

The introduction of modern electronic payment systems is supported by our society due to their potential advantages including, but not limited to:

a) Cost Reduction: Digital payments and transactions do not require provisions for infrastructure such as bank branches, which lowers costs to the providers of financial services. Consumers also prefer to pay less fees when compared to other forms of banking.
b) Improved Financial Inclusion: Mobile payment technology provides access to financial services such as savings, credit, and insurance to the unbanked population. Using mobile payments significantly improved access to credits and savings for the less fortunate in developing countries. This assertion is supported by Chatterjee (2024).
c) Economic Development: Digital payment systems increase productivity and economic activities through fast and secure transactions. For example, Bruno et al. (2019) claimed that the adoption of digital payments could boost global GDP by $3.7 trillion by 2025.
d) Comfort and Protection: Faster transactions, better security through asymmetric encryption, and biometric identification reduces chances of fraud. This allows for greater convenience.

1.1.3 Challenges: The Need for Cultural Adaptation

Cultural barriers are the critical barrier to the widespread adoption of digital payment systems; these barriers change with different demographic groups. A primary drawback to the advancement of many digital payment systems is the cultural gap that exists between their prospective users and the communities that these systems intend to serve. Many pay systems that are digital in nature are created to cater to a global clientele and therefore tend to ignore the cultural intricacies of specific user groups.

a) Cash-Centered Society: In cash-driven societies, cash is not just a currency but a cherished way of life; it is viewed as instant and trustworthy. As an example, South Asian cultures have some practices which do hold great cultural significance, such as Bohni (the first sale of the day) which is symbolic and cannot be replicated through digital transactions.
b) Reduced Literacy Levels: Adoption is hampered due to limited understanding and knowledge of digital payment systems, especially in rural and semi urban population centers.

c) Absence of Trust: In cash societies, users believe the use of physical cash is more secure than the use of digital payment systems.

In order to respond to these issues, digital payment systems must urgently develop culturally flexible approaches. These approaches involve the use of local dialects, adapting to local customs, and building confidence through targeted advertisements and educational initiatives.

1.2 Importance of Cultural Adaptation in Financial Technology

The implementation of digital payment systems relies not only on the development of technologies, but also on how these systems relate to the culture of the people they wish to serve. Cultural adaptation in financial technology (FinTech) is perhaps one aspect that determines if a payment platform can serve its users from various socio-cultural backgrounds. This is particularly true in areas where culture is considered to be an integral part of socio-economic practices and are often observed to be oriented based on values, beliefs, and rituals that go far beyond functional service delivery.

1.2.1 Cultural Barriers to Adoption

Cultural issues may greatly slow down the acceptance of digital payments, particularly in developing nations or regions such as South Asia and Southeast Asian countries. These issues stem from existing cultural norms, social habits, and traditionalized monetary practices, which have been part of society for many years.

In South Asian countries, cash transactions are accompanied by deep cultural and sentimental values such as Bohni, which is the first sale of the day. Bohni is usually performed in cash payment because it is considered auspicious. This attachment to cash has resulted in exceptionally high dominated cash circulation, for instance, in Pakistan and India cash-in-circulation is17% and 12% of GDP respectively, compared to only 5% in more developed nations (World Bank, 2022).

Social expectations and gender norms equally influence the delayed shift towards digital payment methods in certain regions. Bangladesh, for instance, has one of the highest gaps in relation to financial inclusion for women who struggle with adopting modern digital financial instruments. Some of these barriers consist of low digital literacy, confidence in using mobile devices, limited English skills, and the domineering family setup. Notably, almost 99% of digital finance agents in Bangladesh are men (World Bank, 2022).

Cultural beliefs regarding the nature of money and transactions may also inhibit the use of digital payments. In cultures where cash was the primary formative medium of exchange, there is usually an overdependence and trust associated with cash. This cultural bias towards cash makes it difficult for the people to trust and use digital payment services (Qadri, 2023).

The effects of cultural attitudes on the use of digital payments are not restricted to South Asia only. In Southeast Asia, many regions continue to depend on cash transactions even with the increasing adoption of digital wallets and mobile payment applications. For instance, a survey conducted in 2021 showed that a large majority, or 66% of respondents from the Philippines, preferred cash on delivery as a payment option in social commerce (Komoju, 2024).

To tackle these cultural obstacles, there is a need to design and implement programs that respect local customs and ways of life. Such programs may include culturally-oriented digital and financial literacy initiatives that aim at amending the gap between the access and use of digital payment systems. Also, payment service providers and financial institutions working with e-commerce businesses have to allow the development of more holistic digital payment systems that consider the local culture.

In conclusion, while the use of digital payments is being highly encouraged by technological progress alongside government policies, some cultural hurdles persist. Overcoming these obstacles calls for a carefully crafted plan that leverages the advantages of financial inclusion while respecting and integrating local cultural traditions.

1.2.2 Case Studies of Cultural Integration

The successful implementation of these fintech solutions rests on the integration of local culture into the customer's online platform, which requires some degree of belief, value, and ritual adaptation of the targeted market. Such adaptations serve the dual purpose of creating trust and increasing participation in the use of digital payment systems by making them more culturally relevant and familiar.

a) WeChat Pay in China: WeChat Pay is a prime example of cultural adaptation through the incorporation of sending WeChat monetary gifts (red envelopes) during the Chinese New Year holiday. Hongbao, or red envelopes, are considered a gift in the form of cash given to family and friends during major celebrations with the hope of bringing good fortune. With WeChat Pay, this custom was transformed into a digital practice where users could send virtual hongbao as gifts. This innovation allowed WeChat Pay to enhance user participation during significant cultural festivities, while further increasing its dominance as the primary digital payment platform in China (Butt et al., 2024). By offering a modernized approach to traditional festive gifting, WeChat Pay successfully encouraged its use by a larger audience.

b) M-Pesa 's services in Kenya: Kenya is also renowned for M-Pesa, a mobile money service that transformed financial access in the nation. M-Pesa capitalized on local customs by allowing users to receive and send money through mobile phones. This capability was especially useful in a country that relied heavily on cash and had very limited banking services. M-Pesa also leveraged informal money transfer practices and introduced agents in local communities who facilitated these transfers giving hope to a society that primarily operated in cash. M-Pesa was able to build trust and acceptance of cashless transactions by balancing the sociocultural and economic environment. Within no time, M-Pesa was able to grow its user base across the whole region of Eastern Africa despite the initial hesitation to adopt mobile payments (Jack & Suri, 2011).

c) Paytm in India: As one of the top digital payment companies in India, Paytm provides an outstanding example of cultural integration within the context of digital payments. Founded in 2010, Paytm started out as a mobile recharging site, but later incorporated digital wallets, bill payments, and an array of e-commerce services to keep pace with the increasing demand for modern payment systems in India. Paytm's success can be explained by its cultural integration policies which took into consideration India's social and cultural customs and preferences. For example, the platform's functionality, which included regional language provisions and user-friendly design, was aimed at a vast populace, including those lacking digital skills. Paytm also gained the confidence of its customers through local merchant partnerships which enabled cashback offers that aligned with India's strong cash and loyalty culture. In addition, Paytm took advantage of the increasing availability of mobile internet and government support for financial inclusion, especially after the 2016 demonetization when the use of digital payments became necessary. By combining traditional community payment concepts, local merchants and modern payment technologies, Paytm achieved a remarkable level of acceptance into the Indian culture and facilitated widespread use of digital payments (Bhatia-Kalluri & Caraway, 2023).

These case studies show that digital payment systems which align well with local cultural values have a higher likelihood of successful adoption and growth.

1.2.3 Impact of Cultural Adaptation

Culturally customized digital payment systems have a profound influence on user trust, engagement, and ultimately adoption. Users are more willing to engage with technology when they are ethnically appreciated by a financial platform. For instance, Almuhairat (2024) found that users of financial services incorporated localized features like language, cultural symbols, and region-specific rewards that incorporated localized features reported higher retention, some platforms reported forty percent higher retention.

In addition, cultural adaptation fosters trust. This is particularly the case in less developed regions where there is limited experience with digital transactions. Failing to incorporate local customs and practices significantly impedes acceptance of the system. Fintech companies willing to respect such cultures can foster acceptance and ultimately, speed up the move from cash to digital payments. The inclusion of local language, culturally relevant offerings, and alignment with important culture events can enhance adoption and user engagement.

1.3 Review of Literature

The use of digital financial technologies all over the world shows rapid development, which has resulted in a new area of research – the cultural adaptation of digital payment systems. One relevant issue of this adaptation is the incorporation of digital payment systems into community-based financial practices, which differ from one culture to another. An in-depth understanding of these localized practices is fundamental for the effective implementation and acceptance of digital payment systems in foreign markets. This review aims to summarize major contributions on the issue, including contemporary ones focusing on the relationship between digital payments and community-based financial systems.

1.3.1 The Effect of Trust and Culture on Digital Payment Transactions

There has been considerable scholarly attention on the impact of cultural values on attitudes towards digital payments, especially through the lens of Hofstede's (1984) cultural dimensions theory. As Shin (2010) explains, collectivist cultures tend to trust social networks more, which affects the willingness to accept digital payment systems. Financial activities, such as lending circles and P2P transfers, are built on trust and social relationships. When digital payment systems seek to enter into these cultures, they are more likely to adopt systems that prioritize social validation and endorsements, such as WeChat Pay in China (Arner et al., 2020). Social endorsement as a marketing strategy is critical in cultures where people would rather trust systems that accept them based on their social group than the system's technical capabilities (Yoon, 2009).

1.3.2 Community Based Financial Practices and Financial Inclusion

In numerous developing countries, informal community-based financial practices like rotating savings and credit associations (ROSCAs), informal lending, and community-based saving groups are still widely used. These systems are relevant in facilitating financial inclusion, especially where formal banking systems are lacking. Research by Klapper et al. (2016) illustrates these practices as vertically empowering because they allow the most marginalized people in the society to obtain financial resources through community based trusted systems. When tailored to these cultural settings, digital payments can improve the efficiency and effectiveness of these practices. For instance, M-Pesa in Kenya is a mobile money platform that has adapted community based financial systems by allowing people to carry out financial transactions through mobile phones, often using community-based networks for verification (Jack & Suri, 2011).

1.3.3 Barriers to Adoption of Digital Payments in Community-Based Societies

The integration of new technology is often met with some degree of resistance. In particular, community-oriented societies exhibit high levels of barriers such as lack of trust in formal financial systems, inadequate digital skills, and technological phobia. Such phenomena described by Yang et al. (2023), suggest that the older, rural demographic groups are the most averse to digital payment systems because they are technologically challenged and highly depend on interpersonal or face-to-face transactions. Meanwhile, Nurjihadi et al. (2024) argue that cultural norms that are associated with cash payment and physical interaction often induce a lower willingness to accept digital systems that lack a personalized approach or human element. Such resistance is further exacerbated when mobile payment systems fail to integrate into the societal community framework, underlying the importance of paying attention to particular local customs and values.

1.3.4 Digital Payments and Socio-Cultural Norms

The impact of socio-cultural norms on the adoption of digital payment systems is one of the most studied topics in recent literature. Opoku-Okuampa (2024) claims that community oriented financial activities are fundamentally performed within a system of cultural norms that manage social relations to a greater extent than they manage financial transactions. For example, in several Southeast Asian societies, there are informal borrowing and saving (referred to as "susu" in some places) circles within the socio-economic culture.

Unless these social norms are modified, the shift to digital payment systems may disrupt these cultural practices. Alipay is known to have a more culturally sensitive approach by allowing group transactions, which is more aligned with the community-oriented nature of Chinese financial transactions (Huang et al., 2020).

1.3.5 Recent Developments and Localized Strategies

The modification and development of modern cultures along with region-specific new technologies has fostered a rise in demand for region specific systems, leading to the restructuring of various digital payment systems. Aysan et al. (2025) explain that local payment methods (LPMs) are coming into existence, permitting firms to target and expand to particular areas around the globe. Domestic payment systems facilitate inexpensive payments, boost economic activity, and ensure economic autonomy for the country. Research conducted in India indicates that technology adoption for using digital payments differs significantly between users due to the availability of smartphones, especially Google Pay and Paytm. The existing trends in the digital ecosystem are characterized by the expansion of Fintech, the growing number of active smartphone users, and the enhanced user experience in payments towards convenience and security that clearly shows the importance of capturing foreign markets.

Furthermore, Dasarraju (2024) states that there have been localized approaches in the recent innovations for cross border remittance that enhance the inclusiveness of the services and respond to the specific characteristics of the migrant communities. In particular, mobile money and digital platforms have become the most important channels for cross-border remittances within and between regions such as Africa and South East Asia. These platforms provide facilities for quick acces to funds which subsequently improves the transaction fees, which in turn boosts financial activities within these communities. These new types of business models are changing the remittance market for borderless transfers. They include mobile money cross border remittances, online and internet platforms, peer-to-peer services, and even blockchain solutions. The introduction of new technologies such as QR codes, Blockchain, and artificial intelligence is revolutionizing the payments and remittances across the borders. New services being introduced are improving the effectiveness, safety, and transparency of these transactions.

1.3.6 Challenges in Regulatory and Policy Frameworks

Digital payment systems differ from region to region which is often culture-bound. Policies which aim to foster financial inclusion in different parts of the world may actually worsen the deployment of digital payment systems if local context is not considered (Arner et al., 2015). For example, in informal economies, there may be legal restrictions on how informal practices can be accommodated into modern digital financial systems. There is a real challenge in integrating informal practices into existing legal frameworks.

The reviewed literature highlights the intricate issue of the continuation of the culture preservation and the innovation phenomenon of the digital payment systems in relation to the local legal and financial boundaries. This equilibrium is important for the proper utilization and assimilation of digital payment systems in different communities. The construction of digital payment systems from the cultural context of the community is a complex issue, considering the local socio-cultural ecosystem. Recent research shows that successful incorporation of digital payments into these systems necessitates substantive knowledge of cultural and socioeconomic factors.

1.4 Research Gap and Need for Culturally Inclusive FinTech Solutions

The interplay between the cultural context and the dynamics of adopting digital payment systems remains underexplored, given the context of innovation-driven growth. There has been considerable focus on such technical components as security, interface, and system scalability of fintech innovations, but not much attention has been paid to the culture, emotion, and sociology of finance. Most digital payment methods are developed to cater to a global audience without regard to the specific customs, practices, and traditions of different societies. Consequently, people from some geographical areas where cash payments and cultural related customs are prevalent may not identify with these systems.

1.4.1 Existing Gaps in Research and Practice

a) Limited Research on Integrating Traditional Practices into Digital Platforms

Although fintech solutions have transformed financial services around the globe, the integration of traditional practices with digital payment platforms is still being explored. Numerous digital payment systems ignore the impact that cultural practices such as 'Bohni' (first sale of the day) in South Asia, have on user behavior. Financial rituals and practices are often traditional and cultural, therefore incorporating these elements into the design of fintech platforms is essential to achieve acceptance. Further research is required on how these practices can be digitized or integrated into platforms to increase adoption within culturally sensitive user groups.

b) Insufficient Focus on Rural and Semi-Urban Populations

Digital payment systems have been adopted the most in urban cities where infrastructure intel is more available. Rural and semi-urban populations, who are more culturally conservative regarding finances, have difficulties adopting such payment systems. Traditional values and cash economies are deeply rooted in these areas. Furthermore, the absence of appropriate digital payment services for these populations widens the gap between urban and rural places in terms of financially accessible places. It would be more beneficial to devote more attention and funding to research and development that seeks to address the culturally significant barriers facing these neglected societies.

c) Absence of Partnership Between Fintech Companies and Local Communities

One of the reasons that create the gap in the adoption of inclusive culture-oriented fintech solutions is the partnership gap between the communities and the fintech companies. Most fintech platforms are built to be user-friendly without consulting intended users on their desires, procedures, and beliefs which is essential for proper ethnographic market research. Such an approach often ignores certain cultural aspects such as the tendency to pay or the social interaction involved in these payments. Consulting the targeted communities while the solutions are being developed and deployed will make it easier for the fintech companies to design solutions that will be suitable for the given culture.

1.4.2 The Need for Culturally Inclusive FinTech Solutions

The absence of financial technologies that guarantee cultural diversity directly affects the levels of acceptance of electronic payment methods, especially in areas where cash is still the most popular payment method. Research suggests that many micro-business owners, who are important to the subsistence economy, are often reluctant to adopt digital payment methods because of their traditional orientation as well as issues of trust, security, and convenience. As an example, the ubiquity of smartphones in South Asia is not accompanied by the widespread use of digital payment systems, as only 30% of small vendors use them regularly (Carlos et al., 2024). This represents a disturbing disconnect between the acceptance of technology and the corresponding culture that underscores the necessity of cultural- sensitive fintech solutions.

Adoption of digital payments might be slow for older people and even for some merchants and people in the countryside as they might find no direct value in it or might find it culturally important. For them, cash payments are more personal and the perception of digital payments can seem to be quite distant. In addition, the absence of proper online safety, lack of cyber literacy, and unfamiliarity with online systems broaden the gap. Therefore, these problems can only be addressed using a functional approach that incorporates emotion and culture.

1.4.3 Opportunities for Culturally Inclusive FinTech

The tasks involved in defining culturally inclusive solutions offers practitioners a difficult problem to solve and also presents a wide potential opportunity. The positive outcomes that are expected from responsive inclusive fintech solutions are the following:

a) Enhanced Participation of Targeted Users in Financial Services:

Adoption of local culture norms in practice of digital payment systems particularly in economically and socially marginalized rural and rural areas can be very supportive towards high adoption ratios.

These interventions can reach niche markets, women, and other underserved populations that consider digital financial services irrelevant or difficult to access.

b) Increasing Productive Economic Engagements

There is great potential to enhance economic engagement and growth through culturally-tailored fintech applications. In economies where businesses are particularly struggle to receive payments, payment solutions that are tailored to local customs can significantly enhance the business value. Moreover, the combination of mobile wallets with local saving and lending cooperatives may enable small entrepreneurs to access funding, which would promote small business activities.

c) Market Expansion for Fintech Companies

Culturally tailored solutions could provide the necessary opportunity for underserved communities to be served by FinTech's. By implementing local cultural practices and customs, community finance can be better positioned to develop solutions with higher user acceptance and loyalty. This strategy also serves to differentiate firms in crowded markets by providing relevant local solutions. For instance, in India, Paytm has been successful because of the local payment culture and the usage of the app during traditional festivals, which made it more acceptable to users (Bhatia-Kalluri & Caraway, 2023).

d) Building Trust and User Retention

Trust, an important integration component in payment adoption, can be built through culturally adapted systems. Users can relate to adaptively and culturally relevant services, which according to Almuhairat (2024) improves user retention by nearly 40%. Trust is essential in financial dealings, and when such dealings are placed in a particular cultural context, there is bound to be a better and stronger relationship between customers and fintech companies.

1.5 Objectives

1. To analyze the role of cultural factors in the adoption of digital payment systems.
2. To identify the challenges and opportunities in integrating traditional financial practices into digital platforms.
3. To propose culturally adaptive models for digital payment systems that promote financial inclusion.
4. To provide global case studies showcasing successful cultural adaptation in digital finance.
5. To offer policy and strategic recommendations for stakeholders to foster inclusive digital economies.

1.6 Research Methodology

A mixed approach has been taken in this chapter. It includes a thorough review of scholarly literature, industry reports, and cultural dimensions on cultural adaptation in the FinTech space. This is augmented by an extensive analysis of these payment systems and their cultural contexts through cases that discuss the successful and unsuccessful systems in place. In addition to this, the successful, culturally adaptive financial models is studied, including their innovative, strategic, and key success factors as well as the lessons that can be drawn for future implementations. Discussions were carried out with experts in FinTech, policymakers, and leaders in the communities where they discuss the perspectives of the challenges and opportunities of digital finance that are culturally responsive and set inclusive of the community's needs.

2. CULTURAL INFLUENCE ON FINANCIAL BEHAVIOUR

Culture influences people's financial actions such as their decisions on spending, saving, lending, borrowing, and their interaction with money. This includes how people adjust to new advancements in finance such as digital payment systems. These cultural factors are shaped by deep-rooted traditions, norms, beliefs, and societal expectations. In most cases, financial decisions are not only rational choices made out of an economic context, but are also influenced by cultural and social factors. Evaluating such cultural factors is important in designing technologies for finance that are effective and culturally acceptable.

2.1 A Cultural Approach to Financial Decision-Making

Different cultural groups tend to approach different financial decisions differently and that is why Hofstede's cultural dimensions (2001) are relevant. Hofstede's cultural dimensions include power distance, individualism versus collectivism, uncertainty avoidance, long versus short term orientation, and indulgent vs restrictive. Each of these factors determine how people view and interact with money, financial systems, and how they make decisions regarding saving, spending, investing, and borrowing.

2.1.1 Power Distance and Financial Decision Making

In societies that have a high-power distance, such as head of households or business owners in a family, financial decisions tend to be made by few powerful individuals. This level of hierarchy dictates how particular resources are divided or distributed among families or communities. For example, in a number of South Asian or some African cultures, household heads or male bread winners perform a majority of the financial decisions autonomously with minimum input from family members. This has important implications for fintech solutions. In case digital payment systems are to be successful in such environments, they ought to ensure that important decision-makers are in full support of using these systems. Moreover, the digital payments systems must be sophisticated yet straightforward to the level where these decision makers can easily dominate and comprehend them (Hofstede, 2001). In cultures with lower power distance such as in many Western countries, financial decisions are much more decentralized and there is greater discretion in how an individual can manage his/her finances. This can impact the way that individual control and choice are designed into the fintech platform interfaces.

2.1.2 Individualism Vs. Collectivism in Financial Practices

The approaches taken towards finances can be significantly altered by the level of individualism or collectivism exhibited within a culture. The United States and Western European countries are examples of highly individualistic societies which greatly value independence and self-reliant finances. In these scenarios, digital payment systems advertise extreme convenience, personalization, and autonomy in finances to their customers. For example, individualistic cultures tend to appreciate self-directed investment plan, personalized financial planning services, and individual savings accounts (Bitar & Tarazi, 2022).

On the other hand, collectivist societies, which include India, China, and much of Africa, focus on group wellbeing, family responsibilities, and community cohesion. In these cultures, economic choices are usually made for the good of the family or the community. Thus, digital payment systems in these contexts should accommodate group centered actions. For these users, some important aspects include group saving features, family accounts, and cooperative budgeting accounts. Furthermore, these collective financial rituals, in particular, communal savings groups (susu in West Africa, chit funds in India, tanda system in Latin America and the Caribbean, hui in China, and stokvels in South Africa), need to be incorporated into the digital payment platforms to increase their usage and significance in these parts of the world (Triandis, 1995).

2.1.3 Uncertainty Avoidance and Risk in Financial Decisions

Different cultures treat risks and uncertainty in various ways. Uncertainty, however, refers to the extent in which a certain individual is comfortable with the possibilities of ambiguities in his or her environment. For example, Japanese and many Europeans show greater levels of constraint and try to avoid certain risks financially (Reis, 2022). Adopting new payment systems digitally may severely jeopardize such societies unless there are favorable clear outcomes to do so. For risk wary users, security measures such as fraud protection, encryption, and user supportive customer service must be focused on and employed. Also, trade banking or cash based financial system may appeal more predictable and safer than the new digital platforms, therefore, making them more preferred. In such societies, fintech companies have a great challenge of earning trust and protecting people's private information.

However, cultures with low degrees of uncertainty avoidance like Israel and the U.S. are more willing to take and endure risks as it leads to enabling the adoption of new innovative payment options regardless of how uncertain they are (Hofstede, 2001).

2.1.4 Short-Term versus Long-Term Considerations in Financial Decision-Making

In cultures with a long-term orientation like China and India, there is an emphasis on working hard, saving money and planning. People in these cultures tend to think long-term by saving towards future goals such as furthering their education or retirement at a younger age. Such individuals, if offered credit or investment opportunities, may be more willing to use digital payment platforms. Additionally, digital payment platforms in these cultures can aid economically by providing features such as savings accounts, retirement accounts, and investment recommendations.

Short-term oriented cultures such as parts of the U.S. and Latin America place greater importance on fulfilling immediate needs. Financial decisions in these cultures revolve around meeting one's immediate desires rather than worries of long-term security. Digital payment platforms aimed at these short-term oriented users may be more effective if they offer easy and instant credit, quick transfers, or rewards (Hofstede, 2001).

2.2 Role of Traditions and Social Norms in Financial Practices

Financial practices are influenced by social norms and traditions, which have a significant impact on their management, expenditure, and the perception of financial institutions. These norms determine how people interact with their finances and how they interface with payment systems.

2.2.1 Cash-Dependent Economies

Across the globe, particularly within developing nations, cash payments continue to be the dominant form of payment. This is especially true for rural areas or lower socio-economic populations with limited access to banking facilities and low faith in automation processes. Cash payments are believed to be more real, as well as to provide a higher degree of personal safety and autonomy (Bandura & Ramanujam, 2021). Residents from most Asian, African, and Latin American nations prefer cash transactions because they are simpler to deal with. In these circumstances, where the tangible aspect of cash is preferred, the acceptance of digital payment systems becomes extremely challenging.

Social interaction and cash are often intertwined. In countries such as India and Kenya, cash payments are made personally, enhancing connection and relationships. This social element, however, becomes very difficult to recreate in digital spaces which are considered more neutral. Resistance to adopting digital payment systems is made worse by fears of identity theft, losing control over transaction processes, and ignorance about how such systems operate.

2.2.2 Symbolic Practices in Financial Transactions

Paying or receiving money may have deeper meaning in various cultures. For instance, in several African cultures, money exchanged via a wedding or funeral ceremony is seen as a key way of sustaining social ties and meeting cultural expectations. In the same way, in China, giving out money contained in hongbao (red envelopes) during the Lunar New Year is not only about the movement of cash, but also good fortune, prosperity, and family unity (Saxena, 2024). Adoption of digital payment systems will enable People in other cultures to understand and make use of these platforms when they are designed to cater to these principles. Understanding and embedding cultural symbols, rituals, and social practices into the payment systems can help the users appreciate and embrace these digital systems.

2.2.3 Community-Oriented Savings and Lending Practices

Numerous regions in the world particularly South Asia, Africa, and Latin America, practice community-focused financial activities such as chit funds in India or susu in Western Africa. These informal financial schemes facilitate members of a community to save money and enable access to loans in resource-pooled manner. Digital payment services seeking to enter these markets need to consider these self-help financial methods. For example, a service that enables collective saving, peer-to-peer lending, or other joint financial projects with set objectives would be more acceptable to users familiar with community-centered financing (Jack & Suri, 2011). Furthermore, the communal aspect of such activities, which includes trust and shared accountability, needs to be incorporated into digital systems in order to ease integration and usage.

2.3 Barriers to Digital Payment Adoption Due to Cultural Factors

Even with the benefits that can be derived from using digital payments in advancing inclusiveness and growing the economy, varying cultures exist as obstacles in trying to adopt such technologies in a lot of countries. It is important to know and tackle such issues in order to enable digital payment systems in these relatively advanced, cash driven economies.

2.3.1 Trust and security issues

Trust is central to the adoption of any financial system and of utmost concern for cultures that have relied on face-to-face cash transactions. In developing nations, especially in the countryside, many people are concerned about the security and authenticity of digital payments. These contexts are more sensitive to technology-related concerns about privacy, fraud, and hacking. Digital payments are often dependent on mobile networks and internet connectivity, which could be seen as unstable and non-secure. People from cultures where trust is built through personal interaction are less likely to engage with digital systems that have no personal contact or responsibility (Aker & Mbiti, 2010).

2.3.2 Digital Illiteracy and Technology Issues

Lack of digital literacy poses the most significant challenge for the adoption of digital payments, particularly in lower-income and rural areas. While the availability of mobile and smart phones is ubiquitous, the majority of the population lacks the ability to utilize digital financial platforms proficiently. Sometimes, older people who are not so tech-savvy tend to adopt new systems with greater reluctance. Moreover, users living in areas with poor internet or mobile coverage may find it difficult, if not impossible, to use digital payment systems. It is essential to overcome these obstacles by creating digital payment options with as few technological constraints as possible (Anwar et al., 2024).

2.3.3 Cultural Preference Towards Cash Payments

One of the challenges facing the shift to online payments is the cultural inclination that favors cash transactions. In numerous societies, cash goes beyond mere currency; it represents a form of identity as it denotes independence, affluence, and power. For example, in most rural settings, people regard cash transactions as a sign of trust since they facilitate a direct form of exchange. The trust and emotional attachment that cash transactions engender (Pascal & Edith, 2023) is hard to achieve with digital payments that are viewed as cold and immaterial. As a result, online platforms have to work towards achieving a user experience that imitates the immediate and tangible feel of cash transactions to appeal to these societies.

3. CHALLENGES IN CULTURAL ADAPTATION OF DIGITAL PAYMENTS

3.1 Trust and Security Concerns in Cultural Adaptation of Digital Payments

Trust and security issues constitutes one of the major challenges in the cultural adoption of digital payment systems. Digital payment systems are premised on the electronic transfer of monetary value through mobile wallets, banking applications, and online payment systems. While the systems provide considerable ease and speed, they also pose significant consequences to the users, including fraud, cyberattacks, and data breaches that heavily impact trust.

3.1.1 Trust as a Cultural Concept

In developed countries, people are highly uncertain and tend to avoid change because of the new technology's propensity to misuse their sensitive information and money (Hofstede, 1980). Culturally, people from these economies might need comprehensive legal and security frameworks along with an efficient resolution of conflicts as prerequisites for allowing the use of digital payments systems (Gefen et al., 2003).

Furthermore, the level of trust perceived for particular institutions such as banks or payment service providers varies from region to region. Users from low-trust countries are wary of sharing personal financial information or using services associated with traditional banking, and the same behavior can be seen in regions affected by fraud or corruption, which results in a low level of trust and lack of engagement with the digital system (Lu et al., 2011).

3.1.2 Security Concerns

In comparison to cultural attitudes towards payment systems, security fears pose enormous barriers to the adoption of such systems in the economy. security and safety remain of utmost concern in the level of the economy, especially in the context of growing identity theft, forged documents, spyware, and financial scamming. It is possible to lose data consisting of ID cards, banking accounts, mobile phones, and passports, which leads to hardships beyond financial loss. Likewise, evidence shows certain consumers are scared of revealing sensitive information such as ID or credit card numbers, and hence restrain themselves from using digital payment systems (Sleiman et al., 2021).

For instance, individuals from nations with weak cybersecurity are likely to challenge the capability of the service provider to protect their information. Even technologically advanced countries have users skeptical of online payment platforms due to prominent data breaches, such as Target's 2013 breach and Equifax's 2017 breach. Furthermore, insufficient information on how personal data will be processed can worsen these fears because people may be concerned about being monitored, their data being abused, or even sold without their permission (Bajwa et al., 2023).

3.1.3 Cultural Influences on Perceptions of Security

Cultural attitudes are also important when analyzing how different communities respond and interact with digital security challenges. In collectivist cultures like those in Asia, Africa, or Latin America, the primary basis of confidence tends to come from relationships and community support. In social groups where a certain platform is widely accepted as trustworthy, it is likely to be adopted by other members of that group even when they are not fully aware of its technical security features (Creese et al., 2021). In addition, these cultures exhibit a rather lethargic attitude towards self-protection strategies, such as providing personal passwords, relying instead on institutions or service providers to take responsibility for everything.

Compared to collective cultures, individualistic cultures such as those in North America, and Western Europe, are highly autonomous, which makes personal responsibility a priority when assessing security. These users tend to analyze technical protections such as encryption, user controls, and privacy policy transparency to determine whether trust can be established (Shin, 2010). Furthermore, skepticism towards institutions makes people demand more control over their data, which can be seen in the regulation instruments of the GDPR (General Data Protection Regulation) in the European Union.

Perceptions and attitudes towards security may further be affected by power distance and uncertainty avoidance. Power distance and uncertainty avoidance contribute towards an individual's perception of security. In cultures with high power distance, such as the Middle East and some areas in Asia, citizens are more willing to accept information and trust platforms that use endorsements from authorities or governments as sources of information (Luo et al., 2010). However, cultures that focus on high uncertainty avoidance such as Japan and Greece prefer policies that include visible security features and those that provide a detailed explanation of how risks will be controlled. On the other hand, countries with lower uncertainty avoidance, such as the United States and Singapore, are more willing to take the risk of adopting new technology without fully understanding the risks, as long as the benefits are evident.

Systems and platforms should be created and designed with careful consideration of these cultural differences. Community endorsement should be the focal point of the platforms in collectivist cultures as it increases trust, while in individualistic cultures, strong technical safeguards and user privacy should be the emphasis. Engaging with such cultural values builds trust and acceptance in users around the globe.

3.1.4 Mitigating Trust and Security Concerns

Digital payment service providers can take the following initiatives to manage the trust and security issues:

a) Enhanced Data Privacy Compliance: Being compliant with global laws such as GDPR is a prerequisite, as the data of users must be handled with care and clear communication should be provided.
b) Enhanced Security Measures: With Multi-factor Authentication/AI Security Systems, users gain more assurance regarding their information due to the high level of encryption provided.
c) Trust Building Through Education: Eliminating fears regarding the identification of scams is achieved through educating users on how to use digital payments systems.
d) Political Support: Trust can be fostered by augmenting the cultural norms of a locality with low digital literacy through the aid of agents facilitating digital payments in person.
e) Strategy formulation by regulators: Establishing compliance as well as other industry standards becoming consumer protective entities is crucial for the government and other regulatory bodies.

3.2 Digital Literacy and Technological Barriers

The skill of operating the internet has now become an essential requirement of the world we live in (Spante et al., 2018). However, this seamless technique faces several restrictions in its use and adaptation due to modern technological barriers. Some of these constraints may be lack of adequate infrastructure, poor access to essential commodities, absence of necessary skills, mental or cultural blocks. This barrier fundamentally impedes the large-scale adoption of a particular technology, and is likely to affect marginalized groups who traditionally have restricted access to education, employment, and the digital economy. To successfully target these problems, an understanding of such issues is paramount, as it is critical to fostering an all-encompassing digital society.

3.2.1 Access and Infrastructure Barriers

The absence of access to reliable and affordable digital technologies remains the biggest barrier to attaining basic digital literacy skills. The digital gap, defined as the divide between those who own digital devices and those who do not, is very much a function of social and spatial inequality. Van Dijk (2020) points out that residents of rural regions or the peripheral parts of poor metropolitan areas have very little access to broadband infrastructure, thus constituting a major obstacle to their attainment of digital skills. This can be exemplified through sub–Saharan Africa and parts of South Asia, where electricity supply is incredibly unreliable along with low internet penetration rates making access to digital tools very difficult (Jacome, 2023). In the absence of such tools, people are barred not just from optimized learning, but participating in the economy as well.

Furthermore, the absence of digital tools worsens the issue of accessibility. Digital devices, for example computers, tablets, smartphones, internet subscriptions, and even steam subscriptions are way beyond the financial scope of a majority of families in low or middle classes. Even in developed countries, members of lower socioeconomic communities can find themselves hindering access to advanced devices or high-speed internet due to the devices cost (Hilbert, 2016). This gap creates a new cycle of inequality, where individuals who do not have access to modern technologies, still cannot keep up with the rapid pace of technology development.

3.2.2 Skill Gaps and Digital Competence

Achieving digital literacy is not only dependent on access to technology, but skill gaps can also be a challenge. Skills involved in the use of technology range from basic things like operating a computer or smartphone to upper-level skills like online information evaluation, data protection, and use of digital tools for work productivity (Helsper & Eynon, 2013). For many people, especially those who are less educated, the absence of technology is heavily prohibitive to obtaining these competencies.

These skill gaps are accentuated by differences brought about in each generation. Terms like “digital natives” are often used to refer to the younger generation who, due to their prolonged exposure to technology, are relatively more skillful at operating advanced technological equipment. Nonetheless, their adeptness in technology is usually confined to social and entertainment purposes, which creates gaps in advanced capabilities like content creation or even digital security (Karjaluoto et al., 2020). On the other hand, older generations, also referred to as digital immigrants, are usually met with cognitive as well as psychological hurdles when trying to learn any digital skill. Such barriers consist of apprehension towards complex interfaces, anxiety over making mistakes, and low self-efficacy towards learning new technologies (Bandura, 1997). These challenges make it difficult for older adults to positively engage with available digital spaces and fosters feelings of disconnection in an increasingly digitized society.

3.2.3 Cultural and Psychological Barriers

Culture and psychology, as factors, greatly influence how digital literacy is perceived. For instance, the attitude towards technology differs from one society or region to another. People from high context cultures, where interpersonal communication and societal hierarchy are highly valued, may not readily adopt digital tools because they consider them highly intrusive (Hall, 1976). For example, elders in such cultures prefer in person communication; viewing everything else as too impersonal, hence resist adopting digital means of communication.

Psychological factors like self-efficacy and perceived usefulness affect digital literacy. Bandura's (1997) self-efficacy theory outlined that individuals who have low confidence with technology tend to be unresponsive towards it. This self-doubt can emerge from previously experiencing failure, having limited exposure, or not engaging due to anxiety. Furthermore, the context of where these tools are used has an impact on how useful they are, for instance, in farming or local trading, people in rural areas may not see digital literacy as relevant at all.

3.2.4 Overcoming the Technological Challenges of Digital Literacy

In order to overcome these challenges, there needs to be collaboration between the government, businesses, and educational bodies. Initially, there is the need to solve issues relating to accessibility and affordability. The government and NGOs can take the lead by improving digital infrastructure in these regions. Providing public Wi-Fi hotspots, distributing devices to school children, and subsidizing Internet access can greatly ease accessibility problems (Rahtio, 2024).

Second, the development of skills through trainings designed for various sociographic groups is a must. Older individuals, for instance, can be trained digitally through workshops and community activities that build confidence and competence to use digital tools. Educational institutions, as a rule, should embrace and teach digital literacy, so that younger generations are not limited to only elementary skills. Free or inexpensive online training programs are also important for widening access to digital knowledge (Helsper & Eynon, 2013).

Lastly, solving the cultural and psychological barriers requires that these problems are approached in a local and culturally appropriate manner. Community leaders and other influential persons can also play a pivotal role in implementing campaigns advocating for the importance of digital literacy in highly conservatively accepting areas. Campaigns that endorse the concrete benefits of being digitally skilled, for example, better healthcare services, financial aids, and educational opportunities, will help change people's attitude towards technology. In addition, providers of the technology should emphasize intuitive design and user-friendly interfaces that would make technology accessible for the new users with minimal cognitive effort.

3.3 Regulatory and Policy Issues

The development and implementation of digital payment systems is often hampered by considerable regulatory and policy issues that cultural differences aggravate. The regulation of consumer protection, financial privacy, and other issues differ from one country to another because of different cultural and societal settings. This diversity poses challenges to the worldwide expansion of digital payment systems and makes cultural adjustment more difficult.

3.3.1 Data Privacy and Security Regulations

Lack of uniformity in data security and privacy poses one of the most critical challenges. For example, the European Union General Data Protection Regulation imposes hefty constraints with regard to user consent as well as user data collection and retention practices, which illustrates the degree of respect for privacy in Western societies (Voigt & Von dem Bussche 2017). Many Asian and African countries, however, adopt a lighter model of privacy regulation that emphasizes the provision of digital financial services. These varying approaches force international providers of digital payment systems to adapt their systems to the specific privacy laws of each country, increasing both the cost and the complexity of maintaining the systems.

3.3.2 Anti-money laundering (AML) and know-your-customer (KYC) regulations

The distinction in the AML/KYC policies per country is determined within the specific region's culture and inclination to take on financial risks. For example, India and Kenya are more lenient with identification policies due to the financial inclusion objective, whereas, the United States and Germany need to implement stricter policies. Klapper et al. (2016) argues that these countries aim at preventing fraud by using thorough identity verification systems. This leads to the conclusion that a balance has to be established between ease of the user and strict compliance by the provider in regards to digital payment systems.

3.3.3 Government intervention in financial systems

The misconception by the general public about the government's intervention in the financial system also influence regulatory frameworks. Nations such as China or those in the Middle East with high power distance cultures tend to have active government participation in the management and promotion of digital payment systems. The digital payments centralization under Alipay and WeChat-pay in China is a good example of how much the government controls an economy's activities (Arner et al., 2020). However, low power distance cultures such as the United States and the United Kingdom have taken a more laissez faire approach where regulations are predominantly self-imposed, which encourages competition, but tends to create fragmentation in the system.

3.3.4 Cross-border interoperability

Another concern of cross border interoperability is the differences in regulations regarding limits of transactions, payments of taxes, and policies for currency exchange that inhibit the effective operation of digital payment systems across borders. For instance, much of Asia and Africa still lack good regulation because of their regulatory inconsistency and diversity in culture and economy, while Europe on the other hand has made great strides in Digital payments, especially with the creation of the Single Euro Payments Area (SEPA). This greatly enabled them (Song et al., 2024; Kouty, 2024).

3.3.5 Emerging Technologies Regulation

Finally, the regulation of blockchain technology and cryptocurrencies presents a set of challenges that are distinctly different from those faced earlier. Some cultures are more innovative and ready to experiment, such as the U.S. and Singapore, while others are more conservative mainly due to fears of fraud, misuse, and technological instability. For instance, India's supervision of cryptocurrency regulation is much more cautious and restrained in comparison to El Salvador's aggressive adoption of Bitcoin as legal tender. This represents different cultural attitudes in innovation adoption and risk sensitivity (Mittal & Singh, 2024).

Most of these regulatory and policy hurdles will only be resolved by understanding the local culture and customs. Decision makers need to control the degree of security and innovation provided, while global digital payment providers need to spend towards regulation compliance. Some problems can be resolved through international cooperation like regulatory and public-private partnerships which enhance the possibilities of operational expansion of the digital payment systems for the cultures and regulatory environments.

4. OPPORTUNITIES FOR CULTURALLY ADAPTIVE DIGITAL PAYMENTS

The integration of new forms of digital payment can pose a challenge which can be overcome by recognizing and accommodating local financial practices and cultural values. The essence of culture-sensitive digital payment systems is to foster their ease of adoption in both developed and developing nations. In many societies, informal financial practices like community savings, peer-to-peer lending, and physical transactions continue to thrive due to their trust-based nature and deeply ingrained social structures. The adaptation of digital payment systems to these community-embedded practices increases user trust, adoption rates, and supports financial inclusion (Klapper et al., 2016). For instance, M-Pesa in Kenya and WeChat pay in China have successfully fused these modern digital products with more traditional practices making their services a lot easier to use (Jack & Suri, 2011; Yang et al., 2019). It is these adaptive systems that integrate the old world and the new, thus enabling economic growth and aiding financial education in the process (Arner et al., 2020). Thus, the opportunity is in creating payment technologies that are sensitive to socio-cultural issues while enhancing economic opportunities for people.

4.1 Using Traditions to Integrate into Digital Space

Cultural adaptation of digital payment systems can revolve around integrating traditional financial practices into a digital platform. Many societies, particularly those in developing countries, place great value in informal community-based financial practices such as rotating savings and credit associations (ROSCAs), community lending, and peer-to-peer transactions. These activities tend to stem from a deep cultural attitude which revolves around trust and shared community ideals. The ability for digital payment systems to incorporate these cultural practices will aid in the systems' widespread adoption and enhance financial inclusion.

4.1.1 Combining Vernacular Informal Financial Practices with Digital Platforms

Mobile payment systems like M-Pesa have revolutionized community-based financial practices through mobile by enabling users to send, receive, and even make payments for group savings or loans within their social set, all from their mobile devices. Such mobile payment platforms have achieved great success in Kenya by incorporating the existing vernacular informal financial practices into the system. For instance, in Kenya where there is no formal banking infrastructure, ROSCAs and community savings groups fill the gap, arguably providing mobile payment platforms the opportunity to market their services (Jack & Suri, 2011). Digitizing these communal financial systems improves the level of transparency, efficiency and accessibility created while embedding the trust these communities have in their systems.

4.1.2 Cultural Adaptation Through Social Influence and Peer Validation

Digital payments are able to utilize relationships formed on trust which are present in many societies. In a collectivist culture, social influence is of utmost importance for decision-making. The use of digital payment systems in any society can develop user's endorsement or social recommend features to fit the cultural characteristics of that society. This can be seen in the case of China where WeChat Pay or Alipay users can approve transactions, partially disclose their transaction histories, and have peers review their transactions for endorsement (Arner et al.,2020). In such cultures, acceptance and use of new technologies is usually dictated by family or associates and sometimes even influential people within a particular locality. Hence, the creation of a digital payment system within a social framework that serves as the basis for motivation allows for greater acceptance of these forms of payment.

4.1.3 Rural and Community-Based Financial Solutions

Rural populations along with those from still developing regions face barriers and restrictions in gaining access to established financial institutions- but this only helps in strengthening their reliance on community based financial structures. However, digital payments represent an innovative opportunity to deliver real financial services to those areas which lack traditional financial services. If rural banking providers reconsider the structure of their platforms to adequately attract rural users, micro lending, mobile money, and local insurance services can easily be availed (Klapper et al., 2016). While the ability to borrow and lend still exist in traditional community framework, the secure digital modalities and remote access leverage makes it uniquely advantageous. This has the potential to trigger poverty alleviation by allowing economic activities in the neglected regions.

4.1.4 Localized Payment Solutions That Reflect Cultural Practices

Digital cultural payment integration possesses wider scope and deeper relevance than just the inclusion of relevant region payment systems. For instance, there are cultures that have unique ways to mint out financial transactions like gifting and trading which do not involve cash. Even barter trade which is common in some rural and tribal societies can easily be modernized by the use of tokens or credit notes that represent the physical exchange of goods. Furthermore, cash on delivery (COD) is widely accepted in a lot of Asian countries, and therefore, payment systems are easily deployed by providing COD and cash responsive offerings (Saha & Nag, 2019). Payment systems can be developed to incorporate local preferences and culturally sensitive transactions to enhance the ease of adoption and acceptance of digital systems.

4.1.5 Financial Literacy and Education Through Community Networks

Education concerning finances greatly matters within digital payment frameworks, more so in societies that largely grapple with low economic literacy. Users can be trained and encouraged to use the digital payment system through established community networks. Local celebrities, community heads, or even local bank managers can be used in as a motivational tool for acceptance and trust in those features of digital payment systems. The collaboration between Paytm and local vendors to train consumers on digital wallets is paying off in rural areas (Ravindra & Sreedhar, 2017). In this light, community-based education is a subtle marker of both the culture as well as the technology present in the locality.

4.2 Community-Based Digital Financial Models

In regions devoid of physical banking institutions, these community-based financial models have continually enhanced economic development. Such models are predicated on social capital and networks that promote trust and responsibility, thereby facilitating saving, lending, and investing activities. The increased adoption of digital payment technology has a tremendous potential to automate and improve these traditional practices. This facilitates the creation of digital community finance models that integrate local cultures with modern technology. Such models enable and empower communities by allowing them to retain the underlying socio-economic frameworks alongside a sophisticated financial structure.

4.2.1 Digitizing Informal Savings and Lending Practices

Digitization of rotating savings and credit associations or ROSCAs along with other savings associations is one of the most outstanding ones. Community-based savings model ROSCAs, which is common in many counties in Africa, Asia, and Latin America, have group members contributing a certain amount after predefined intervals, and collated sums are divided among members on a regular basis. Digital platforms offer new and effective ways to manage ROSCAS through secure methods of depositing and disbursing funds. M-Pesa, for instance, used in Kenya and Tanzania, provide users with mobile money services which have greatly assisted in the electronic organization of ROSCAS, thereby facilitating informal saving systems while diminishing the risk of fraud (Jack & Suri, 2011). Similarly, other apps such as Chama Apps in Kenya support community savings groups with automation for monitoring and notifications, which helps to enhance trust and accountability among savers (Chidziwisano et al., 2020).

4.2.2 Cultivating Peer-to-Peer Lending Using Digital Systems

The phenomenon of P2P lending is an additional element to the community based financial systems that have emerged as a result of digitization. In the past, communities have depended on interpersonal lending to cater to their short-term financial requirements. Digital platforms that strive to mimic the trust-based system are likely to improve the level of access and security while maintaining the communitarian attitude underlying it. Kiva is a platform that has successfully globalized P2P lending by allowing people from developed nations to lend modest amounts to people in developing areas. These systems adopt systems of social accountability and embed culture at the community level to assure repayment and trust among participators (Ly & Mason, 2012).

4.2.3 Integration of Microfinance and Crowdfunding

The combination of microfinancing with crowdfunding has resulted in growth in community-based digital financing models. Crowdfunding has become an effective tool for raising funds for small businesses and local projects in areas with high levels of economic interdependence. GoFundMe and Kickstarter have inspired Ketto in India, which addresses culturally specific needs in funding such as health care, education, and business. Microfinance institutions have used technology to provide low-interest loans to the disadvantaged while taking into consideration cultural practices such as group lending, which is common in many societies (Armendáriz & Morduch, 2010).

4.2.4 Cultivating trust by employing culture relevant functionality

One of the most important factors for the success of community-oriented digital financial models is the development of trust within the cultural context of the target group. In many cultures, trust is cultivated through peoples' social relationships, which means that social interaction is a facilitator of trust, and thus, can be easily adopted. In China, users can transfer money through group messaging and sending virtual red packets as gifts with WeChat Pay (Yang et al., 2019). This shows how digital financial platforms are able to combine traditional ways and modern-day technology in a way that is appealing to users.

4.2.5 Empowering Women and Marginalized Groups

Digitized financial models that are deployed by the community possess the greatest value in their capability to empower women and other marginalized groups. Women around the world, especially those hailing from more traditional backgrounds, serve as key managers of household finances, while also participating in informal savings groups. By digitizing these practices, platforms provide women with better access to resources which enables them to participate in the development of the community and the household. It has been established that the inclusion of women's digital finance improves their welfare and the economy overall (Demirgüç-Kunt et al., 2017). Microfinance models that empower women and help them escape poverty, pioneered by Bangladeshi Grameen Bank and India's Self-Employed Women's Association (SEWA), illustrate culturally attuned digital programs.

4.3 Significance of Local Languages and Cultural Symbols in User Participation

A key success factor for a digital payment system is the acceptance and use of the system by the intended users which is determined by their culture and language. Local language and cultural symbols are instrumental in building trust, aiding accessibility, and promoting adoption in different communities. They go beyond aiding the use of the platforms to developing a familial and emotional bond which is critical towards maintaining engagement, especially in places where technology trust, literacy, and use is low.

4.3.1 Local Language: Bridging Accessibility Gaps

Understanding language as a barrier for many people and augmenting gaps in accessibility can be achieved by incorporating local languages into digital payment platforms. Because users engage with technology through the medium of language, offering platforms in native languages can greatly ease cognitive load, thereby enhancing user experience. Research indicates that incorporating local languages in digital platforms enhances trust and confidence particularly among novices and users with little formal education (Wiratno & Rozi, 2024). For example, Paytm in India is a mobile payment application that allows its users to interact with the platform in over ten regional languages. Such a feature has significantly boosted adoption by rural, non-English speakers (Spillers & Gavine, 2018).

Furthermore, assistive voice technology in vernacular languages is becoming increasingly important for enhancing increased access. For instance, Paytm users in India are able to transact using voice commands in Hindi, Tamil, and many other regional languages. Such features help users with low literacy skills, making digital financial services more accessible to all (Chaudhuri & Roy, 2021).

4.3.2 Cultural Symbolism: Enhancing Emotional Connection

Cultural symbolism can be in the form of icons, motifs, or themes that are culturally relevant, and can enhance user interaction. In collectivistic societies where financial choices are a communal or family concern, these types of platforms will appeal more to users. For instance, China's WeChat Pay has features like "red envelopes" which imitate the custom of giving money during festivals and other special events. This form of cultural modification not only helps in the adoption of the platform, but also helps integrate it into the daily life and social customs of the users (Yang et al., 2019).

User engagement increases on platforms that embed local cultures and symbols of the festivals. Payment software in Muslim-majority countries adjusts its designs and marketing efforts by including crescent moons and charity sponsorships during the holy month of Ramadan. This kind of culturally themed graphics during Ramadan incites user participation because of its feeling of relevance and connectedness.

4.3.3 Building Trust Through Cultural Sensitivity

Using local languages and cultures to design interfaces and carry out digital payments is an effective strategy to build trust in regions that are distrustful. Ethnic identity is a strong predictor of the willingness of the users to engage with the technology. The adoption rates of new technologies tend to be higher when the designs are sensitive to the user's culture and traditions (Shin et al., 2021). For example, the incorporation of branding using Swahili and Kenyan culture on M-Pesa increases the confidence of the rural and semi-urban populations in Kenya (Jack & Suri, 2011).

4.3.4 Overcoming Resistance Through Localization

Local languages blended with cultural symbols aid in overcoming the unfamiliarity, which helps in reducing the gap and distance towards accepting digital payments. Digital payments are still very foreign to many individuals, especially those who are not familiar with advanced technological countries. Certain platforms failing to adjust to local biases may become estranged and harmful to their market growth. PayPal is an example of a globally accepted platform that has no localized features and as a result does not penetrate the remoter regions of India and Indonesia, which are more rural (Spillers & Gavine, 2018).

5. CULTURAL DIMENSIONS OF DIGITAL PAYMENTS

Different cultures around the world have a direct effect in the adoption and use of digital payment systems. Hofstede's cultural dimensions of individualism and collectivism, as well as uncertainty avoidance, are well-known as having a significant impact on user behavior and preferences. In cultures emphasizing individualism, there is a stronger concern on private and security issues regarding digital payments and users prefer systems that ensure data protection (Chang, 2024). However, collectivist cultures seem to focus on community trust and group associated relationships in payment systems (Deufel et al., 2019). Countries with strong uncertainty avoidance, like Japan and other European countries, are usually more traditional in financial issues and are less likely to use digital payment options unless they are guaranteed a strong level of security. Research has shown that consumers' purchasing decisions, especially in comparison between offline and online shopping, are greatly influenced by the uncertainty avoidance bias (Alkhwaldi et al., 2024). Such behavioral patterns not only affect particular user preferences, but also determine the configuration and the implementation of digital payment systems in different countries. Therefore, the patterns underline the importance of adopting culturally sensitive policies in the design and deployment of payment systems, illustrating the need for culturally sensitive approaches in the development and implementation of fintech strategies (Zhao & Pan, 2023).

5.1 M-Pesa in Kenya: Adopting the Traditional and Modern Financial Services

M-Pesa, Kenya's leading mobile money service, has transformed the economy by merging cultural elements with banking procedures and services. This has significantly changed the socio-economic and financial landscape of Kenya. M-Pesa's success can be primarily credited to the collectivistic nature of Kenya. Socially collectivistic cultures tend to prefer payment methods that foster social interaction. M-Pesa capitalized on the already existing cultural traditions in which users support and help members of their extended families by allowing payments to be made to relatives in rural regions. This goes hand in hand with the Kenyan custom of strong kinship ties which expects a considerable amount of interaction and aid to one's relatives after migrating to urbanized regions (Hinz, 2014).

The platform's influence on financial inclusion illustrates its relation to cultural aspects. Access to formal financial services in Kenya before the launch of M-Pesa in 2007 was limited to 26.7% of the population, while 41.3% were completely cut off. As of 2019, 82.9% of the population had access to formal financial services, with those excluded and informally accessing services drastically reduced to 11% and 6.1% respectively (Ndung'u, 2021). The drastic increase in financial inclusion shows the effectiveness of M-Pesa in bridging the divide between the traditional informal financial services and modern digital finance.

M-Pesa's success is equally attributed to capitalizing on prevailing informal systems. The exchange of airtime as a form of currency by Kenyans prior to the launch of M-Pesa enabled the swift acceptance of mobile money. This is in line with the cultural dimension of uncertainty avoidance since people were used to mobile technology being utilized for monetary transactions.

The incorporation of M-Pesa's mobile money service into Kenya's social structure, which was rooted in a vigorous familial attachment, represented a significant cultural innovation. The service harnessed the existing custom of urban workers remitting money to their families in the countryside and, thereby, fulfilling an enormous cultural interest. This unison with collectivist principles aided M-Pesa in achieving rapid acceptance and adoption (McBride & Liyala, 2021).

The platform's influence goes beyond economic exchanges to the redefinition of social relations and cultural behavior. Digital transactions have reconfigured cultural rites and norms; many people do not physically attend funerals and weddings but send money electronically instead. This proves the case that digital finance does not only adapt to change cultures, but actively defines its evolution. Further, M-Pesa's integration with the financial ecosystem of the country illustrates how the platform bridges the gap between traditional and digital finance. The growth of mobile money has sparked a wider revolution in financial services in Kenya, with later generations of innovation such as banks discovering the e-money platform, credit being offered on the platform, cross-border and international remittance payments being facilitated, and partnering with other fintech services (Ndung'u, 2021).

M-Pesa has undoubtedly had an impact on the economy. Mbiti & Weil (2015) note that M-Pesa users experienced a reduction in competing money transfer services. The prices for transactions under 35,000 shillings fell from about 14% to 4%. This illustrates the ability of digital finance to stimulate competition and enhance productivity within the financial services sector.

M-Pesa's success in Kenya is a case of how digital financial platforms can serve as effective intermediaries between traditional and modern finance systems by incorporating and responding to cultural parameters. The platform's endorsement of disadvantaged social groups and support for financial inclusivity, within the wider collectivist framework, demonstrates that fintech innovations can respect and redefine cultural values while facilitating economic growth.

5.2 WeChat Pay in China: The Convergence of Social and Financial Transactions

In China, WeChat functions as a social networking site and has incorporated WeChat Pay which has become the most popular mobile payment service in the country. This has created a unique ecosystem where financial transactions and social engagements intertwine, and it has been working well because of Chinese sociocultural practices. This also allows for the reduction of platforms used to achieve multiple goals. WeChat Pay adoption is said to be influenced by a multitude of factors. For example, Wang and Gu (2017) showed that perceived social influence together with perceived networking capability impacts the perceived ease of accomplishing relationship maintenance activities, which affects the use intention of WeChat Pay. This adoption was highlighted as it showcases the platform's success with factors existing surrounding the social interactions.

The convenience and effortless integration of the platform into daily life also contribute to its success. Vatsa et al. (2024) pointed out that social and hedonic views are of great concern to users' continuance intention with WeChat, which suggests that the platform's popularity rests on its ability to make the users' life enjoyable and easy. This corresponds to the "scenarios of use" theory, in which WeChat Pay enables the creation of new contexts in which the application can be utilized, such as shopping, bill payments, and many other aspects of users' routines.

Cultural elements significantly influence the adoption of WeChat Pay. Lisha et al. (2017) suggested integrative guanxi-based constructs in the context of a technology acceptance model. They found that "g nqíng" (emotional attachment) positively affects a users' continuance intention to use WeChat whereas "mìanz " (face) negatively impacts the intention. This indicates the readiness and willingness to appreciate and understand local cultures in the design of mobile payment systems is crucial.

The growth of WeChat Pay's market share is clear evidence of its rapid growth. By the end of 2019, WeChat was estimated to have one billion users with a penetration rate of 92.4% (Torre & Xu, 2020). The growing acceptance of WeChat Pay has greatly transformed the Chinese market by enabling new modes of business-customer interactions, contributing to economic growth and expansion as well as advancing financial inclusiveness in poorer regions.

However, obstacles still remain. Vatsa et al. (2023) studied the use of mobile payments in China and revealed that demographic and socioeconomic factors do play a role. The adoption of mobile payments is negatively correlated with age, and males are less likely than females to take advantage of mobile payment systems. There is a chance that some demographic groups may be more receptive and, therefore, easier to encourage to adopt mobile payment systems than others.

In conclusion, the case of WeChat Pay in China demonstrates how a society integrates social and financial services to adapt for culture. WeChat Pay is now known to be a part of the cyber culture in China through social networks, local traditions, and other innovative uses of technology. This offers an insight to other cultures on how the system of digital payments could be constructed, highlighting the need for social cohesion and cultural diversity in the application of FinTech's advanced systems.

5.3 Paytm in India: Cultural Localization of Digital Payments

As the top digital payment provider in India, Paytm has transformed the country's cultural intricacies into opportunities endorsing innovation in the fintech domain. Its approach has greatly facilitated the cultural localization of digital payments in a multi-demographic and multi-regional India.

Localization is one of the tactics employed by Paytm to extend its services to a wider market. Acknowledging India's linguistic tallies, Paytm launched its multilingual interface in November 2016, supporting ten regional languages alongside English. As a result, Paytm experienced a tremendous increase in user adoption from tier II and tier III cities, which now constitute 50% of Paytm's users. The addition of support for vernacular languages increased traffic from smaller towns by five times, indicating that the relative importance of language accessibility in the adoption of digital payments in a culturally rich country like India (Bhatia-Kalluri & Caraway, 2023).

The achievements of Paytm's cultural adaptations go beyond language, as it pivots to the specificities of rural and semi urban areas. The firm has oriented itself to build an inclusive system that serves a wide range of ethnic groups in India. Paytm has collaborated with local suppliers and enterprises to help close the gap in digital financial literacy while explaining the advantages of expansion and growth through the use of digital payment systems. This bottom-up strategy has been exceptionally helpful in improving financial transactions for rural areas and small businesses owners in tier three and four towns. A research study conducted in the year 2023 assessed the role of mobile banking services, specifically Paytm, in advancing financial inclusion in the rural and semi-urban regions in India. The study established that nearly 72% of the households surveyed said that dealing with financial transactions using Paytm and other related platforms has become easier. This brought about an increase in revenue of 35% for local businesses and a reduction of 40% in cash transactions (Patel et al., 2023). These results prove the ability of Paytm in overcoming the digital financial literacy divide and in encouraging the use of digital payment systems for business activities in rural and semi urban demographics.

The implementation of the Paytm Soundbox technologically and without regard to culture is a noteworthy example of cultural innovation by Paytm. The claim is substantiated by the recent publication of a study in 2024 by researchers at ResearchGate titled "Impact of Soundbox by Third Party Application Providers (UPI) on Retailers with Special Reference to Greater Chennai Corporation." The study investigates various features and functions of Paytm Soundbox including its adoption and impact. The introduction of the Paytm Soundbox which features audio output of transaction the breakdown in vernaculars is an example of how technology is used to mitigate cultural obstacles. This feature speaks the transaction details in local dialects, thereby enabling users who find it difficult to read text appreciate the digital payments interface and develop trust over time. Such innovations show Paytm's attempts to respond to local payment systems and cultures (Denny, 2024).

Paytm's efforts to localize its culture have also tackled the gender disparity in financial inclusion. Data shows that an astounding 95% of women who use digital payments express that such forms of payment have lessened their reliance on family to take care of cash and bill payments (GOI, 2024). The financial independence empowered by digital payments for women is a remarkable cultural change that has been brought about by payment platforms such as Paytm.

The integration of the company's business strategy with cultural practices has some hurdles to overcome. Some customers have reported experiencing problems with issue and technical reliability, especially in South India where the need for effortless use of apps and customer services is at its highest (Bose, 2024). These problems underscore the reality that there is still a need for the digital payment platforms to improve their services in order to satisfy the needs of users from different cultures.

Paytm's success in local cultural adaptation provides significant lessons to the rest of the digital payment industry. Wang and Gu (2017) suggested that social influence and individual networking capabilities have considerable impact on the use of digital payment services. Paytm's social and financial cross-function integration strategy fits these results by developing a payment system that is culturally and socially responsive.

Thus, Paytm's experience in India reveals the necessity of cultural adaptation in the digital payments industry. Through localization of languages, customization of services at the state level, and other culturally appropriate measures, Paytm has managed to brand itself as a champion of the digital payment revolution in India. With the changes in the digital payment ecosystem, strategies of cultural adaption employed by Paytm will probably serve as a basis for novel approaches in India's and other culturally rich economies' financial technologies.

6. PROPOSED FRAMEWORK FOR CULTURALLY ADAPTIVE DIGITAL PAYMENT SYSTEMS

The implementation of different payment systems through technology has grown in a short period of time. This has shown that cultural adaptation should take place to ensure users can access and utilize these systems affordably and efficiently. A suggested structure for evolving payment systems through the use of technology addresses the needs and challenges of various user groups because cultures directly affect payment practices and the acceptance of technology.

As technology continues to advance, cultural factors such as trust in financial institutions, deep-seated norms, and historical context act as barriers to adoption. As APEXX Global (2023) suggests, there is a stark contrast between what is adopted in different countries, such as the preferred use of mobile wallets in China as opposed to the SEPA direct debit and bank transfer preference in Germany (Padee, 2023).

It has been established that the adoption of digital payment tools relies on a myriad of factors such as, but not limited to, easeness of access, security, and trust alongside government efforts. A 2024 survey suggested that the adoption of digital payments among younger adults is primarily driven by ease of use, trust in security features, and promotional activities (Sundararaj & Meera, 2024). Still, data privacy and digital literacy issues remain challenges that must be addressed in a culturally appropriate manner.

In developing a framework for culturally adaptive digital payment systems, aspects such as language translation, graphic user interface, payment security, and local financial institution compliance have to be addressed. The success of Paytm in India shows that vernacular audio content is an innovation that can greatly increase the trust and ease with which users who are not comfortable with text interfaces relate to digital payments (Eelu & Nakakawa, 2018).

In addition, the framework must incorporate the recently observed cross-country evidence around the non-linear relationship between trust and cash usage. As Wischnewsky (2024) notes, social trust enables societies to move away from cash dependency and adopt digital economies. Addressing these cultural aspects, it is possible to create a model for culturally adaptable digital payment systems that could facilitate the shift to cashless societies inclusively and satisfactorily across all cultural boundaries.

6.1 Culturally Sensitive User-Centric Design

Culturally sensitive user-centric design is integral to constructing a digital payment system that meets the needs of potential users from multi-regions and cultures. This design strategy requires that the preferences, actions, and expectations of users be contextually and culturally relevant which might be highly different from other people.

Cultural sensitivity in user-centric design involves the understanding more complex issues than localization of language, which include payment cultures, levels of trust towards financial institutions, and deeply set cultural context. As APEXX Global (2023) pointed out, payment methods are often country dependent; for example, China predominantly uses mobile wallets, while Germany relies more on SEPA direct debit and bank transfers (Padee, 2023).

6.1.1 Psychological self-assurance of users

The user's psychological self-assurance as a vital construct in culturally sensitive user- centric design. This process includes communication regarding security, unobstructed presentation of interfaces, and assurance of secured transaction processes. For Apple Pay, the Face ID and Touch ID authentication system not only provides security but also assures users that their transactions are secure (Vidjikant, 2024).

6.1.2 Cultural Differences

Digital payment experiences vary with different cultures, particularly in the balance of security and convenience. Most Western societies, known for their individualistic nature, tend to be more concerned about privacy and security in payment systems, while having a lower risk propensity toward payment platform community trust (Chang 2024).

6.1.3 Accessibility

Culturally responsive user-centric design also implicates accessibility. Facilitating usability and accessibility for payment systems is achievable if digital payments are compatible with screen readers, offer different ways of input, and give visual and sound feedback. Square payment solutions are an example of this approach because they provide easy-to-read text, high contrast displays, and are visually impaired friendly (Vidjikant, 2024).

6.1.4 Comprehension of local contexts

User friendly designs with cultural consideration of the region being served requires thorough comprehension of local contexts. As illustrated in a project in Indonesia, human-centered design methodologies tend to be powerful in revealing and addressing the actual needs of the users while conforming to the known culture. This has highlighted the need to focus on the larger social environment within which payments are made, which people tend to be more conscious of due to the likelihood of their payment decisions having a relational impact on other people (Edwards et al., 2014).

Thus, user-centric design with cultural considerations is critical in the design and development of digital payment systems that work effectively and, at the same time, appeal to the user's cultural practices and norms. Such an approach is likely to increase acceptance, user experience, and ultimately, the effectiveness of digital payment systems in various regions of the world.

6.2 Public-Private Partnerships for Inclusive FinTech

PPP partnerships are one of the most remarkable strategies aimed at enhancing inclusiveness in FinTech and increasing the scope of outreach. The public sector institutions, private business firms, and sometimes third sector organizations and NGOs work in collaboration in order to make use of the assets from each sector, and construct solutions for financial issues more efficiently and sustainably.

6.2.1 Advantages of PPPs in Inclusitive Finance Technology

a. Greater Intelligence and Knowledge Integrating

Integration of information and training between the governments and the private sector is enhanced through PPPs. Financial institutions provide transactional information and behavioral data, and government institutions offer relevant information that is policy driven. All these elements combine to form a far better comprehensive understanding of the issues and chances regarding financial inclusion (Sokrovolska et al., 2022).

b. Enhanced Infrastructure and Service Provisioning

PPPs indeed facilitate the development of the infrastructure needed to support digital financial services in unserved regions. The Center for Strategic and International Studies published a document in March 2024 that explains these concepts further. The report outlines the ways in which PPPs can facilitate the digitalization of services while guaranteeing that essential government services are accessible during difficult periods. In Ukraine, for example, Google has provided technical infrastructure support and USAID has developed the Diia e-government application. These illustrate how PPPs can harness public funds together with private sector efficiency to enhance the availability of digital financial services. Furthermore, the analysis noted that there is a greater dependence on public-private partnerships to foster investment towards achieving the Digital Services' UN Sustainable Development Goals provision.

Even partnerships that have blended finance components, can widen the scope of mobilizing other investors and help to bridge the funding gap, particularly in poorer countries. By tailoring development strategies to local contexts, these low-income people can have better access to digital financial services (Bandura et al., 2024).

c. Timeliness and Cost Effectiveness in Delivery of the Project

The private sector's involvement in infrastructure development, including in FinTech's projects, has shown to be more effective and productive. A Report of Study published from ResearchGate in 2024 indicates that the propensity of the private sector to take on such infrastructure projects seems to be positively associated with the efficiency and effectiveness of service delivery. The study also notes that such private sector involvement tends to lead to more rapid project completion and cost-effective delivery because of their project management and evaluation skills (Miyamoto et al., 2005).

6.2.2. Challenges in Implementing PPPs for Inclusive FinTech

a. Legal and Regulatory Barriers

Implementing Public Private Partnerships (PPP) initiatives can be hampered by the existence of various legal systems and privacy laws across different jurisdictions that have varying levels of information privacy. This issue is especially problematic in developing countries where the legal framework is poorly articulated and the institutional structures are at best substandard. Batjargal and Zhang (2021) observe that the differences in organizational goals, insufficient political and legal support, and lack of appropriate risk sharing and allocation are some of the key obstacles to the execution of PPPs. There is need to build strong regulatory frameworks and encourage communication among partners to enhance trust to deal with such issues.

b. Operational and Technical Challenges

The integration of diverse public and private sector systems and processes in collaborative ventures poses operational and technical challenges that include achieving interoperability and protecting sensitive data. This complexity was highlighted by a study published in 2024, which studied the challenges associated with implementing sustainable building projects through Public-Private Partnerships. The research asserts that public-private partnerships (PPPs) frequently face conflicts of different partner organizational cultures and goals, lack adequate institutional environments, and suffer from ineffective risk sharing and responsibility allocation practices. These reasons help explain the challenges in achieving integration and interoperability of public and private systems while protecting the sensitive data (Asuamah Yeboah, 2024).

6.2.3 Strategies for Effective PPPs in Inclusive FinTech

a. Focused and Time Bound Objectives

Partnerships must have clearly defined targets that aim at having a tangible impact to local communities, especially to forcibly displaced persons and their hosts (Hansen & Pistelli, 2024).

b. Public Policy Conversations and Collective Studies

The organization of policy dialogues, roundtables, and public forums enables active participation from all stakeholders. Undertaking collective research on the effects of inclusive policies and initiatives helps find gaps and areas of improvement opportunities (Hansen & Pistelli, 2024).

c. Application of FinTech in administration of benefits

Benefit application and management processes in PPPs can be enhanced through the use of FinTech. For instance, Code for America's collaboration with the White House and the U.S. Department of Treasury showcased how the use of technology can be leveraged to make public benefits accessible with the launch of the bilingual CTC management tool (Cabusora et al., 2022).

To conclude, the use of public-private partnerships (PPPs) to advance the development of inclusive FinTech is promising, but achievement will depend on initial design work that achieves a target within a given regulatory environment. These forms of partnership, when properly structured, have the power to address financial and economic imbalances through the optimized interaction of public and private sectors.

6.3 Policies and Regulations Supporting Cultural Adaptation

The integration of electronic payment systems into culture has specific policies and regulations facilitating it, which makes it possible to implement these policies within systems of different cultural features. Such policies seek to create an environment that is not only focused on financial inclusion, but also sensitive to cultural and social factors and norms.

6.3.1 Consumer education and protection

In relation to the payment systems, the focus on education and consumer protection among the digitally excluded or underserved populations remains an important issue towards ensuring inclusivity and accessibility. The Better Than Cash Alliance's responsible digital payment principles developed in 2016 encourage the engagement of cultural and ergonomic specialists for the design of these systems. This corroborates the works of Shende and Kathiriya (2024) who have suggested the necessity of consumer awareness campaigns to teach users about safe participation in digital transactions and detection of fraud cases. Such efforts go beyond the basic protection of consumers because they build trust which leads to increased adoption of digital payment systems.

6.3.2 Institutional policy provisions and frameworks

These frameworks are also important in the order of establishing the cultural change process. The Centers for Disease Control and Prevention design and publish guidelines for practitioners adopting evidence-based interventions. These documents seek to help enhance the adoption of these guidelines at the cultural context without losing key elements that comprise the best evidence (Alper et al., 2020).

6.3.3 Evaluation of Scope and Context of the Population in Question

Policies that encourage cultural customization of digital payments in developing nations frequently stress the need to conduct thorough investigations of the population and context in question. This is especially important for countries that have a strong cash culture as changes in cultural attitude towards digital payments will take considerable time and effort. Qadri (2023) cites evidence from developing countries like Pakistan and Sri Lanka, where mobile payment services are not widely adopted due to poor digital infrastructure, limited accessibility to banking, and low confidence in technology. The research underscores the influence of cultural beliefs - for instance, an overreliance on physical currency makes it difficult to adopt digital payment services. Moreover, Omoniwa (2021) argues that national cultural factors, to a great extent, moderate the relationships among the determinants of technology use and acceptance, and therefore, impact intention and use behavior of digital financial services in developing countries. These conclusions stress the need for context-specific educational strategies and motivational policies to enhance the shift from cash to digital payments in culturally diverse settings.

6.3.4 The Role of Cultural Adaptation in Financial Inclusion Strategies

Policies that enable cultural adaptation in digital payments tend to merge these factors in broader financial inclusion policies. This means that payment systems adapted to digital technology consider payments as a means of addressing inequities among different social groups. Women's sub-segment and other user segments that need special attention are emphasized in the Responsible Digital Payments Policies Guidelines of the United Nations (BTCA, 2024). This is consistent with Anyangwe et al. (2022), who argue that there is need for more adoption of digital payment systems with culturally sensitive design approaches for different user groups. Additionally, Qadri (2023) captures the need to appreciate cultural attitudes and serve the local context with digital payment solutions especially for developing countries where cash is the dominant culture.

Different studies have established that culture impacts the use of payment systems and adoption of digital payment systems. Patil et al. (2017) studied social influence and perceived enjoyment as two aspects that mitigate the perceived risk of mobile payment systems. This demonstrates that cultural values and social norms need to be integrated in the design and implementation of digital payment policies.

Policies and regulations that promote the cultural localization of digital payments allow for the systematic adaptation of these systems so that they are more relevant to and effective within the target culture. These policies foster the creation and use of culturally adapted digital payment systems by emphasizing consumer education, outlining adaptation procedures, and integrating cultural factors into higher level policies.

7. POLICY AND STRATEGIC RECOMMENDATIONS

The Policy and strategic recommendations for cultural localization of digital payment systems are important in regard to the functionality and successful application of the systems across different cultures. Different payment cultures will influence the rapidly evolving digital payment technologies. There is a need for policies and practices that define the scope of different cultural characteristics for different user groups to be developed.

Different studies have proved that even culture has an impact on the acceptance and use of digital payment systems. The research of Zhao and Pan (2023) regarding the acceptance of mobile payments in China and Korea, proposes that the specific cultural attributes of a nation significantly impact the behavioral intentions of its citizens towards the adoption of modern technology. Therefore, efforts to create behavioral change must integrate local customs and practices.

In formulating policies, there is need to consider the challenges faced by different user groups. Patil et al. (2018) have noted that some users, particularly the younger generation, are more likely to use digital payments due to a perceived ease of use, trust in the systems' security features, and marketing incentives. However, there still remain problems concerning data privacy and low digital literacy that require an educative and informative solution. Policy reforms in developing countries that have low or non-existent digital infrastructure, as mentioned by Qadri (2023), require an understanding of the local culture along with cultural sensitivity to these regions. Such regions are still reliant on cash as a dominant form of payment; therefore, there is still a lack of integration of digital payment structures into the financial systems of these places.

Easy and effective digital payment policies will need the inclusion of these cultural issues alongside the research so they can effectively serve the people.

7.1 For Governments: Creating Supportive Regulatory Environments

Governments need to extend the scope of their policies and regulations to enable consumer protection and innovation in order to facilitate the adoption of digital payments within community-based financial activities. This, in essence, means that there is a need to formulate policies and regulations aimed at enhancing economic innovation and growth while protecting public interest and maintaining financial stability.

An example of a system that enables the development of such policies and regulations is the introduction of digital payments through merchant accounts that allow businesses and other economic actors the ability to readily accept payment for goods and services via new digital payment systems. These policies balance the need for competition and innovation within a framework of oversight that is necessary in order for them to be effective.

Equally, Governments can play a complementary role towards ensuring adequate pricing by refraining from intervening too much and instead, offering adoption incentives and dampening the benefits associated with informality. This can help facilitate adoption of digital payment systems by merchants while taking into account the informal economies' practices.

Within the scope of community-based financial services, governments can facilitate better collaboration and interoperability among payment service providers (PSPs). The United Nations' Responsible Digital Payments Guidelines have highlighted the need for the alignment of the private sector's interests with government objectives in service to vulnerable users (BTCA, 2024). Defining financial service providers as co-creators of solutions rather than subjects to control for the decision-making process is an easier method to foster their buy-in.

The need for regulatory endorsement of financial literacy initiatives is an equally important aspect. Governments could support the development of digital payment systems for excluded groups by providing user-friendly voice messages in the local languages. This underscores the role of cultural sensitivity in the promotion of digital financial inclusion.

In addition, governments could take the lead in national, regional, and global initiatives directed at the regulation of responsible artificial intelligence use in digital payments, considering imbalances of power and protecting at-risk populations. This also involves assessing algorithmic impacts and interrogating inputs and assumptions in Areas of Action like customer profiling to find and rectify racist assumptions.

With the adoption of these strategies, governments will be able to foster the development of regulatory frameworks that encourage cultural adaptation of digital payments whilst their enabling and promoting community-based financial activities. Such policies would contribute to the development of more inclusive, robust financial systems that integrate and adapt local socio-cultural practices while fostering innovation and financial inclusiveness.

7.2 For FinTech Companies: Integrating Cultural Research in Product Development

For FinTech firms, the integration of cultural research in product development is pertinent in achieving the cultural integration of digital payments with community-oriented financial activities. This guarantees that digital payment services are not only sophisticated but also relevant and usable by different cultural groups.

The most critical way to apply cultural research is to employ a cross-cultural user design approach. Strebkov (2024) pointed out that more and more fintech firms shift towards more customer centric models rather than focusing only on product driven growth strategies. Such changes require understanding local customs, languages, and laws, which is in most cases not homogenous in many countries.

Cultural research can help people to develop features that support local custom and practices in payment systems. FinTech companies, for example, may look into ways to join digital services with community-based financial systems in informal self-help groups for saving or cooperative lending groups. This may assist in changing the traditional ways of doing financial transactions, which are considered to be on paper, to electronic ways which are more advanced.

The advancement of cultural research can be achieved while utilizing AI and machine learning technologies. For example, AI solutions such as behavioral chatbots can analyze the sentiments, preferences, and needs of clients which aid in crafting business understanding. This, in turn, helps inform product development, feature prioritization, and resource allocation (Strebkov, 2024).

Additionally, engagement with local organizations can help deepen understanding of community-centered finances fostered by social integration in FinTech companies. Sibley (2024) describes partnerships with NGOs, charities, and community organizations as pivotal for the realization of financial inclusion. Such partnerships enable the provision of relevant cultural context, which allows to design FinTech solutions to specific community needs.

The incorporation of cultural research attempts also involves consideration of the non-adoption barriers. The Better Than Cash Alliance talks about the need to account for transparency concerning the protection of funds and, digital payment accountability. Subsequently, cultural research should then be used by FinTech organizations to identify and address trust, security, and privacy issues that some communities or cultures experience.

To conclude, incorporating detailed cultural study in a product development cycle enables FinTech firms to design sophisticated digital payment systems that are culturally sensitive. Adoption of this strategy would improve user interfaces, enhance usage, and contribute towards deeper financial inclusion in various populations in the diverse societies.

7.3 For Local Communities: Building Trust and Digital Literacy

Trust and digital literacy are critical for communities to adequately adjust to digital payments particularly with respect to community centric financial services.

It requires a combination of teaching, building trust and integrating the culture with technology.

An important fundamental for improving digital literacy is the implementation of basic educational programs. Digital Financial Literacy (DFL) programs, as noted by the Alliance for Financial Inclusion (2021), should have intervention components of multi-dimensional scope:

1. Digital Financial Service (DFS) Knowledge: Clients need to be informed about the existing DFS, their functions, and even classification of products.
2. DFS Awareness: Teaching learners about dangers such as phishing, data breaches and other online risks.
3. Financial Cyber Protection: Teaching users how to secure their information such as using strong passwords and multi-factor authentication.
4. Knowledge of Redress Procedures: User Rights and Sociolegal Solutions in Case of Issues or Fraud

Community-centered initiatives foster trust and improves digital literacy. Sibley (2024) insists on the need to collaborate with non-governmental organizations (NGOs), charity and community groups aimed at managing financial exclusion. Such collaborations would facilitate the adaptation of digital payment systems to the local culture and context.

In order to manage any forms of trust gap, a balance between digitalization and human contact needs to be maintained. Research by the European Investment Fund (EIF) indicates that while intuitive user journeys and real time data reporting are the focus of FinTech's, Microfinance Institutions (MFIs) tend to favor on-site educational events and story-telling. This can aid in assisting communities to migrate towards digital payment systems while allowing them to retain their traditional finance systems (European Investment Fund, 2024).

For cash-intensive communities, a phased approach could be more effective. The Federal Reserve Bank of Atlanta (2023) recommends temporarily continuing the use of cash while simultaneously granting easier access to more modern forms of payment. This includes the establishment of cash-in/cash-out (CICO) systems which allow users to exchange cash for digital currency.

Finally, it is crucial to understand that digital financial literacy is a continuum. The GSMA calls for the cooperation of all stakeholders within mobile money ecosystems to implement national DFL programs. This collaborative approach maintains that digital payment systems are responsive to the people's needs and their cultural practices (GSMA, 2023).

Addressing these issues enables communities to boost trust and digital literacy, thus facilitating successful digital payment adoption while safeguarding community-centered financial practices.

8. CONCLUSION

8.1 Summary of Key Insights

The introduction of digital payment systems has drastically changed how people conduct financial transactions. However, there remain concerns about what main variables lie behind the use of cash and how to properly design central bank digital currency (CBDC) policies. There are new studies that point out the influence of cultural and behavioral variables on the patterns of the usage of digital payment systems and cash. Social trust, for instance, has been found to underlie the global transition from being dependent on cash to being integrated into the digital economy and posits a particular non-linear relationship to cash usage (Wischnewsky, 2024).

The advent of digital payment systems has resulted in economic growth and development. Its adoption has resulted in higher financial inclusion, lower transaction expenses, and advancement of novel financial products and services. By erecting appropriate infrastructure and formulating relevant policy, governments make a significant contribution toward the development of digital payment systems. These systems enable financial institutions to expand their customer base while reducing costs and improving efficiency for businesses (Mishra & Sharma, 2024).

8.1.1 Cultural Influence on Financial Behavior

Cultural elements modify how people save, invest, and spend money. One important cultural factor that determines financial behavior is the difference between collectivist and individualistic societies. Group harmony and interdependence are characteristic of some Asian countries that are collectivist. This is often characterized by relatively higher tendencies towards family saving and spending compared to other cultures. In Western cultures, there is more emphasis placed on individualistic achievements and autonomy, coupled with a relatively self-centric approach to financial planning and investment. Different cultures' view on time also influences the economic behavior of people. Some cultures are more present-oriented and tend to consume more, which results in higher expenditure and lower savings. On the other hand, future oriented cultures focus more on saving and investment planning (Berghofer, 2023). Research suggests that there is a significant correlation between socialization and financial literacy (socializing with pocket money and free banking at a young age) and time preferences, especially delay of gratification. Financial socialization is the main mediator of financial literacy across cultural groups (Brown et al., 2018).

8.1.2 Issues Related to Cultural Integration of Digital Payments

There are several challenges with the cultural contextualization of the digital payment systems. In developing nations, the resistance to the digital payment systems is rooted in a plethora of issues, including the broken technological infrastructure in rural parts. Many users are unable to accept digital payments due to the absence of dependable internet connections and payment networks. Additionally, people's security concerns significantly influence the use of payment systems because cyberattacks, fraud, and data breaches are always a threat. More often than not, people from emerging markets tend to not want to adopt digital payments for this reason, instead of learning different ways to protect themselves. In order for emerging markets to become major digital hubs, there is a need for people to be able to trust technology and understand data security. Another challenge is the absence of customer awareness. Education on using technology is vital for people to adopt digital methods, especially for the aged and rural folk who may not be familiar or even know about the process of making payments through digital means. Initiative programs from the government within the ecosystem to promote users' digital literacy in payments will, in the end, lead to the acceptance of digital payments (Aurazo & Gasmi, 2024).

8.1.3 Prospects of Culturally Adaptive Digital Payments Systems

While challenging, the prospects of culturally relevant digital payments remain promising. A significant approach is to analyze and customize payment options. Companies need to dedicate some efforts to examining the preferred payment options available in every target region and ensure such methods are incorporated into their systems. When new and better techniques are to be used, training becomes important. Preferred methods will always have clear features, both in the level of communication provided and the security offered. Working with local partners, for instance, banks, and other financial institutions can aid understand the local market and offer guidance that adds value and trust. It is also important to provide additional options because not everyone will be satisfied. There are numerous preferences from clients paying from different parts of the world, so it is important to enable payment gateways for different types of payments. In nations like Brazil, government programs for financial inclusion, aid efforts to provide access to digital financial services, is widely accepted. PIX, for instance, is meant for every Brazilian, even those who do not have access to conventional bank accounts. Such innovations in

Fintech like mobile applications and payment using QR codes are rapidly changing the practices of many small businesses and consumers (Saiyid, 2023).

8.1.4 Cultural Dimensions of Digital Payments like M-Pesa, WeChat, Paytm

Systems of digital payment such as M-Pesa, WeChat Pay, and Paytm have successfully integrated into particular cultural settings. For example, mobile wallets such as WeChat Pay and Alipay are prevalent in China, while Germans prefer SEPA direct debit and bank transfers. And in the United States, credit cards are the most preferred option. These differences do not solely stem from technology; they are equally a function of cultural practices, trust in financial institutions, and socioeconomic history (Deufel et al., 2019). These examples indicate that policymakers and business leaders need to focus on the local culture while implementing any platform. For example, M-Pesa's success in Kenya is a direct result of the service's ability to link it with the country's informal financial systems while catering to the unbanked population's needs (Mas & Radcliffe, 2010). Social media embeds with WeChat Pay resonates with social preferences of Chinese people towards social interactions during financial expenditures (Wang & Gu, 2017). Paytm's success in India lie in its cultural - transforming strategy that favors cash payment, enable regional languages, and supports government projects like demonetization and UPI which are aimed at making digital payments easier.

8.1.5 Proposed Framework for Culturally Adaptive Digital Payment Systems

A framework for systems with culturally adaptive digital payments must have multifaceted considerations. Security must be an utmost priority in order to gain acceptance in various cultures as it helps in building trust and also ensures that payment systems conform to internationally accepted security regulations. The user experience must be equally prioritized, as systems should be straightforward and easy to navigate with minimal steps, and maximum clarity at each stage. It must also be proactive and responsive to change, in order to meet and take advantage of new market opportunities. Actively interacting with international clientele and soliciting their opinions on the existing payment systems is imperative for the enhancement of these systems. The framework should also acknowledge that the pace at which different countries assimilate to the use of cashless payment systems differs due to cultural and socio-economic as well as technological reasons. Strategic understanding and contextualization are essential for effective transitions (Eelu & Nakakawa, 2018).

8.1.6 Policy and Strategic Suggestions for Governments, FinTech's, and Societies

The following are the most pertinent policy and strategic recommendations which aim to politically integrate cultures to the digital payment systems for governments and communities as well as fintech firms based on the findings of the study.

a. Initiatives towards Financial Inclusion

Government agencies should give priority to financial inclusion programs in collaboration with private firms aimed at widening the scope of financial service utilization. This may involve setting up mobile wallet micro- services, where clients can easily apply for mobile loans. Orange for example, has provided micro loans in several African countries with 1.3 million micro loan clients in Cote d'Ivoire alone. These loans which rely on mobile money credit history are granted even to people who do not have formally assessable credit history (Kane, 2024).

b. Advanced Digital Payment Infrastructure Expansion

Focus on the formation of public-private partnership at a national level and the providing of subsidies and incentives for the expansion of internet infrastructure in regions where the majority of the population lives, especially in rural areas, need to be emphasized by payment service providers together with policy makers. These are necessary steps in catering the needs of rural and marginalized regions of society (Falch & Henten, 2010).

c. Comprehensive Approach to Digital Payments Adoption

Fintech organizations must appreciate that the adoption of digital payments goes beyond just technology. Research results show that there is a high acceptance of, and willingness to use, digital payment systems by Microfinance Institutions (MFIs) and consumers because these systems are user friendly and well designed. In addition, there is a growing popularity of Automated Teller Machines (ATMs), internet payments, and the use of debit and credit cards, as well as mobile applications (Saravanan & Chithra, 2024).

d. Emerging Payment Technologies

Fintech and government agencies must take into account new technologies on payments. Some suggestions include focusing on the development of digital currencies, which need to be initiated by the central banks and controlled at the OIC Level, and creating a unified Digital Payment System Platform which would enable transactions to be carried out internationally. Besides, a new legal framework has to be designed for the purpose of establishing an intergovernmental regulatory authority to coordinate the implementation of multilateral cooperation in the region for payment systems (Fung & Halaburda, 2016).

e. Cultural Sensitivity and Adaptation

The acceptance of cashless payment systems varies considerably among different groups or cultures. Although some customers are more accustomed to using credit cards and other electronic means of payment, there are those who use cash or even checks. This may stem from their sentiments, cultural influences, or the absence of relevant digital payment infrastructure in some parts of the world. All marketing strategies aimed at electronic payment systems integration should take inclusiveness and accessibility into account. There are, however, some customers, like the elderly or physically challenged, who may find it difficult to make use of online payments. Ensuring that electronically accessible options are available to every customer is necessary for universal acceptance (Saravanan & Chithra, 2024).

f. Building Trust through Education

The government and fintech firms should prioritize the implementation of digital literacy campaigns on a national scale focusing on rural regions where financial literacy is considerably low. Also, there is a need to periodically update and synchronize the laws with the development of the digital financial services industry. In addition, broadening the scope of innovations covered in the regulatory sandbox and easing the licencing framework for new technologies should be considered (Parenti, 2020; Pandey, 2023).

g. Collaborative Approach

Encouraging continuous innovation and regulatory support for new entrants in the digital payment ecosystem will promote healthy competition in the fintech industry. There is need for the government and fintech companies to seek ways of promoting financial inclusion through the use of sophisticated digital payment systems that are uncomplicated for the unbanked and underbanked. They should also focus on digital literacy and basic technical skill training to ensure that there is a sustainable future workforce in the banking sector. In addition, there is a need to increase the cooperative action among central banks of the member countries to strengthen cybersecurity measures, as well as create a working group composed of digital payment specialists to keep pace with advancements in payment systems (Nwafor & Ayodele, 2024).

Addressing cultural aspects thoughtfully can support the efforts of governments, fintech companies, and communities toward crafting effective and more accessible digital payment systems.

8.2 The Future of Culturally Inclusive Digital Finance

The future of digital finance with cultural inclusivity has the capacity to transform traditional financial services in developing countries, as it seeks to incorporate local customs and practices. This approach seeks to integrate modern technology and community financial approaches that are already in existence.

8.2.1 Community-Based Financial Organizations (CBFOs)

Community-based financial organizations are essential in the provision of primary financial services to neglected groups in society, especially those living in remote region. These organizations can take the following forms:

a) Small community-based time limited saving groups: These include rotating savings and credit associations (ROSCAs), village savings and loan associations (VSLAs), and village banks (VBs). Such groups are usually informal and self-sustained (Ritchie, 2007).
b) Self Help Groups (SHGs): Common in countries such as India, SHGs are independent groups ranging between 10-20 members who save and share money among members (Sultania, 2024)
c) Large community based cooperative groups: These organizations have social goals in addition to providing a wider range of financial services (Ritchie, 2007).

8.2.2 The Digital Transformation of Community-Based Finance

Financial ecosystems of the future are likely to be enabled through:

a) Digitalization of former practices: New digital applications can be designed to automize the work processes of ROSCAs or SHGs, improving record keeping and fund management while preserving the social interactions of its members (Feyen et al., 2021).
b) Personalization through AI: Development of advanced AI and big data analytics will allow the crafting of unique financial products tailored to the distinct

requirements of individuals and small businesses in the society (Omogbeme et al., 2024).

c) Adoption of cryptocurrency and blockchain technology: These innovations would increase the level of openness as well as security of transaction in community-based finances, especially in the case of international remittances (Feyen et al., 2021).

8.2.3 Culturally Inclusive Digital Finance

a) Klong Pia Savings Group (Thailand): This grassroots-based group began 24 years ago with only 57 members, but has grown to over 7000 members today. They provide a variety of loans to cover health, education, occupation, and housing expenses. They also assist children, youth, women, and folk artists which shows the evolution of community-based finance and its responsiveness to local needs (Worakul, 2006).

b) Village Savings and Credit Organizations (VSCOs) in Sri Lanka: The Sri Lanka Gemidiriya framework features large VSCOs of 300 or more members, linked to multipurpose village companies. This model demonstrates how broader community development can be integrated with Digital Finance by providing infrastructure and livelihood support services to community members (Munshi et al., 2007).

c) Native Women Lead (United States): This organization serves Native American women entrepreneurs by operating a variety of funds and coordinating networking opportunities. It is an example of how finance that seeks to be inclusive can empower certain communities and meet specific cultural needs (Hill & Hoss, 2018).

8.2.4 Cultural Considerations in Digital Finance

In order to ensure inclusivity, future approaches to digital finance should be able to:

a) Adapt to local customs: The “Bohni” ceremony that is observed among South Asian markets is highly symbolic, and digital payment systems should cater to such customs.

b) Target women as discrete clients: Women have been ignored as clients in more traditional financial systems. New digital financial products should not just be designed for women, but rather be designed specifically for them (Goodwin Groen & Klapper, 2020).

c) Encourage users of any age to actively learn and participate: Financial concepts can be simplified through gamification and interactive engagements to encourage a culture of financial empowerment.

The vision of an inclusive digital finance system rests on the combination of modern technologies and community-centered financial systems. Through respect to local cultures, addressing social gaps, and encouraging collaborations from different sectors, digital finance can effectively aid in financial inclusion and economic empowerment for developing countries.

As we approach the year 2034, it is anticipated that money will shift from being a tangible object to a service integrated within the digital sphere. Additionally, AI will radically transform financial services making them hyper personalized and automated benefiting everyone, but positively impacting the last mile stragglers the most; the poor in developing economies, women, and those residing in the rural areas (Thompsett, 2024).

Embracing these changes demands strong inter-institutional collaboration, enabling legislation, cooperation between fintech companies and traditional banking, NGOs and international institutions each with unique sets of requirements, to foster a flexible system. If progress is met with these conditions and attention to cultural inclusivity is prioritized, then the future of digital financing is bound to encourage deeper financial inclusivity.

REFERENCES

Aker, J. C., & Mbiti, I. M. (2010). Mobile phones and economic development in Africa. *The Journal of Economic Perspectives*, *24*(3), 207–232. DOI: 10.1257/jep.24.3.207

Alkhwaldi, A. F., Al-Qudah, A. A., Al-Hattami, H. M., Al-Okaily, M., Al-Adwan, A. S., & Abu-Salih, B. (2024). Uncertainty avoidance and acceptance of the digital payment systems: A partial least squares-structural equation modeling (PLS-SEM) approach. *Global Knowledge, Memory and Communication*, *73*(8/9), 1119–1139. DOI: 10.1108/GKMC-07-2022-0161

Alliance for Financial Inclusion. (2021). Digital Financial Literacy. Retrieved from https://www.afi-global.org/publications/digital-financial-literacy/

Almuhairat, A., Alti, A., & Alswailim, M. (2024). Improving cultural awareness and trust towards m-banking apps in Jordan. *Bulletin of Electrical Engineering and Informatics*, *13*(5), 3695–3704. DOI: 10.11591/eei.v13i5.7824

Alper, A., Pinar, O., & Marin Espinosa, A. G. (2020). Electronic Payments Acceptance Incentives–Literature Review and Country Examples. World Bank Group. https://documents.worldbank.org/en/publication/documents-reports/documentdetail/793041588860799013/Electronic-Payments-Acceptance-Incentives-Literature-Review-and-Country-Examples

Anwar, S., Alburaiki, G., Ibrahimi, A., & Qadri, L. (2024)... *Digital Banking Issues Experienced by Older People*, *1–5*, 1–5. Advance online publication. DOI: 10.1109/ICOA62581.2024.10754510

Anyangwe, T., Vanroose, A., & Fanta, A. (2022). Determinants of financial inclusion: Does culture matter? *Cogent Economics & Finance*, *10*(1), 2073656. DOI: 10.1080/23322039.2022.2073656

Armendáriz, B., & Morduch, J. (2010). *The Economics of Microfinance*. MIT Press.

Arner, D. W., Auer, R., & Frost, J. (2020). Stablecoins: risks, potential and regulation. SSRN *Electronic Journal*. https://dx.doi.org/DOI: 10.2139/ssrn.3979495

Arner, D. W., Barberis, J., & Buckley, R. P. (2015). The evolution of fintech: A new post-crisis paradigm. *SSRN*, *47*(4), 1271–1319. DOI: 10.2139/ssrn.2676553

Asuamah Yeboah, S. (2024). Navigating the Nexus: Overcoming Challenges in Public-Private Partnerships for Sustainable Building Initiatives in Developing Countries. Munich Personal RePEc Archive. https://mpra.ub.uni-muenchen.de/122667/

Aurazo, J., & Gasmi, F. (2024). Digital payment systems in emerging economies: Lessons from Kenya, India, Brazil, and Peru. *Information Economics and Policy*, *69*, 101113. DOI: 10.1016/j.infoecopol.2024.101113

Aysan, A. F., Ozturk, O., & Selim, N. H. (2025). Local power, global reach: The rise and impact of domestic payment networks in the financial landscape. *Journal of Science and Technology Policy Management*, *16*(1), 143–162. DOI: 10.1108/JSTPM-12-2023-0227

Bajwa, I. A., Ahmad, S., Mahmud, M., & Bajwa, F. A. (2023). The impact of cyber-attacks awareness on customers' trust and commitment: An empirical evidence from the Pakistani banking sector. *Information and Computer Security*, *31*(5), 635–654. DOI: 10.1108/ICS-11-2022-0179

Bandura, A., & Wessels, S. (1997). *Self-efficacy*. Cambridge University Press.

Bandura, R., McLean, M., & Girgvliani, S. (12 March 2024). *Advancing digital transformation and digital public infrastructure: The role of the private sector*. Center for Strategic & International Studies. Retrieved from https://www.csis.org/analysis/advancing-digital-transformation-and-digital-public-infrastructure-role-private-sector

Bandura, R., & Ramanujam, S. R. (21 September 2021). *Developing inclusive digital payment systems*. Center for Strategic & International Studies. Retrieved from https://www.csis.org/analysis/developing-inclusive-digital-payment-systems

Batjargal, T., & Zhang, M. (2021). Review of key challenges in public-private partnership implementation. *Journal of Infrastructure. Policy and Development*, *5*(2), 1378. DOI: 10.24294/jipd.v5i2.1378

Berghofer, G. (26 September 2023). *How clients' culture influences their financial behavior: Tips and tricks to improve your advisory*. Behavior quant. Retrieved from https://behaviorquant.com/en/understanding-how-clients-culture-influences-their-financial-behavior-tips-and-tricks-to-improve-your-advisory/

Better Than Cash Alliance. (30 November 2016). *Responsible Digital Payment Guidelines*. Retrieved from https://www.betterthancash.org/explore-resources/responsible-digital-payments-guidelines

Better Than Cash Alliance. (2024). *UN Principles for Responsible Digital Payments*. Retrieved from https://responsiblepayments.org/

Bhatia-Kalluri, A., & Caraway, B. R. (2023). Transformation of the digital payment ecosystem in India: A case study of Paytm. *Social Inclusion (Lisboa)*, *11*(3), 320–331. DOI: 10.17645/si.v11i3.6687

Bitar, M., & Tarazi, A. (2022). Individualism, formal institutional environments, and bank capital decisions. *Journal of Corporate Finance, 76*, 102244. DOI: 10.1016/j.jcorpfin.2022.102244

Bose, I. (November 6, 2024). *Regional divide: Paytm and PhonePe dominate north India, Google Pay leads in the south.* Storyboard18. Retrieved from https://www.storyboard18.com/how-it-works/regional-divide-paytm-and-phonepe-dominate-north-india-google-pay-leads-in-the-south-46712.htm

Brown, M., Henchoz, C., & Spycher, T. (2018). Culture and financial literacy: Evidence from a within-country language border. *Journal of Economic Behavior & Organization, 150*, 62–85. DOI: 10.1016/j.jebo.2018.03.011

Bruno, P., Denecker, O., & Niederkorn, M. (2019). *Global payments report 2019: amid sustained growth, accelerating challenges demand bold actions.* McKinsey Global Banking Practice, September. Retrieved from https://www.mckinsey.com/~/media/mckinsey/industries/financial%20services/our%20insights/tracking%20the%20sources%20of%20robust%20payments%20growth%20mckinsey%20global%20payments%20map/global-payments-report-2019-amid-sustained-growth-vf.ashx

Butt, A. H., Ahmad, H., Muzaffar, A., Irshad, W., Mumtaz, M. U., & Zubair Ahmad Khan, T. (2024). WeChat gamification: Mobile payment impact on word of mouth and customer loyalty. *Spanish Journal of Marketing - ESIC.* Advance online publication. DOI: 10.1108/SJME-01-2023-0021

Cabusora, H., Cordova, E., Kirk, A., & Watson, I. (18 January 2022). *Can Fintech Resolve Public Benefit Obstacles?* Change-Machine. https://change-machine.org/can-fintech-resolve-public-benefit-obstacles/

. Carlos, M. S., Guiao, J. M., Lagman, J. G., Pangilinan, E. B., Torres, S. K. M., & Akiate, Y. W. D. (2024). Factors Affecting the Willingness of Micro Enterprises to Adopt E-Wallets. *International Journal of Innovative Science and Research Technology.* https://doi.org/DOI: 10.38124/ijisrt/IJISRT24APR594

Chang, J. (5 August 2024). How cultural differences influence security and convenience in digital payment experiences. UX Across Cultures. Retrieved from https://www.uxmatters.com/mt/archives/2024/08/how-cultural-differences-influence-security-and-convenience-in-digital-payment-experiences.php

. Chatterjee, P. (2024). The Rise of Mobile Payment Systems: How Information Technology Shapes the Fintech Ecosystem. *International Journal of Engineering and Computer Science.* https://doi.org/DOI: 10.18535/ijecs/v12i08.4712

Chidziwisano, G. H., Wyche, S., & Kisyula, R. (2020). *SmartChama: Designing Collaborative Digital Financial Systems for Chama Groups in Kenya*. Information and Communication Technologies and Development., DOI: 10.1145/3392561.3394641

Creese, S., Dutton, W. H., & Esteve-González, P. (2021). The social and cultural shaping of cybersecurity capacity building: A comparative study of nations and regions. *Personal and Ubiquitous Computing*, *25*(5), 941–955. DOI: 10.1007/s00779-021-01569-6 PMID: 33994905

Dasarraju, V. K. (2024). The Evolution of Cross-Border Payments and Remittances: Integrating QR Codes, Blockchain, and AI. *Journal of Artificial Intelligence & Cloud Computing*, *3*(1), 1–3. DOI: 10.47363/JAICC/2024(3)364

Demirgüç-Kunt, A., Klapper, L., Singer, D., Ansar, S., & Hess, J. (2017). The Global Findex Database 2017: Measuring financial inclusion and the fintech revolution. *World Bank Group*. https://openknowledge.worldbank.org/server/api/core/bitstreams/4c546107-6b3e-5069-b8d1-b2cc25ea2aa6/content

Denny, B. (2024). Impact of Soundbox by Third Party Application Providers (UPI) on Retailers, with Special Reference to Greater Chennai Corporation. *Journal of Corporate Finance Management and Banking System*, *4*(43), 16–26. DOI: 10.55529/jcfmbs.43.16.26

Deufel, P., Kemper, J., & Brettel, M. (2019). Pay now or pay later: A cross-cultural perspective on online payments. *Journal of Electronic Commerce Research*, *20*(3), 141–154. http://jecr.org/sites/default/files/2019vol20no3_Paper1.pdf

Edwards, B., Stallings, H., & Cheng, I. (30 July, 2014). *Using human-centered design for e-payment systems in Indonesia*. CGAP. Retrieved from https://www.cgap.org/blog/using-human-centered-design-for-e-payment-systems-in-indonesia

Eelu, S., & Nakakawa, A. (2018). Framework towards enhancing adoption of electronic payment in a developing economy: A case of Uganda. *The African Journal of Information Systems*, *10*(3), 222–245. https://digitalcommons.kennesaw.edu/ajis/vol10/iss3/5

Euromonitor International. (25 April 2024). Asia-Pacific accounts for 52% of global paperless transactions: Euromonitor International. Retrieved from https://www.euromonitor.com/press/press-releases/april-2024/asia-pacific-accounts-for-52-of-global-paperless-transactions-euromonitor-international#:~:text=SINGAPORE%2C%20Singapore%20%E2%80%93%20Asia%2DPacific,to%20findings%20from%20Euromonitor%20International

European Investment Fund (EIF). (2024). EIF working paper 2024/97, *Addressing financial and digital illiteracy: Insights from microfinance institutions and fintech organisations.* https://www.eif.org/news_centre/publications/EIF_Working_Paper_2024_97.htm

Falch, M., & Henten, A. (2010). Public private partnerships as a tool for stimulating investments in broadband. *Telecommunications Policy*, *34*(9), 496–504. DOI: 10.1016/j.telpol.2010.07.010

Federal Reserve Bank of Atlanta. (June 2023). *Shifting the focus: Digital payments and the path to financial inclusion.* https://www.atlantafed.org/-/media/documents/promoting-safer-payments-innovation/publications/2023/06/13/SCOPI_Report.pdf

Feyen, E., Frost, J., Gambacorta, L., Natarajan, H., & Saal, M. (2021). *Fintech and the digital transformation of financial services: implications for market structure and public policy. BIS papers*, Bank for International Settlements, number 117. https://www.bis.org/publ/bppdf/bispap117.pdf

Fung, B. S. C., & Halaburda, H. (2016). *Central bank digital currencies: a framework for assessing why and how* (No. 2016-22). Bank of Canada Staff Discussion Paper. https://dx.doi.org/DOI: 10.2139/ssrn.2994052

Gefen, D., Karahanna, E., & Straub, D. W. (2003). Trust and TAM in online shopping: An integrated model. *Management Information Systems Quarterly*, *27*(1), 51–90. DOI: 10.2307/30036519

Goel, N., & Nath, V. (2020, March). An exploratory study on digital payment systems and its impact on trust and continuance intention in newly remonetized and digitized era. In *Proceedings of the International Conference on Innovative Computing & Communications (ICICC).* https://doi.org/DOI: 10.2139/ssrn.3562948

GOI. (February 2024). Impact of DIGIDGAN mission on India's digital payment ecosystem. Retrieved from https://www.meity.gov.in/writereaddata/files/Impact-Assessment-of-DigiDhan-Mission.pdf

Goodwin Groen, R., & Klapper, L. (July 2020). Advancing Women's Digital Financial Inclusion. Better Than Cash Alliance. *Women's World Banking, and the World Bank Group.* https://www.gpfi.org/sites/gpfi/files/sites/default/files/saudig20_women.pdf

Grand View Research. (2023). *Digital Payment Market Size, Share & Growth Report, 2030. Retrieved from* https://www.grandviewresearch.com/industry-analysis/digital-payment-solutions-market

GSMA. (2022). *State of the Industry Report on Mobile Money.* GSMA Intelligence. Retrieved from https://www.gsma.com/sotir/wp-content/uploads/2022/03/GSMA_State_of_the_Industry_2022_English.pdf

GSMA. (23 March 2023). *The digital financial literacy toolkit: Addressing the gap in low and middle-income countries*. Retrieved from https://www.gsma.com/solutions-and-impact/connectivity-for-good/mobile-for-development/gsma_resources/the-digital-financial-literacy-toolkit-addressing-the-gap-in-low-and-middle-income-countries/

Hansen, L. M. P., & Pistelli, M. (9 July, 2024). *Effective public-private partnerships to advance financial inclusion and financial well-being of forcibly displaced persons and their hosts- Aligning the business cases*. e-MFP. Retrieved from https://www.e-mfp.eu/post/effective-public-private-partnerships-advance-financial-inclusion-and-financial-well-being

Helsper, E., & Eynon, R. (2013). Distinct skill pathways to digital engagement. *European Journal of Communication, 15*(4), 739–758. DOI: 10.1177/0267323113499113

Hilbert, M. (2016). The bad news is that the digital access divide is here to stay: Domestically installed bandwidths among 172 countries. *Telecommunications Policy, 40*(6), 567–581. DOI: 10.1016/j.telpol.2016.01.006

Hill, M., & Hoss, M. A. K. (2018). Reclaiming American Indian women leadership: Indigenous pathway to leadership. *Open Journal of Leadership, 7*(3), 225–236. DOI: 10.4236/ojl.2018.73013

Hofstede, G. (1980). Culture and organizations. *International Studies of Management & Organization, 10*(4), 15–41. DOI: 10.1080/00208825.1980.11656300

Hofstede, G. (1984). *Culture's consequences: International differences in work-related values*. Sage Publications.

Hofstede, G. (2001). *Culture's Consequences: Comparing Values, Behaviors, Institutions, and Organizations Across Nations*. Sage Publications.

Huang, Y., Wang, X., & Wang, X. (2020). Mobile payment in China: Practice and its effects. *Asian Economic Papers, 19*(3), 1–18. DOI: 10.1162/asep_a_00779

Huo, W., Xiohui, W., Zulfiqar, M., Chand, A., & Ullah, M. R. (2024). Communication dynamics: Fintech's role in promoting sustainable cashless transactions. *Humanities & Social Sciences Communications, 11*(1), 1–15. DOI: 10.1057/s41599-024-03729-4

Jack, W., & Suri, T. (2011). *Mobile Money: The Economics of M-Pesa. NBER Working Paper Series*, 16721. https://doi.org/DOI: 10.3386/w16721

Jacome, V. (2023). Limits of energy access research on unreliability. *Journal of International Development*, *35*(8), 2626–2644. Advance online publication. DOI: 10.1002/jid.3790

. Kane, A. (27 March 2024). *Mobile money and the next phase of financial inclusion*. AFI. Retrieved from Mobile money and the next phase of financial inclusion - Alliance for Financial Inclusion

Karjaluoto, H., Shaikh, A. A., Leppäniemi, M., & Luomala, R. (2020). Examining consumers' usage intention of contactless payment systems. *International Journal of Bank Marketing*, *38*(2), 332–351. DOI: 10.1108/IJBM-04-2019-0155

Khando, K., Islam, M. S., & Gao, S. (2022). The emerging technologies of digital payments and associated challenges: A systematic literature review. *Future Internet*, *15*(1), 21. DOI: 10.3390/fi15010021

Klapper, L., El-Zoghbi, M., & Hess, J. (2016). *Achieving the Sustainable Development Goals: The role of financial inclusion. CGAP*. Retrieved from https://www.cgap.org/sites/default/files/Working-Paper-Achieving-Sustainable-Development-Goals-Apr-2016_0.pdf

Komoju (January 17, 2024). Southeast Asia's payment methods: The local payment culture by country. Retrieved from https://en.komoju.com/blog/payment-method/southeast-asia/

Kouty, M. (2024). Boosting the intra-African digital trade in the AfCFTA context: Does regulatory framework matter? *Digital Economy and Sustainable Development*, *2*(1), 9. Advance online publication. DOI: 10.1007/s44265-024-00036-3

Lisha, C., Goh, C. F., Yifan, S., & Rasli, A. (2017). Integrating guanxi into technology acceptance: An empirical investigation of WeChat. *Telematics and Informatics*, *34*(7), 1125–1142. DOI: 10.1016/j.tele.2017.05.003

Lu, Y., Yang, S., Chau, P. Y. K., & Cao, Y. (2011). Dynamics between the trust transfer process and intention to use mobile payment services: A cross-environment perspective. *Information & Management*, *48*(8), 393–403. DOI: 10.1016/j.im.2011.09.006

Luo, X., Li, H., Zhang, J., & Shim, J. P. (2010). Examining multi-dimensional trust and multi-faceted risk in initial acceptance of emerging technologies: An empirical study of mobile banking services. *Decision Support Systems*, *49*(2), 222–234. DOI: 10.1016/j.dss.2010.02.008

Ly, P., & Mason, G. (2012). Individual preferences over development projects: Evidence from microlending on Kiva. *Voluntas*, *23*(4), 1036–1059. DOI: 10.1007/s11266-011-9255-8

Mas, I., & Radcliffe, D. (2010). *Mobile payments go viral: M-PESA in Kenya*. World bank Document. https://documents.worldbank.org/curated/en/638851468048259219

Mbiti, I., & Weil, D. N. (2015). Mobile banking: The impact of M-Pesa in Kenya. In *African successes, Volume III: Modernization and development* (pp. 247-293). University of Chicago Press. https://doi.org/DOI: 10.7208/9780226315867-009

McBride, N., & Liyala, S. (2021). Memoirs from Bukhalalire: A poetic inquiry into the lived experience of M-PESA mobile money usage in rural Kenya. *European Journal of Information Systems*, •••, 1–22. DOI: 10.1080/0960085X.2021.1924088

Mishra, J., & Sharma, S. (2024). A Study on impact of adaptability and accessibility of digital payment systems and its impact on growth and development of economy. *Academy of Marketing Studies Journal*, *28*(1), 1–5. https://www.abacademies.org/articles/a-study-on-impact-of-adaptability-and-accessibility-of-digital-payment-systems-and-its-impact-on-growth-and-development-of-economy-16173.html

Mittal, H., & Singh, A. (2024). Crypto in India: Time for SEBI to take the lead. DNLUSLJ, 1, 20-25. https://dnluslj.in/wp-content/uploads/2024/04/DNLU-SLJ-SPECIAL-VOLUME-4.pdf

Miyamoto, K., Sato, Y., & Kitazume, K. (2005). Private-sector participation in infrastructure projects and value for money: Economic and financial impacts. *Transportation Research Record: Journal of the Transportation Research Board*, *1932*(1), 16–22. DOI: 10.1177/0361198105193200103

Munshi, M., Ritchie, A., & Nair, A. (2007). *Community-managed microfinance: a new model from Sri Lanka*. World Bank. Retrieved from https://www.semanticscholar.org/paper/Community-managed-Microfinance-A-New-Model-from-Sri-Munshi-Ritchie/38b23fd5a76703e00b621f180a99e19c71c29483

Ndung'u, N. (2021). *A digital financial services revolution in Kenya: The M-Pesa case study*. African Economic Research Consortium., Retrieved from https://www.researchgate.net/publication/349548752_A_Digital_Financial_Services_Revolution_in_Kenya_The_M-Pesa_Case_Study

Nurjihadi, M., Ibrahim, O., & Pasaribu, S. M. (2024). University education and innovation resistance: A multigroup analysis of digital payment barriers among Indonesian farmers. *International Journal of Academic Research in Progressive Education and Development*, *13*(4), 2199–2216. DOI: 10.6007/IJARPED/v13-i4/23812

. Nwafor, K. C., & Ayodele, E. A. (2024). Regulatory challenges and innovations in financial technology: Safeguarding against fraud while maximizing ROI. *International Journal of research Publication and Reviews, 5*(10), 4983-4994. https://doi.org/DOI: 10.55248/gengpi.5.1024.3125

. Omogbeme, A. O., Phil-Ugochukwu, A. I., Nwabufo, I. J., & Onyebuchi, J. (2024). The role of artificial intelligence in enhancing financial inclusion: A review of its impact on financial services for the unbanked population in the United States. https://doi.org/DOI: 10.30574/wjarr.2024.23.2.2489

Omoniwa, B. (2021). Nearly cashless: Digital transformation or cultural transmission? EconStor. Retrieved from https://www.econstor.eu/bitstream/10419/289862/1/1886219311.pdf

Opoku-Okuampa, I. Y. (2024). The Influence of Cultural Factors on Financial Decision-Making in Ghana. *International Journal of Finance and Banking Research, 10*(6), 118–125. http://www.intjfobr.com/article/10.11648/j.ijfbr.20241006.12. DOI: 10.11648/j.ijfbr.20241006.12

Padee, V. (31 October, 2023). Overcoming cultural barriers in payment systems. Apexx Global. Retrieved from https://apexx.global/blog/overcoming-cultural-barriers-in-payment-systems/

Pandey, D. (2023). Empowering rural communities through digital literacy initiatives in India. [IARE]. *International Assulta of Research and Engagement, 1*(1), 1–12. DOI: 10.5281/iare.v1i1.31

Parenti, R. (2020). Regulatory sandboxes and innovation hubs for FinTech. *Study for the Committee on Economic and Monetary Affairs, Policy Department for Economic, Scientific and Quality of Life Policies, European Parliament, Luxembourg, 65*. Retrieved from https://www.europarl.europa.eu/supporting-analyses

Pascal, I. N., & Edith, O. (2023). Cross-cultural approaches to managing the adoption of cashless policies in the workplace. *International Journal of Management, Social Sciences. Peace and Conflict Studies, 6*(2), 345–353. https://ijmsspcs.com/index.php/IJMSSPCS/article/view/538/587

Patel, A. S., Rao, V. K., & Radhakrishnan, M. K. (2023). Impact of Mobile Banking Platforms Paytm and Google Pay on Financial Inclusion in Rural and Semi-Urban Areas in India. *Journal of Finance and Accounting, 7*(5), 113–122. DOI: 10.53819/81018102t4205

Patil, P. P., Dwivedi, Y. K., & Rana, N. P. (2017). Digital payments adoption: an analysis of literature. In *Digital Nations–Smart Cities, Innovation, and Sustainability: 16th IFIP WG 6.11 Conference on E-Business, E-Services, and E-Society, I3E 2017, Delhi, India, November 21–23, 2017, Proceedings 16* (pp. 61-70). Springer International Publishing. https://doi.org/DOI: 10.1007/978-3-319-68557-1_7

Patil, P. P., Rana, N. P., & Dwivedi, Y. K. (2018). Digital payments adoption research: A review of factors influencing consumer's attitude, intention and usage. In *Challenges and Opportunities in the Digital Era: 17th IFIP WG 6.11 Conference on e-Business, e-Services, and e-Society, I3E 2018, Kuwait City, Kuwait, October 30–November 1, 2018, Proceedings 17* (pp. 45-52). Springer International Publishing. https://doi.org/DOI: 10.1007/978-3-030-02131-3_6

Qadri, E. N. (2023). Critical Factors that Affect the Adoption of Mobile Payment Services in Developed and Developing Countries. Master's Degree Project. Retrieved from https://uu.diva-portal.org/smash/get/diva2:1800492/FULLTEXT01.pdf

. Rahtio, H. (2024). Digitalization as a Driver of ASEAN Inclusivity After the Global Pandemic. *Edunity Kajian Ilmu Sosial dan Pendidikan, 3*(1), 126-132. https://doi.org/DOI: 10.57096/edunity.v3i1.220

Reis, A. C. S. (September 2022). The impact of culture on risk aversion in financial decisions: evidence from a TV show. Retrieved from https://repositorio-aberto.up.pt/bitstream/10216/146593/2/597064.pdf

Ritchie, A. (2007). *Community-based financial organizations: a solution to access in remote rural areas?* World Bank, ARD. Retrieved from https://www.rfilc.org/wp-content/uploads/2020/08/1175091210909_community_based_FO_rural_solution.pdf

Saiyid, M. (2023). Pix: Brazil's Successful Instant Payment System. *IMF Staff Country Reports*, 53-61. Retrieved from https://www.elibrary.imf.org/view/journals/002/2023/289/article-A004-en.xml

Saravanan, S., & Chithra, M. (2024). Adaptation of digital payment system among rural consumers in Enhancing the performance of the MFI. *International Journal of Arts. Science and Humanities, 11*(1), 59–65. DOI: 10.34293/sijash.v11iS1-June.7791

Shende, A., & Kathiriya, S. (2024). Consumer Protection and Payment Compliance in the Digital Age: Challenges, Innovations, and Regulatory Responses. 9. Retrieved from https://www.ijnrd.org/papers/IJNRD2401114.pdf

Shin, D.-H. (2010). The effects of trust, security, and privacy in social networking: A security-based approach to understand the pattern of adoption. *Interacting with Computers*, *22*(5), 428–438. DOI: 10.1016/j.intcom.2010.05.001

Sibley, J. (16 July 2024). *Redefining community finance: Practical strategies for payment leaders to drive financial inclusion*. The Payments Association. Retrieved from https://thepaymentsassociation.org/article/redefining-community-finance-practical-strategies-for-payment-leaders-to-drive-financial-inclusion/

Sleiman, K. A. A., Juanli, L., Lei, H., Liu, R., Ouyang, Y., & Rong, W. (2021). User trust levels and adoption of mobile payment systems in China: An empirical analysis. *SAGE Open*, *11*(4), 21582440211056599. DOI: 10.1177/21582440211056599

Sokrovolska, N., Prodius, O., Zhuk, O., Yamnenko, G., & Fedyna, V. (2022). Public-private partnership in microfinance. *Financial & Credit Activity: Problems of Theory & Practice*, *4*(45), 93–103. Advance online publication. DOI: 10.55643/fcaptp.4.45.2022.3842

. Song, X., Qin, X., & Feng, X. (2024). Creating financial inclusion in "Belt and Road" countries in Europe, Asia and Africa: regulation, technology and financial literacy. *Journal of Financial Regulation and Compliance*. https://doi.org/DOI: 10.1108/JFRC-11-2023-0180

Spante, M., Hashemi, S. S., Lundin, M., & Algers, A. (2018). Digital competence and digital literacy in higher education research: Systematic review of concept use. *Cogent Education*, *5*(1), 1519143. DOI: 10.1080/2331186X.2018.1519143

Spillers, F., & Gavine, A. (2018). Localization of mobile payment systems: cultural and temporal rhythms in user adoption. *Nordic Conference on Human-Computer Interaction*, 862–867. https://doi.org/DOI: 10.1145/3240167.3240276

Strebkov, T. (28 October, 2024). The presence and future of fintech UX & product design: Survey results. Pixels and Sense. Retrieved from https://www.pixelsandsense.com/insights/the-presence-and-future-of-fintech-ux-product-design-survey-results

Sultania, A. (2024). Are We Underestimating the Success of Self Help Groups in India? A Systematic Review. *Open Journal of Business and Management*, *12*(4), 2640–2661. DOI: 10.4236/ojbm.2024.124137

Sundararaj, J., & Meera, R. (2024). Adoption and User Behaviour Towards Digital Payments by Millennial. *South Eastern European Journal of Public Health*, •••, 822–828. DOI: 10.70135/seejph.vi.2126

Sunil, S. V., & Nalwaya, N. (2023). Digital Payment Systems–An Overview of Categories and Extant Opportunities and Challenges. [IJRASET]. *International Journal for Research in Applied Science and Engineering Technology, 11*(4), 2321–9653. DOI: 10.22214/ijraset.2023.50019

Thompsett, L. (4 October, 2024). The future of finance: Emerging technologies deep-dive. FinTech Magazine. Retrieved from https://fintechmagazine.com/articles/the-future-of-finance-emerging-technologies-deep-dive

Torre, D., & Xu, Q. (2020). Digital payments in China: adoption and interactions among applications. *Revue d'économie industrielle*, (172), 55-82. Retrieved from https://shs.hal.science/halshs-03313693/file/torre-xu-def.pdf

Triandis, H. C. (1995). *Individualism and collectivism*. Westview Press. Triandis, H. C. https://psycnet.apa.org/record/1995-97791-000

Van Dijk, J. (2020). *The Digital Divide*. Routledge Press.

Vatsa, P., Ma, W., & Zheng, H. (2024). Mobile payment adoption in China: Do demographic and socioeconomic factors matter? *MDE. Managerial and Decision Economics, 45*(3), 1428–1434. DOI: 10.1002/mde.4086

Vidjikant, S. (31 July, 2024). User experience in digital payments – A guide through best practices. Softjourn. Retrieved from https://softjourn.com/insights/user-experience-in-digital-payments-a-guide-through-best-practices

Voigt, P., & Von dem Bussche, A. (2017). The eu general data protection regulation (gdpr). *A Practical Guide, 1st Ed., Cham: Springer International Publishing, 10*(3152676), 10-5555. https://link.springer.com/book/10.1007/978-3-319-57959-7

Wang, J., & Gu, L. (2017). Why is wechat pay so popular? *Issues in Information Systems, 18*(4), 1–8. https://iacis.org/iis/2017/4_iis_2017_1-8.pdf

. Wiratno, E. P., & Rozi, F. (2024). Prabowo Gibran's Local Language Campaign On The Tiktok Account Fillixpratama. *JKOMDIS : Jurnal Ilmu Komunikasi Dan Media Sosial*. https://doi.org/DOI: 10.47233/jkomdis.v4i2.1903

Wischnewsky, A. (2024). Nearly Cashless: Digital Transformation or Cultural Transmission? *Available at SSRN* 4794163. https://dx.doi.org/DOI: 10.2139/ssrn.4794163

Worakul, W. (2006). Community-Based Microfinance: An Empowering Approach Towards poverty Alleviation and Community Self-Reliance. *UNDP Thailand*. Retrieved from https://www.undp.org/sites/g/files/zskgke326/files/migration/th/Microfinancebook.pdf

World Bank. (2022). South Asia's digital opportunity. World Bank Group. Retrieved from https://openknowledge.worldbank.org/server/api/core/bitstreams/778a3dc7-9da1-5722-a128-50b6b2aa38d7/content

Yang, C. C., Yang, S. Y., & Chang, Y. C. (2023). Predicting older adults' mobile payment adoption: An extended TAM model. *International Journal of Environmental Research and Public Health*, *20*(2), 1391. DOI: 10.3390/ijerph20021391 PMID: 36674145

Yoon, C. (2009). The effects of national culture values on consumer acceptance of e-commerce: Online shoppers in China. *Information & Management*, *46*(5), 294–301. DOI: 10.1016/j.im.2009.06.001

Zhanga, A., Yue, X., & Kong, Y. (2011, June). Exploring culture factors affecting the adoption of mobile payment. In *2011 10th International Conference on Mobile Business* (pp. 263-267). IEEE. https://doi.org/DOI: 10.1109/ICMB.2011.32

Zhao, Y., & Pan, Y. H. (2023). A Study of the Impact of Cultural Characteristics on Consumers' Behavioral Intention for Mobile Payments: A Comparison between China and Korea. *Sustainability (Basel)*, *15*(8), 6956. DOI: 10.3390/su15086956

Chapter 13
Ethical and Legal Dimensions of Drones and Robots in Smart Business:
Navigating the Autonomous Innovations in Logistics

Ananya Pandey
https://orcid.org/0000-0002-2419-6314
Christ University, Bangalore, India

Jipson Joseph
https://orcid.org/0009-0005-1795-5051
Christ University, Bangalore, India

ABSTRACT

Use of drones and robots are slowly but surely making their way into people's personal and professional lives. International companies are spending millions of dollars on drone research and development for logistics. Companies like Google, Amazon, DHL, and Walmart are actively experimenting on autonomous systems and robotics. Drones and robots are used in the business strategies of these organizations to ensure last-mile delivery, which involves transporting packages straight from a depot to the ultimate customer, maintaining inventory, loading and unloading of trucks, order processing, and the like. But the usage of these advancements poses some critical ethical and legal issues. Authorities must pay attention to these challenges and difficulties in order to ensure their meaningful usage. This chapter, in this

DOI: 10.4018/979-8-3373-2240-7.ch013

perspective, overviews the most significant legal and ethical concerns pertaining to the use of autonomous vehicles in smart businesses and logistics and provides recommendations for their effective utilization in future.

INTRODUCTION

A significant area of interest for transnational and multinational corporations is smart logistics. While increasing sales, it helps them to become more efficient. Smart logistics technologies such as drone and robot usage are used to track and manage all items, goods, and processes as they move through different supply chain phases (Ding et al., 2021). Accurate and pertinent data is provided by smart logistics helping to boost productivity and reduce errors. Logistics management is the process of acquiring, distributing, and creating materials and items in the right amounts and locations based on their intended use (Sergi et al., 2021). Originally used to describe the upkeep, storage, and movement of army personnel and supplies, logistics has now been superseded by supply chain management. Planning, putting into practice, and overseeing effective strategies to sustain forward and reverse flow are all part of the logistics management process (Mangan & Lalwani, 2016). Additionally, it makes easier to store products and services and submit relative data between the site of origin and the point of consumption as needed by clients. A multidimensional approach to raw material management, supply-chain management makes sure that all client requirements are properly integrated into the production process up until the product is delivered to the end user (Habib, 2011).

Supply chains are changing due to the incorporation of mobile robots and drones that present previously unheard-of chances for creativity and efficiency. As these technologies develop further, their strategic implementation will be essential to preserving competitiveness and resilience in an always changing market (Javaid et al., 2021). The manufacturing sector is leading the way in the widespread use of mobile robots and drones. Since automated guided vehicles (AGVs) and industrial robotics have been around for almost 50 years, manufacturing industries—especially the automotive and high-tech electronics sectors—are in a good position to spearhead the adoption of these technologies (Gunasekaran et al., 2017). Moving goods from point A to point B using a variety of payloads and vehicle sizes is the most frequent use case. Although they are lagging behind manufacturing, logistics and warehousing are also embracing robotics (Domingo Galindo, 2016). The range of use cases is wider here, as businesses utilize a variety of robots to support picking, sorting, packaging, and shipping processes in addition to unloading the trucks. With hundreds of thousands in operation globally, collaborative picking robots and mobile goods-to-person systems are especially well-liked (Villa, 2022). Autonomous vehicles such as drones and robots provide moral and legal conundrums that require careful consideration. For instance, self-driving systems must emphasize avoiding injury while making decisions in scenarios where an accident is inevitable. It is difficult to decide how autonomous cars should respond to these moral conundrums and make sure that the choices they make are consistent with society's ideals (Poszler et al., 2023). Keeping in view the above, this chapter explores the ethical and legal dimensions of the usage of drones and robots in logistics and supply chain. It also provides suitable policy recommendations for their responsible adoption.

AUTONOMOUS INNOVATIONS IN LOGISTICS

Over the past ten years, Unmanned Aerial Vehicle (UAV) technology—also known as drones—has advanced so quickly that it is now widely used in many different industries (Kim et al., 2020). Drones, in particular, have become increasingly popular in the logistics sector as their developing capabilities have turned them from experimental instruments into essential parts of contemporary logistics systems, particularly in delivery services (Cornell et al., 2023). To increase speed and efficiency in both urban and rural areas, some nations, for example, have integrated drones into last-mile delivery procedures (Kim et al., 2020).

Applications of Drones and Robots in Logistics

Drones and robotics use is being researched, tested, or, depending on the application, already used in logistics for transportation and warehousing. The following are some of the applications of drones and robots in logistics.

Delivery

Drone delivery basically means using UAVs to deliver goods to customers and businesses, including food, medical supplies, and parcels. This technology is the logistics industry's next big thing (Zieher et al., 2024). The main reasons for the increasing popularity of this distribution method are its claims of efficiency and speed. Traditional distribution techniques are frequently hampered by local issues like transportation congestion, inadequate infrastructure, and geographic limitations (Wiesbrod & Fitzroy, 2011). Drones can swiftly overcome hurdles and reach clients. Rapid commerce has emerged as a result of the dramatic change in customer expectations, and robots and drones are poised to take center stage. The delivery scene is changing dramatically, from the steady trundle of robots on our sidewalks to the quiet hum of delivery drones darting across the sky (Anderie et al., 2024).

Robots and drones for package delivery have advanced technologically. Autonomous robots and drones will be able to negotiate complicated routes, effectively avoid obstacles, and make deliveries more quickly with the support of enhanced AI and machine learning. One may anticipate other advancements that will broaden the use of robots and drone delivery services, such as increased payload capacity and a longer route range (Dhaliwal, 2021). For example, Amazon Prime Air has successfully advanced toward the future of delivery services in spite of the intricate web of logistical and regulatory obstacles. The service, which was started with the goal of getting products to clients in 30 minutes or fewer, has delivered about 100 parcels in two U.S. cities. The company is still investing the possibilities to improve the capabilities of its drone, even though it did not meet the original target of 10,000 deliveries by the end of 2023 (Berke et al., 2023). Also, an Irish drone delivery business called Manna Aero has now made over 100,000 drone deliveries in different parts of Ireland and is on track to do over 1,000 deliveries every day in crowded urban areas (Tarr et al., 2021).

Product Storage

The storage of goods is one of the most popular uses of robotics in logistics facilities. Automating storage tasks entails using state-of-the-art technology to carry out facility motions. Forklifts are replaced with stacker cranes, a form of automated handling machinery used for both pallets and boxes (Klumpp & Ruiner, 2018). These devices swiftly and independently place unit loads in their designated locations. They relocate the items to the appropriate location after picking them up from one end of the storage aisle (Gagliardi et al., 2012). Similarly, in accordance with directives from the warehouse management system (WMS), they take products out of the racks and move them to an outgoing conveyor or a pick and deposit (P&D) station (Wang et al., 2012).

Internal Transportation

Both handling equipment and operators, who devote a large portion of their time to these tasks, are disrupted by the flow of items. Automating internal product movements is another useful robotics application for high-throughput factories (Derossi et al., 2023). Large amounts of cargo are moved via on-site transport systems such as autonomous mobile robots (AMRs), automated guided vehicles, electric monorails, and pallet conveyors. Conveyors for pallets and boxes are the most common internal transport solution in warehousing out of all the available options (Meincke, 2022). To cut expenses and guarantee quicker delivery, they are utilized in distribution centers to move goods across the various sections of the building. AMRs are among the most promising conveyance technologies (Javaid et al., 2021). Without human assistance, these machines are capable of carrying out activities and moving around the warehouse. The size of the worldwide AMR market "is expected to expand at a compound annual growth rate (CAGR) of 15.5% from 2023 to 2030" (Simpson et al., 2019). These robots move products to the working areas as needed by the activities, connecting the various zones of a warehouse, logistics center, or production center.

Loading and Unloading of Trucks

Automatic truck loading and unloading systems allow businesses to automate the flow of items into and out of their facilities. This method, also referred to as automatic truck loading systems (ATLS), automates the process of loading and unloading pallets from trucks (Anderson, 2020). It uses skids, extended forks, and rollers or chains to transport objects with little assistance from the operator (Wawrla et al., 2019). Since the loading and unloading may be completed in a few minutes, this technique increases productivity. Additionally, automation reduces the possibility of product damage by preventing rough handling of goods (Rejeb et al., 2023).

Order Processing

One of the most expensive and intricate tasks in any warehouse is order processing. Additionally, picking using the goods-to-person technique is made easier and more efficient by automation (Bormann et al., 2019). By using this selection approach, workers can assemble orders without having to travel as automated systems deliver the products to them. The Shuttle System is one of the most successful goods-to-person robots' applications (Villa, 2022). Workstations are supplied with the goods needed to fulfill orders by the conveyors, elevators, and transfer cars that are incorporated into an automated storage and retrieval system (AS/RS) for boxes (Rejeb et al., 2023). Robotic arms, which are programmable devices that mimic or substitute human arms during production and shipping operations, are another option available to businesses with high order volumes (Dhaliwal, 2021). Heavy items can be carried by these arms more quickly and without fatigue. These fully autonomous robots are perfect for temperature-controlled environments (Javaid et al., 2021).

Inventory Management

One may greatly increase the data accuracy of the inventory count by using drones. In order to count the designated inventory items, the drones fly to inventory locations and scan the barcodes that are visible from floor level to the topmost rack (Ali et al., 2024). They can even detect empty spaces and scan barcodes that are broken or difficult to read. When the drone returns to the charging station, all of the information from each inventory scan is immediately updated on the dashboard. This greatly lowers the possibility of human error and provides one with instant access to the inventory status (Yoon et al., 2023).

Warehouse Automation

One practical usage for an autonomous drone in a warehouse is inventory counting. An aerial drone can fly through the warehouse and capture images of every area where merchandise is kept (Wawrla et al., 2019). A drone, as opposed to a mobile robot on the ground, can fly up and capture an image of a pallet that is kept at any height.

Humanitarian Aid

Transportation issues can make it difficult for medical professionals to provide children with high-quality care in remote and geographically challenging places, and they can also hinder early diagnosis (Rejeb et al., 2023). Supply chain management also depends on efficient transportation, especially when it comes to transporting temperature-sensitive goods, replenishing necessary supplies and medications, and providing life-saving emergency supplies in humanitarian situations (Rodman, 2004). Technologies and services based on UAVs or drones are showing promise in delivering life-saving supplies and producing significant societal benefits in the process. One way to cut down on transit time and guarantee immediate therapy to needy children is the use of drones (Mohsan et al., 2022). In addition to providing dependable and prompt patient care for disease diagnosis or effective epidemic management, on-board digital monitoring systems can assist with inventory tracking, temperature control, and other critical aspects of temperature-sensitive cargo (Ali et al., 2024). In many humanitarian situations where life-saving vaccines, biological samples, transfusion plasma, or organs are needed, drone delivery of emergency medical kits and supplies can speed up the process. For instance, UNICEF and the Malawian government started an air corridor in June 2017 to evaluate the possibility of using UAVs, or drones, for humanitarian purposes (Dean, 2015).

BENEFITS AND CHALLENGES OF AUTONOMOUS TECHNOLOGY IN LOGISTICS

One important development in recent years has been the rise of autonomous vehicles in logistics and transportation. Often referred to as self-driving automobiles or driverless cars, autonomous vehicles are capable of functioning with little to no human intervention (Hancock et al., 2019). To sense their environment, maneuver, and make judgments, they make use of cutting-edge technologies such as sensors, cameras, radar systems, and AI. Even though autonomous vehicles have many advantages, there are certain issues and worries that need to be resolved (Sadaf et al., 2023). Among these difficulties are a number of safety issues and technological constraints.

Benefits of Using Drones and Robots in Logistics and Supply Chain

Autonomous vehicles have gained popularity in the logistics and transportation sector as drones and robots have the potential to completely transform the way people and products are moved. Some of the significant benefits of using drones and robots are discussed below.

Technological Advancement

The emergence of autonomous vehicles has been fueled by the quick advancement of technology, especially in the areas of AI, sensor technology, and processing power (Rejeb et al., 2023). Vehicles can properly sense their environment with the support of sophisticated sensors like LiDAR, cameras, and radar systems. In order to assess the surroundings and make judgments in real time, AI algorithms analyze the sensor data, enabling safe and effective vehicle navigation (Vargas et al., 2021). These systems offer real-time insights, automate risky jobs, and establish safer workplaces, ultimately saving lives and minimizing accident-related project delays.

Safety, Efficiency and Productivity

One of the main drivers of autonomous vehicles adoption is safety. Autonomous vehicles have the ability to reduce or perhaps eliminate human error, which is a major contributor to traffic accidents (Shetty et al., 2021). Autonomous vehicles can improve road safety by carefully following traffic laws, maintaining ideal speeds, and responding to possible hazards more quickly than human drivers, who are prone to distractions, exhaustion, or poor judgment (Wang et al., 2020). Increased productivity and efficiency in logistics and transportation operations are anticipated with autonomous vehicles. With the use of sophisticated algorithms and real-time data processing, they can decrease traffic issues, optimize routes, and use less fuel (Dikshit et al., 2023). Additionally, they may work around the clock, doing away with the need for driver breaks or shifts, which would maximize vehicle utilization and boost industrial output (Liu, 2019).

Cost Savings

Transportation and logistics costs could be significantly reduced by autonomous cars. Businesses can cut labor expenses related to driver pay, benefits, and rest periods by doing away with the requirement for human drivers (Willems, 2021). Autonomous vehicles can also optimize fuel efficiency, minimize insurance premiums because of better safety records, and save maintenance expenses with predictive maintenance systems. Businesses can cut expenses by as much as 60% with the use of drones for the last mile (Grote et al., 2024).

Last Mile Delivery

The last-mile delivery industry has been primarily impacted by the emergence of autonomous cars. In order to manage the complex process of delivering items to customers' doorsteps, businesses are investigating the use of autonomous delivery robots and drones (Said et al., 2023). These cars can handle the difficulties of last-mile delivery and provide better client experiences by navigating congested metropolitan areas, streamlining delivery routes, and operating faster and more efficiently (Mohammad et al., 2023). Logistics networks now prioritize getting shipments into customers' hands during the last-mile of delivery, which entails moving goods from warehouses to residences or places of business. The need for seamless last-mile delivery has increased dramatically due to the continued explosion in popularity of online shopping, placing a significant burden on already overworked employees (Sawik, 2024). Given that the majority of shipping businesses estimate that last-mile delivery presents the largest problem in the entire order-fulfillment process, this vital sector of the logistics industry faces significant obstacles (Sorooshian et al., 2022).

Challenges of Using Drones and Robots in Logistics and Supply Chain

Even with the large investments and technological developments in recent years, many autonomous vehicles companies continue to face obstacles that keep them from expanding more quickly. The below mentioned points analyze some of the challenges they face.

Theft and Misidentification

Conventional delivery methods employ delivery workers to deliver the package to the recipient in person or deposit it in a secure mailbox. However, as drone technology can only drop a package in a relatively wide location, it cannot provide this convenience (Chen et al., 2021). For example, the package cannot just be dropped in front of the door at an apartment. Businesses can use sophisticated technology like facial recognition, but they don't offer the same flexibility as traditional approaches (Raposo, 2024).

Urban Delivery

A successful drone delivery in a remote area is demonstrated in the encouraging drone delivery films from Google, Amazon, and Dominos, suggesting that drones will be the delivery method of the future (Bine et al., 2024). It is important to remember, nevertheless, that the vast majority of people in the US and other countries reside in cities with high-rise structures and other architectural landmarks. Given the numerous barriers they must navigate, using drones to populate the skies in an urban environment can be dangerous (Szolomicki et al., 2019). The future of drone technology currently hinges on whether an unmanned pilot can easily handle drones in an urban environment, while fully autonomous drones are prohibited.

Weather Conditions

Traditional delivery methods that use trucks and vans will not have many problems with the changes in weather. It is adequately protected from wind, rain, and other bad weather (Koiwanit, 2018). However, how would tiny drones perform in inclement weather? Severe weather might harm both the drone and the transported cargo (Garg et al., 2023). Every state and region have seasonal weather patterns, with occasional bad weather occurrences. Addressing this unfavorable weather condition would be a problem for drone technology. The question, whether drones can only operate in ideal weather conditions—which is not very reliable—would also come up (Gao et al., 2021).

Insurance and other Expenses

Drones are incredibly expensive because the technology is still in its infancy. Insurance companies do not like the thought of an unmanned vehicle transporting precious goods hundreds of feet above the ground and delivering them (Rejeb et al., 2023). Additionally, it is more difficult for the insurance firm to consider insurance of drones, much alone estimate a premium, because such a notion is largely untested (Shah et al., 2021). Besides, millions of dollars in compensation may be awarded if drones result in bodily injury. Insurance companies will apply a hefty insurance premium in certain situations. Ultimately, it can raise the price of deploying drone technology for deliveries considerably, making it more costly than conventional delivery methods.

ETHICAL AND LEGAL DIMENSIONS OF USING DRONES AND ROBOTS

A new set of ethical and legal issues arises with this new and growingly popular technology of drones and robotics. Some of the significant issues within the ambit of drone delivery and the use of robots that require careful examination and ethical legal consideration are mentioned below.

Safety and Privacy Concerns

Safety and privacy concerns are two main difficulties raised along with the commercial use of drones (Koh et al., 2023). Since drones may be equipped with cameras that can record both audio and video, privacy concerns are particularly important when it comes to drone technology (Cai & Yu, 2024). Data security and privacy issues have become crucial factors as more companies deploy autonomous inventory drones to optimize warehouse operations. These drones produce and handle a lot of sensitive data, such as inventory records, operational metrics, and maybe customer-specific information, even while they provide unmatched precision, efficiency, and cost savings (Akca et al., 2020). The integrity of warehouse operations depends on making sure that this data is handled and protected securely, just like any other digital technology. Large volumes of data are collected and transmitted by autonomous drones, such as those used for inventory management, to guarantee precise tracking, reporting, and operational effectiveness (Sodiya et al., 2024). This information comprises inventory records i.e., the details regarding the stock, where it is kept in the warehouse, and how it moves. It includes operational performance metrics i.e., the information about how well a drone performs, such as flight routes, battery life, and task completion duration (Wawrla et al., 2019). It also comprises connectivity with WMS (Warehouse Management Systems) where the drones frequently communicate with enterprise resource planning (ERP) and work management systems, which may hold private company data.

Labor Market Disruption

As robotics and AI develop further, more occupations could be automated. Cutting-edge technologies are being created that can carry out operations that were previously exclusive to humans, like manufacturing items, processing data, and operating vehicles (Ongsulee, 2017). Businesses benefit from automation's increased productivity, but it also threatens many current jobs and changes the nature of employment. Protecting the workers during this disruptive time will require modification of the labor laws (Filippi et al., 2023). Many of today's jobs are considered to be at high danger of automation in the years to come. According to some studies, by 2030, technologies like AI, robotics, 3D printing, and processing power could replace almost 30% of occupations in OECD countries (Dwivedi et al., 2021). Since they entail predictable physical tasks that are well suited for automation, routine employment in industries like manufacturing, logistics, transportation, and food preparation are particularly vulnerable (Uzorka & Kalabuki, 2024). However, as AI improves its knowledge-based skills, it also poses a threat to professional positions like paralegals, tax preparers, and financial clerks. Many workers may be displaced as a result of widespread automation since some jobs may simply disappear (Lane, 2021). Many will find it challenging to retrain or modify current employees to take on new responsibilities. For automated workers transitioning, labor laws now offer few protections, such as funding for retraining programs or access to unemployment insurance (Illanes et al., 2018). However, if entire industries are decimated, unemployment insurance might no longer be a sufficient remedy. Other ideas, such as paying living expenses for multi-year retraining, offering wage subsidies for new, lower-paying jobs, and offering early retirement benefits to older workers, are being investigated (Costa et al., 2024).

Accountability

Unprecedented efficiency and innovation have been brought to logistics through the integration of autonomous technologies like robots and drones. But it also brings up important moral questions, especially those of responsibility for mishaps brought on by malfunctioning or broken systems (Misaros et al., 2023). Due to their reliance on intricate algorithms, machine learning, and large data inputs, autonomous systems may exhibit unpredictable behavior in unforeseen circumstances. The issue of responsibility gets complex when such systems result in injury, whether it be injuries, property damage, or interruptions to operations (Buchanan & Keohane, 2015). Should the user, the software developer, the manufacturer, or the organization implementing the system be blamed? Because traditional frameworks for culpability frequently overlook the multi-layered nature of autonomous decision-making, this ambiguity presents ethical conundrums (Bench-Capon, 2020). Operators may point to insufficient system alerts or safeguards, while manufacturers may contend that unanticipated circumstances led to the malfunction. Furthermore, transparency is complicated by AI's "black-box" nature, which makes it challenging to identify the underlying cause of an accident (Cabral, 2020). Clear accountability must be established in order to promote public trust in these technologies as well as for legal and financial restitution (Grau, 2023). To guarantee responsible development and implementation, ethical standards and legal frameworks must change. This includes procedures for allocating culpability that are founded on a careful evaluation of control and accountability. In order to reduce risks and guarantee that these technologies serve society in an ethical manner, it is critical to proactively address accountability as autonomous systems continue to grow in logistics (Naik et al., 2022).

Environment Impact

Drones may lessen traffic on the roads, but they also pose new environmental problems due to their energy consumption and noise pollution (Koiwanit, 2018). For example, the instance of drone delivery services in Rwanda, which have raised worries about the disruption of local wildlife habitats notwithstanding their effectiveness (Umlauf & Burchardt, 2022). Especially for last-mile and short-distance deliveries, drone delivery is frequently hailed as an environmentally benign substitute for conventional automobiles. However, one cannot just replace one's vehicles with drones and hope for a reduction in carbon footprint without also taking into account other factors (Kreier, 2022). When incorporating drones into the current system, it is critical to adopt a comprehensive approach. A number of aspects, including infrastructure, production, and noise as a whole affect the environment (Shilling, 2007). One major issue is noise pollution; drones make a characteristic buzzing noise, therefore flying routes and timings in residential areas must be carefully examined (Schäffer et al., 2021). In addition to the energy and materials needed for drone production, disposing of drones at the end of their useful lives—particularly their batteries—can have adverse environmental effects (Mitchell et al., 2023). There may then be a need for new infrastructure, such as drone spots, which have an environmental cost. For drone delivery to be beneficial, all of these issues must be taken into account (Poikonen & Campbell, 2021).

Economic and Market Concerns

There are economic considerations as well. What will happen to those employed in the conventional delivery industry if drone delivery becomes widely used? How can logistics firms prepare for the possibility of job losses in the sector? (Nurgaliev et al., 2023) Is it ethically required of the industry to offer assistance and training programs to truck drivers in order to help them, for instance, locate employment in the drone industry? (Wang et al., 2021) The so-called "drone divide" is another important factor. This is comparable to the "digital divide," which is the difference between those who profit from digital technology like the internet and personal computers and those who do not (West & Bowman, 2016). Similar experiences, but in the air, are described by the drone divide. Drone technology is available to certain groups and regions, but others run the risk of falling behind. The drone divide is caused by a number of variables (Fox, 2022). It is heavily influenced by economic inequality, as wealthy nations are better able to invest in the technology and make profit. Infrastructure is also important since drones require DroneSpots and recharge stations in order to scale effectively (Alonso et al., 2020). Regulators, drone manufacturers, and other stakeholders must acknowledge the drone divide and take aggressive measures to bridge it, even though drone delivery has the potential to revolutionize isolated and difficult-to-reach locations (Dolata & Schwabe, 2023).

Bias and Fairness in Decision Making

The negative or unjust results that arise from the design, use, or deployment of drones are referred to as drone biases and inequalities. Depending on the application's setting, these have an impact on ecosystems, communities, or individuals (Tubis et al., 2024). For example, drones that monitor particular groups without their agreement or following the proper procedures because of their gender, race, religion, or other traits (Resnik & Elliott, 2019). Drones can also be used to distribute goods or services; however, they may exclude or disadvantage some populations or places due to factors like accessibility, infrastructure, or income. Furthermore, because of the quality, source, or interpretation of data, drones may gather and process information that contains biases, errors, or inaccuracies (Mohsan et al., 2023). Finally, depending on the drone's behavior, design, or feedback, it may engage with people without respecting their autonomy, dignity, or desires. Drone biases and inequalities are crucial because they can have major consequences on the rights, well-being, and opportunities of people and the environment (Khogali & Mekdid, 2023). Along with posing hazards to stakeholders and drone operators in terms of law, ethics, and society, they can also erode the validity, confidence, and acceptability of drone applications. Therefore, for anyone who wants to use drones for good, tackling drone biases and inequalities is both a moral obligation and a competitive advantage (Jha, 2014).

Regulatory and Cybersecurity Challenges

Drone operators must be meticulous in adhering to the regulations because of the perceived and actual risks connected to remotely piloted aircraft systems. But when it comes to drones, the regulatory environment is very difficult (Dolata & Schwabe, 2023). Regulations frequently struggle to keep up with emerging technologies like beyond visual line of sight (BVLOS), and these technologies are so new that there is not a universal standard in place (Politi et al., 2021). Another difficulty is cybersecurity. Since drones are networked devices, there is a risk of hacking, illegal access, and data breaches. Drone operators are required by law and ethics to protect sensitive client information by implementing best practices such as data encryption and restricted access (Sihag et al., 2023).

Data Protection and Security Laws

Navigating through the various regulatory frameworks imposed by other nations is one of the biggest obstacles in international drone operations. Drone flying regulations vary from country to country and include things like registration procedures, flight height limitations, and no-fly zones (Tsiamis et al., 2019). For firms looking to carry out cross-border operations, this presents serious challenges. International harmonization of these regulations is still being worked on (Ingves, 2007). It is difficult to get an agreement among member governments with different interests and agendas, even when groups like the International Civil Aviation Organization (ICAO) work to create broad principles (Selsky & Parker, 2005). As a result, while organizing international drone operations, operators frequently encounter a labyrinth of bureaucratic red tape. Also, the users of drones, with cameras, are advised by the General Data Protection Regulation (GDPR) to operate them responsibly and safely in order to protect others' privacy (Pauner et al., 2015).

Liability and Insurance

In order to protect the public and maintain safety, recreational drone use is governed by a number of rules and regulations. Drone operators are subject to rules and regulations set forth by the Federal Aviation Administration (FAA), which include operational constraints, airspace restrictions, and registration requirements (Nakamura & Kajikawa, 2018). Furthermore, drone operators may be required to abide by particular rules set forth by state and local governments (Rajgopalan & Krishna, 2018). When determining liability in personal injury lawsuits involving recreational drones, it is crucial to comprehend these regulations. In drone-related personal injury claims, determining the culpability can be difficult (Sehrawat,

2018). The owner of the property, the manufacturer, the store, the drone operator, and in certain cases, the FAA, might all be held accountable. When determining culpability, elements including carelessness, deliberate wrongdoing, product flaws, and property ownership are crucial. Every situation needs to be assessed separately, taking the relevant laws and circumstances into account (Soyer & Tettenborn, 2022).

RECOMMENDATIONS FOR ETHICAL AND LEGAL NAVIGATION

Developments in AI-powered drone technology have the potential to revolutionize a number of industries by improving data-driven decision making, efficiency, and safety (Javaid et al., 2021). Nonetheless, there are important issues that need to be resolved in the areas of data processing, object detection, energy efficiency, autonomous decision-making, security, and regulatory compliance. It will require a coordinated effort comprising technological innovation, strategic planning, and cooperation to overcome these obstacles (Jain et al., 2024). In order to ensure that inventors can safeguard their contributions and to encourage continued research and development, intellectual property protection is essential to this ecosystem (Kalanje, 2006). Some of the significant recommendations for navigating the ethical and legal dilemmas are discussed below.

Stakeholder Engagement

In order to fully address the ethical and legal issues surrounding AI in autonomous systems, the stakeholder engagement process is essential (Stahl & Stahl, 2021). The practical difficulties of incorporating ethical considerations into AI algorithms and system design should be discussed by AI developers and stakeholders (Clarke et al., 2001). Policymakers should share their thoughts on the shortcomings and inadequacies of the laws in place, highlighting the necessity of new legislative frameworks (Tsamados et al., 2021). Ethicists should critically analyze the moral conundrums surrounding AI decision-making, responsibility, and bias. Leaders in the field should engage in real-world examples of striking a balance between ethical norms and technological progress (Osasona et al., 2024). In order to comprehend the societal issues regarding safety, privacy, and accountability, public opinion should be finally gathered (Kim, 2009). This wide-ranging involvement guarantees a sophisticated comprehension of the moral ramifications, which makes it easier to formulate comprehensive suggestions for enhancing the regulation and application of autonomous technologies (Mennella et al., 2024).

Intellectual Property Protection

Gaining a competitive edge and protecting technological developments are two benefits of obtaining patents for creative AI-based hardware designs and procedures (Panda & Sharma, 2020). Through the granting of exclusive rights to use and commercialize the discovery, patents safeguard research and development efforts. This promotes a culture of innovation and ongoing investment (Clark & Kowalski, 2012). Trade secrets provide protection for distinctive techniques that might not be patented or where disclosure is undesirable. Companies can respect intellectual property rights while utilizing each other's capabilities through strategic partnerships and licensing agreements (Zhang et al., 2018). Working together can result in mutual advancement and the pooling of resources to more successfully address difficult problems.

Comprehensive Ethical Framework

It is crucial to set up explicit ethical frameworks and rules that follow global norms. Principles including respect for international law, privacy protection, and openness in drone activities should all be included in these frameworks (Alberstadt, 2014). International cooperation is crucial since autonomous drone technology is a global phenomenon. Establishing shared security and ethical norms requires cooperation from nations, organizations, and stakeholders (Wilson, 2014). International agreements can encourage responsible use and help stop an uncontrolled drone weapons race. To maximize the advantages of autonomous drone technology while avoiding potential hazards and moral conundrums, responsible development and implementation are crucial (Meir, 2016). Stakeholders can guarantee that autonomous drones are utilized as a force for good and favorably impact a range of industries, from public safety and disaster response to transportation and agriculture, by significantly prioritizing ethics, safety, transparency, and accountability (Wang et al., 2023).

Transparency and Accountability

Accountability procedures should be in place, and the development process should be open in the usage of drones and robots in logistics. Transparency makes it possible for the public and other stakeholders to evaluate the goals and potential of autonomous drones (Rejeb et al., 2023). Developers and operators should be held accountable for any drone-related mishaps or misuse. Existing laws and regulations pertaining to autonomous drones must be followed by developers and operators (Bhat et al., 2024). Getting the required licenses, permissions, and approvals for drone operations is part of this regulatory framework. Compliance helps steer clear of ethical and legal issues and guarantees that drones are operated within the law (Nugraha et al., 2016).

CONCLUSION

Modern technologies are transforming last-mile delivery in the cities. Specifically, cutting-edge technology has demonstrated encouraging outcomes in commercial applications, such as delivery drones. In densely urbanized locations, contactless technologies have become even more crucial to consumers' daily lives when paired with the ongoing epidemic. The advent of autonomous vehicles changes the system and functioning of logistics and transportation. Modern technology and safety enhancements transform the movement of people and products. This autonomous journey requires navigating obstacles including safety, regulations, and infrastructural preparation. The effects of autonomous vehicles extend beyond hands-free driving. They pledge to improve accessibility, reduce traffic, make roads safer, and preserve the environment. This will lead to workforce transformation, urban planning, and economic prospects in data analytics and vehicle maintenance. By offering transportation as a service and enabling people with few options for mobility, autonomous vehicles are redefining mobility. The future of autonomous vehicles is bright, despite obstacles such as regulatory frameworks and technical constraints. Widespread integration with transportation networks is anticipated as trust increases and technology develops. Self-driving cars are just one aspect of the future; another is rethinking how one transports people and products. Governments, businesses, researchers, and the general public must work together. By addressing obstacles, encouraging creativity, and implementing autonomous vehicles responsibly, their full potential can be realized, bringing about safer, more effective, and environmentally friendly logistics and transportation. The emergence of autonomous vehicles is a paradigm shift in transportation thought as much as a technology advancement. It is a chance to transform cities, generate employment, and improve accessibility and environmental sustainability in transportation. Accept the potential that autonomous vehicles present and strive toward a more secure, intelligent, and interconnected transportation future.

REFERENCES

Akca, A., Kara, I., & Aydos, M. (2020, July). Privacy, Security and Legal Aspects of Autonomous Vehicles. In *International Conference on Materials Science, Mechanical and Automotive Engineerings and Technology (IMSMATEC'20)* (p. 20).

Alberstadt, R. (2014). Drones under international law. *Open Journal of Political Science*, *4*(04), 221–232. DOI: 10.4236/ojps.2014.44023

Ali, S. S., Khan, S., Fatma, N., Ozel, C., & Hussain, A. (2024). Utilisation of drones in achieving various applications in smart warehouse management. *Benchmarking*, *31*(3), 920–954. DOI: 10.1108/BIJ-01-2023-0039

. Alonso, C., Kothari, S., & Rehman, S. (2020). How artificial intelligence could widen the gap between rich and poor nations. *IMF Blog, 2*.

Anderson, J. (2020). *U.S. Patent No. 10,815,080*. Washington, DC: U.S. Patent and Trademark Office.

Bench-Capon, T. J. (2020). Ethical approaches and autonomous systems. *Artificial Intelligence*, *281*, 103239. DOI: 10.1016/j.artint.2020.103239

Berke, A., Ding, G., Chin, C., Gopalakrishnan, K., Larson, K., Balakrishnan, H., & Li, M. Z. (2023). Drone delivery and the value of customer privacy: A discrete choice experiment with US consumers. *Transportation Research Part C, Emerging Technologies*, *157*, 104391. DOI: 10.1016/j.trc.2023.104391

. Bhat, G. R., Dudhedia, M. A., Panchal, R. A., Shirke, Y. S., Angane, N. R., Khonde, S. R., ... & Gawande, S. H. (2024). Autonomous Drones and their Influence on Standardization of Rules and Regulations for Operating-A Brief Overview. *Results in Control and Optimization*, 100401.

Bine, L. M., Boukerche, A., Ruiz, L. B., & Loureiro, A. A. (2024). Connecting Internet of Drones and urban computing: Methods, protocols and applications. *Computer Networks*, *239*, 110136. DOI: 10.1016/j.comnet.2023.110136

Bormann, R., de Brito, B. F., Lindermayr, J., Omainska, M., & Patel, M. (2019). Towards automated order picking robots for warehouses and retail. In *Computer Vision Systems: 12th International Conference, ICVS 2019, Thessaloniki, Greece, September 23–25, 2019* [Springer International Publishing.]. *Proceedings*, *12*, 185–198.

Buchanan, A., & Keohane, R. O. (2015). Toward a drone accountability regime. *Ethics & International Affairs*, *29*(1), 15–37. DOI: 10.1017/S0892679414000732

Cabral, T. S. (2020). Liability and artificial intelligence in the EU: Assessing the adequacy of the current Product Liability Directive. *Maastricht Journal of European and Comparative Law, 27*(5), 615–635. DOI: 10.1177/1023263X20948689

Cai, K., & Yu, Z. (2024). Research on the antecedents of data privacy concern toward intelligent connected vehicles. *International Journal of Human-Computer Interaction,* •••, 1–14.

Chen, H., Hu, Z., & Solak, S. (2021). Improved delivery policies for future drone-based delivery systems. *European Journal of Operational Research, 294*(3), 1181–1201. DOI: 10.1016/j.ejor.2021.02.039

Clark, K. L., & Kowalski, S. P. (2012). Harnessing the power of patent information to accelerate innovation. *Wiley Interdisciplinary Reviews. Data Mining and Knowledge Discovery, 2*(5), 427–435. DOI: 10.1002/widm.1066

Clarke, A., English, V., Harris, H., & Wells, F. (2001). Ethical considerations. *International Journal of Pharmaceutical Medicine, 15*(2), 89–94. DOI: 10.2165/00124363-200104000-00012

Cornell, A., Mahan, S., & Riedel, R. (2023). Commercial drone deliveries are demonstrating continued momentum in 2023. *URL:* https://www. mckinsey. com/industries/aerospace-and-defense/our-insights/future-air-mobility-blog/commercial-drone-deliveries

Costa, I., Machado, C. S., & Conceição, O. (2024). Automation and the Importance of Reskilling Workers: A Case Study in the Automotive Industry. In *Proceedings of The 19th European Conference on Management, Leadership and Governance.* Academic Conferences International. DOI: 10.34190/ecmlg.20.1.2923

Dean, H. (2015). *Building innovation capability in an intergovernmental aid organization: A case study of UNICEF s Innovation Unit* (Master's thesis).

Derossi, A., Di Palma, E., Moses, J. A., Santhoshkumar, P., Caporizzi, R., & Severini, C. (2023). Avenues for non-conventional robotics technology applications in the food industry. *Food Research International, 173,* 113265. DOI: 10.1016/j.foodres.2023.113265

. Dhaliwal, A. (2021). Reinventing logistics: Use of AI & Robotics technologies. *Business Research and Innovation, 147.*

. Dikshit, S., Atiq, A., Shahid, M., Dwivedi, V., & Thusu, A. (2023). The use of artificial intelligence to optimize the routing of vehicles and reduce traffic congestion in urban areas. *EAI Endorsed Transactions on Energy Web, 10.*

Ding, Y., Jin, M., Li, S., & Feng, D. (2021). Smart logistics based on the internet of things technology: An overview. *International Journal of Logistics, 24*(4), 323–345. DOI: 10.1080/13675567.2020.1757053

Dolata, M., & Schwabe, G. (2023). Moving beyond privacy and airspace safety: Guidelines for just drones in policing. *Government Information Quarterly, 40*(4), 101874. DOI: 10.1016/j.giq.2023.101874

Domingo Galindo, L. (2016). *The challenges of logistics 4.0 for the supply chain management and the information technology* (Master's thesis, NTNU).

Dwivedi, Y. K., Hughes, L., Ismagilova, E., Aarts, G., Coombs, C., Crick, T., Duan, Y., Dwivedi, R., Edwards, J., Eirug, A., Galanos, V., Ilavarasan, P. V., Janssen, M., Jones, P., Kar, A. K., Kizgin, H., Kronemann, B., Lal, B., Lucini, B., & Williams, M. D. (2021). Artificial Intelligence (AI): Multidisciplinary perspectives on emerging challenges, opportunities, and agenda for research, practice and policy. *International Journal of Information Management, 57*, 101994. DOI: 10.1016/j.ijinfomgt.2019.08.002

Filippi, E., Banno, M., & Trento, S. (2023). Automation technologies and their impact on employment: A review, synthesis and future research agenda. *Technological Forecasting and Social Change, 191*, 122448. DOI: 10.1016/j.techfore.2023.122448

Fox, S. J. (2022). Drones: Foreseeing a'risky'business? Policing the challenge that flies above. *Technology in Society, 71*, 102089. DOI: 10.1016/j.techsoc.2022.102089

Gagliardi, J. P., Renaud, J., & Ruiz, A. (2012). Models for automated storage and retrieval systems: A literature review. *International Journal of Production Research, 50*(24), 7110–7125. DOI: 10.1080/00207543.2011.633234

Gao, M., Hugenholtz, C. H., Fox, T. A., Kucharczyk, M., Barchyn, T. E., & Nesbit, P. R. (2021). Weather constraints on global drone flyability. *Scientific Reports, 11*(1), 12092. DOI: 10.1038/s41598-021-91325-w

Garg, V., Niranjan, S., Prybutok, V., Pohlen, T., & Gligor, D. (2023). Drones in last-mile delivery: A systematic review on Efficiency, Accessibility, and Sustainability. *Transportation Research Part D, Transport and Environment, 123*, 103831. DOI: 10.1016/j.trd.2023.103831

Grau, G. I. (2023). An Appraisal of the Proposal for a Directive on Liability for Defective Products. *Journal of European Consumer and Market Law, 12*(5).

Grote, M., Oakey, A., Pilko, A., Krol, J., Blakesley, A., Cherrett, T., Scanlan, J., Anvari, B., & Martinez-Sykora, A. (2024). The effects of costs on drone uptake in multi-modal logistics systems within a healthcare setting. *Transport Economics and Management*, *2*, 58–75. DOI: 10.1016/j.team.2024.03.001

Gunasekaran, A., Subramanian, N., & Papadopoulos, T. (2017). Information technology for competitive advantage within logistics and supply chains: A review. *Transportation Research Part E, Logistics and Transportation Review*, *99*, 14–33. DOI: 10.1016/j.tre.2016.12.008

. Habib, M. (2011). Supply chain management (SCM): theory and evolution. *Supply chain management-applications and simulations, 10*, 24573.

Hancock, P. A., Nourbakhsh, I., & Stewart, J. (2019). On the future of transportation in an era of automated and autonomous vehicles. *Proceedings of the National Academy of Sciences of the United States of America, 116*(16), 7684–7691. DOI: 10.1073/pnas.1805770115

Illanes, P., Lund, S., Mourshed, M., Rutherford, S., & Tyreman, M. (2018). Retraining and reskilling workers in the age of automation. *McKinsey Global Institute*, *8*(1), 1–8.

. Ingves, S. (2007). Regulatory Challenges of Cross-border Banking: Possible Ways Forwardl Conference–2007.

Jain, N., Gambhir, A., & Pandey, M. (2024). Unmanned Aerial Networks—UAVs and AI. In *Recent Trends in Artificial Intelligence Towards a Smart World: Applications in Industries and Sectors* (pp. 321–351). Springer Nature Singapore. DOI: 10.1007/978-981-97-6790-8_12

Javaid, M., Haleem, A., Singh, R. P., & Suman, R. (2021). Substantial capabilities of robotics in enhancing industry 4.0 implementation. *Cognitive Robotics*, *1*, 58–75. DOI: 10.1016/j.cogr.2021.06.001

Jha, W. C. U. (2014). *Drone Wars: Ethical, Legal and Strategic Implications: Ethical, Legal and Strategic Implications*. KW Publishers Pvt Ltd.

Kalanje, C. M. (2006). *Role of intellectual property in innovation and new product development*. World Intellectual Property Organization.

Khogali, H. O., & Mekid, S. (2023). The blended future of automation and AI: Examining some long-term societal and ethical impact features. *Technology in Society*, *73*, 102232. DOI: 10.1016/j.techsoc.2023.102232

Kim, J., Moon, H., & Jung, H. (2020). Drone-based parcel delivery using the rooftops of city buildings: Model and solution. *Applied Sciences (Basel, Switzerland), 10*(12), 4362. DOI: 10.3390/app10124362

Kim, P. S. (2009). Enhancing public accountability for developing countries: Major constraints and strategies. *Australian Journal of Public Administration, 68*(s1), S89–S100. DOI: 10.1111/j.1467-8500.2009.00626.x

Klumpp, M., & Ruiner, C. (2018). Regulation for artificial intelligence and robotics in transportation, logistics, and supply chain management: Background and developments. *Network Industries Quarterly, 20*(2), 3–7.

Koh, L. Y., Lee, J. Y., Wang, X., & Yuen, K. F. (2023). Urban drone adoption: Addressing technological, privacy and task–technology fit concerns. *Technology in Society, 72*, 102203. DOI: 10.1016/j.techsoc.2023.102203

Koiwanit, J. (2018). Analysis of environmental impacts of drone delivery on an online shopping system. *Advances in Climate Change Research, 9*(3), 201–207. DOI: 10.1016/j.accre.2018.09.001

Kreier, F. (2022). Drones bearing parcels deliver big carbon savings. *Nature*. Advance online publication. DOI: 10.1038/d41586-022-02101-3

Lane, M. (2021). The impact of AI on the labour market: is this time different. *OECD. AI, Labour Market Economist,* https://oecd. ai/en/wonk/impact-ai-on-the-labour-market-is-this-timedifferent, *erişim, 25*, 2021.

Liu, Y. (2019). An optimization-driven dynamic vehicle routing algorithm for on-demand meal delivery using drones. *Computers & Operations Research, 111*, 1–20. DOI: 10.1016/j.cor.2019.05.024

. Mangan, J., & Lalwani, C. (2016). *Global logistics and supply chain management*. John Wiley & Sons.

. Meincke, P. A. (2022). Cargo handling, transport and logistics processes in the context of drone operation. *Automated Low-Altitude Air Delivery: Towards Autonomous Cargo Transportation with Drones*, 205-243.

Mennella, C., Maniscalco, U., De Pietro, G., & Esposito, M. (2024). Ethical and regulatory challenges of AI technologies in healthcare: A narrative review. *Heliyon, 10*(4), e26297. DOI: 10.1016/j.heliyon.2024.e26297

Misaros, M., Stan, O. P., Donca, I. C., & Miclea, L. C. (2023). Autonomous Robots for Services—State of the Art, Challenges, and Research Areas. *Sensors (Basel), 23*(10), 4962. DOI: 10.3390/s23104962

Mitchell, S., Steinbach, J., Flanagan, T., Ghabezi, P., Harrison, N., O'reilly, S., Killian, S., & Finnegan, W. (2023). Evaluating the sustainability of lightweight drones for delivery: Towards a suitable methodology for assessment. *Functional Composite Materials*, *4*(1), 4. DOI: 10.1186/s42252-023-00040-4

Mohammad, W. A., Nazih Diab, Y., Elomri, A., & Triki, C. (2023, April). Innovative solutions in last mile delivery: Concepts, practices, challenges, and future directions. In *Supply Chain Forum* [). Taylor & Francis.]. *International Journal (Toronto, Ont.)*, *24*(2), 151–169.

Mohsan, S. A. H., Othman, N. Q. H., Li, Y., Alsharif, M. H., & Khan, M. A. (2023). Unmanned aerial vehicles (UAVs): Practical aspects, applications, open challenges, security issues, and future trends. *Intelligent Service Robotics*, *16*(1), 109–137. DOI: 10.1007/s11370-022-00452-4

Mohsan, S. A. H., Zahra, Q. U. A., Khan, M. A., Alsharif, M. H., Elhaty, I. A., & Jahid, A. (2022). Role of drone technology helping in alleviating the COVID-19 pandemic. *Micromachines*, *13*(10), 1593. DOI: 10.3390/mi13101593

Naik, N., Hameed, B. Z., Shetty, D. K., Swain, D., Shah, M., Paul, R., Aggarwal, K., Ibrahim, S., Patil, V., Smriti, K., Shetty, S., Rai, B. P., Chlosta, P., & Somani, B. K. (2022). Legal and ethical consideration in artificial intelligence in healthcare: Who takes responsibility? *Frontiers in Surgery*, *9*, 862322. DOI: 10.3389/fsurg.2022.862322

Nakamura, H., & Kajikawa, Y. (2018). Regulation and innovation: How should small unmanned aerial vehicles be regulated? *Technological Forecasting and Social Change*, *128*, 262–274. DOI: 10.1016/j.techfore.2017.06.015

Nugraha, R. A., Jeyakodi, D., & Mahem, T. (2016). Urgency for legal framework on drones: Lessons for Indonesia, India, and Thailand. *Indon. L. Rev.*, *6*(2), 137. DOI: 10.15742/ilrev.v6n2.229

. Nurgaliev, I., Eskander, Y., & Lis, K. (2023). The Use of Drones and Autonomous Vehicles in Logistics and Delivery. *Logistics and Transport, 57.*

Ongsulee, P. (2017, November). Artificial intelligence, machine learning and deep learning. In *2017 15th international conference on ICT and knowledge engineering (ICT&KE)* (pp. 1-6). IEEE. DOI: 10.1109/ICTKE.2017.8259629

Osasona, F., Amoo, O. O., Atadoga, A., Abrahams, T. O., Farayola, O. A., & Ayinla, B. S. (2024). Reviewing the ethical implications of AI in decision making processes. *International Journal of Management & Entrepreneurship Research*, *6*(2), 322–335. DOI: 10.51594/ijmer.v6i2.773

Panda, S., & Sharma, R. (2020). Impact of patent rights on innovation: A meta-analysis. *Journal of Public Affairs*, *20*(2), e2000. DOI: 10.1002/pa.2000

Pauner, C., Kamara, I., & Viguri, J. (2015, December). Drones. Current challenges and standardisation solutions in the field of privacy and data protection. In *2015 ITU Kaleidoscope: Trust in the Information Society (K-2015)* (pp. 1-7). IEEE.

Poikonen, S., & Campbell, J. F. (2021). Future directions in drone routing research. *Networks*, *77*(1), 116–126. DOI: 10.1002/net.21982

Politi, E., Panagiotopoulos, I. E., Varlamis, I., & Dimitrakopoulos, G. (2021, April). A Survey of UAS Technologies to Enable Beyond Visual Line Of Sight (BVLOS) Operations. In *VEHITS* (pp. 505-512).

Poszler, F., Geisslinger, M., Betz, J., & Lütge, C. (2023). Applying ethical theories to the decision-making of self-driving vehicles: A systematic review and integration of the literature. *Technology in Society*, *75*, 102350. DOI: 10.1016/j.techsoc.2023.102350

Raposo, V. L. (2024). When facial recognition does not 'recognise': Erroneous identifications and resulting liabilities. *AI & Society*, *39*(4), 1857–1869. DOI: 10.1007/s00146-023-01634-z

Rejeb, A., Rejeb, K., Simske, S. J., & Treiblmaier, H. (2023). Drones for supply chain management and logistics: A review and research agenda. *International Journal of Logistics*, *26*(6), 708–731. DOI: 10.1080/13675567.2021.1981273

Resnik, D. B., & Elliott, K. C. (2019). Using drones to study human beings: Ethical and regulatory issues. *Science and Engineering Ethics*, *25*(3), 707–718. DOI: 10.1007/s11948-018-0032-6

. Rodman, W. K. (2004). Supply chain management in humanitarian relief logistics.

Sadaf, M., Iqbal, Z., Javed, A. R., Saba, I., Krichen, M., Majeed, S., & Raza, A. (2023). Connected and automated vehicles: Infrastructure, applications, security, critical challenges, and future aspects. *Technologies*, *11*(5), 117. DOI: 10.3390/technologies11050117

Said, M., Aeschliman, S., & Stathopoulos, A. (2023). Robots at your doorstep: Acceptance of near-future technologies for automated parcel delivery. *Scientific Reports*, *13*(1), 18556. DOI: 10.1038/s41598-023-45371-1

Sawik, B. (2024). Optimizing Last-Mile Delivery: A Multi-Criteria Approach with Automated Smart Lockers, Capillary Distribution and Crowdshipping. *Logistics (Basel)*, *8*(2), 52. DOI: 10.3390/logistics8020052

Schäffer, B., Pieren, R., Heutschi, K., Wunderli, J. M., & Becker, S. (2021). Drone noise emission characteristics and noise effects on humans—A systematic review. *International Journal of Environmental Research and Public Health*, *18*(11), 5940. DOI: 10.3390/ijerph18115940

Sehrawat, V. (2018). Liability issue of domestic drones. *Santa Clara High-Technology Law Journal*, *35*, 110.

Selsky, J. W., & Parker, B. (2005). Cross-sector partnerships to address social issues: Challenges to theory and practice. *Journal of Management*, *31*(6), 849–873. DOI: 10.1177/0149206305279601

Sergi, I., Montanaro, T., Benvenuto, F. L., & Patrono, L. (2021). A smart and secure logistics system based on IoT and cloud technologies. *Sensors (Basel)*, *21*(6), 2231. DOI: 10.3390/s21062231

Shetty, A., Yu, M., Kurzhanskiy, A., Grembek, O., Tavafoghi, H., & Varaiya, P. (2021). Safety challenges for autonomous vehicles in the absence of connectivity. *Transportation Research Part C, Emerging Technologies*, *128*, 103133. DOI: 10.1016/j.trc.2021.103133

Shilling, J. D. (2007). *The Nexus between infrastructure and environment* (Vol. 1). Evaluation Cooperation Group.

Sihag, V., Choudhary, G., Choudhary, P., & Dragoni, N. (2023). Cyber4drone: A systematic review of cyber security and forensics in next-generation drones. *Drones (Basel)*, *7*(7), 430. DOI: 10.3390/drones7070430

Simpson, C., Ataii, E., Kemp, E., & Zhang, Y. (2019). *Mobility 2030: Transforming the mobility landscape*. KPMG International.

Sodiya, E. O., Umoga, U. J., Amoo, O. O., & Atadoga, A. (2024). AI-driven warehouse automation: A comprehensive review of systems. *GSC Advanced Research and Reviews*, *18*(2), 272–282. DOI: 10.30574/gscarr.2024.18.2.0063

Sorooshian, S., Khademi Sharifabad, S., Parsaee, M., & Afshari, A. R. (2022). Toward a modern last-mile delivery: Consequences and obstacles of intelligent technology. *Applied System Innovation*, *5*(4), 82. DOI: 10.3390/asi5040082

Soyer, B., & Tettenborn, A. (2022). Artificial intelligence and civil liability—Do we need a new regime? *International Journal of Law and Information Technology*, *30*(4), 385–397. DOI: 10.1093/ijlit/eaad001

. Stahl, B. C., & Stahl, B. C. (2021). Addressing ethical issues in AI. *Artificial intelligence for a better future: An ecosystem perspective on the ethics of AI and emerging digital technologies*, 55-79.

Szolomicki, J., & Golasz-Szolomicka, H. (2019). Technological advances and trends in modern high-rise buildings. *Buildings (Basel, Switzerland)*, *9*(9), 193. DOI: 10.3390/buildings9090193

Tarr, A. A., Tarr, J. A., Thompson, M., & Ellis, J. (Eds.). (2021). *Drone law and policy: Global development, risks, regulation and insurance*. Routledge. DOI: 10.4324/9781003028031

. Tsamados, A., Aggarwal, N., Cowls, J., Morley, J., Roberts, H., Taddeo, M., & Floridi, L. (2021). The ethics of algorithms: key problems and solutions. *Ethics, governance, and policies in artificial intelligence*, 97-123.

Tsiamis, N., Efthymiou, L., & Tsagarakis, K. P. (2019). A comparative analysis of the legislation evolution for drone use in OECD countries. *Drones (Basel)*, *3*(4), 75. DOI: 10.3390/drones3040075

Tubis, A. A., Poturaj, H., Dereń, K., & Żurek, A. (2024). Risks of Drone Use in Light of Literature Studies. *Sensors (Basel)*, *24*(4), 1205. DOI: 10.3390/s24041205

Umlauf, R., & Burchardt, M. (2022). Infrastructure-as-a-service: Empty skies, bad roads, and the rise of cargo drones. *Environment and Planning A. Economy and Space*, *54*(8), 1489–1509.

Uzorka, A., & Kalabuki, K. (2024). The transformative impact of technological advancements in educational leadership on student experiences and outcomes. *Education and Information Technologies*, *29*(16), 1–22. DOI: 10.1007/s10639-024-12714-y

Vargas, J., Alsweiss, S., Toker, O., Razdan, R., & Santos, J. (2021). An overview of autonomous vehicles sensors and their vulnerability to weather conditions. *Sensors (Basel)*, *21*(16), 5397. DOI: 10.3390/s21165397

. Villa, A. (2022). Warehouse design in the omnichannel context: challenges, configurations and trends.

Wang, L., Zhong, H., Ma, W., Abdel-Aty, M., & Park, J. (2020). How many crashes can connected vehicle and automated vehicle technologies prevent: A meta-analysis. *Accident; Analysis and Prevention*, *136*, 105299. DOI: 10.1016/j.aap.2019.105299

Wang, N., Christen, M., & Hunt, M. (2021). Ethical considerations associated with "humanitarian drones": A scoping literature review. *Science and Engineering Ethics*, *27*(4), 51. DOI: 10.1007/s11948-021-00327-4

Wang, N., Mutzner, N., & Blanchet, K. (2023). Societal acceptance of urban drones: A scoping literature review. *Technology in Society*, *75*, 102377. DOI: 10.1016/j.techsoc.2023.102377

Wang, T., Ramik, D. M., Sabourin, C., & Madani, K. (2012). Intelligent systems for industrial robotics: Application in logistic field. *The Industrial Robot*, *39*(3), 251–259. DOI: 10.1108/01439911211217071

Wawrla, L., Maghazei, O., & Netland, T. (2019). Applications of drones in warehouse operations. *Whitepaper. ETH Zurich, D-MTEC*, *212*, 1–12.

West, J. P., & Bowman, J. S. (2016). The domestic use of drones: An ethical analysis of surveillance issues. *Public Administration Review*, *76*(4), 649–659. DOI: 10.1111/puar.12506

. Willems, L. (2021). Understanding the impacts of autonomous vehicles in logistics. *The Digital Transformation of Logistics: Demystifying Impacts of the Fourth Industrial Revolution*, 113-127.

Wilson, R. L. (2014, May). Ethical issues with use of drone aircraft. In *2014 IEEE International Symposium on Ethics in Science, Technology and Engineering* (pp. 1-4). IEEE. DOI: 10.1109/ETHICS.2014.6893424

Yoon, B., Kim, H., Youn, G., & Rhee, J. (2023). 3d position estimation of objects for inventory management automation using drones. *Applied Sciences (Basel, Switzerland)*, *13*(19), 10830. DOI: 10.3390/app131910830

Zhang, Q., Zhang, J., Zaccour, G., & Tang, W. (2018). Strategic technology licensing in a supply chain. *European Journal of Operational Research*, *267*(1), 162–175. DOI: 10.1016/j.ejor.2017.11.036

Zieher, S., Olcay, E., Kefferpütz, K., Salamat, B., Olzem, S., Elsbacher, G., & Meeß, H. (2024). Drones for automated parcel delivery: Use case identification and derivation of technical requirements. *Transportation Research Interdisciplinary Perspectives*, *28*, 101253. DOI: 10.1016/j.trip.2024.101253

Chapter 14
A Technique for Collecting and Analyzing Physico-Chemical Parameters of Soils Based on Wireless Sensor Network Technology

Sergey Melnikov
http://orcid.org/0000-0002-1669-8947
North-Caucasus Federal University, Russia

Svetlana Govorova
http://orcid.org/0000-0002-3225-1088
North-Caucasus Federal University, Russia

Egor Govorov
http://orcid.org/0009-0006-7046-8217
North-Caucasus Federal University, Russia

ABSTRACT

Taking into account soil and climatic conditions, based on measuring the physico-chemical parameters of the soil, meteorological parameters and processing statistical data on climatic conditions, makes it possible to ensure the completeness of the implementation of plant biopotential due to the rational choice of parameters and timing of agrotechnical measures. The controlled parameters vary significantly in the frequency of measurements and the accuracy of the representation. The results

DOI: 10.4018/979-8-3373-2240-7.ch014

of the analysis of the physico-chemical parameters of soils and climatic conditions are summarized in a summary table, the contents of which allow calculations based on the volume of transmitted traffic and the speed of the wireless sensor network channels. Various standards of low-power wireless networks were considered. Based on the specifications of this standard, the effective data transfer rate, current consumption and operating time of the devices were determined at the declared and calculated data transfer rates.

1 GENERAL STATEMENT OF THE SCIENTIFIC TASK OF DEVELOPING THE METHODOLOGY

Modern high-performance agriculture is based on digital (precision) farming approaches (precision agriculture, satellite farming or site specific crop management). These approaches are aimed at taking into account the intra-field heterogeneity of crop productivity, measured in units and tens of meters, clarifying the causes of such heterogeneity and implementing targeted measures to solve the problems that have arisen (Ladoni et al., 2009; Montgomery, ; Nield et al., 2007).

Information on the quality of land is used in decision-making both at the level of an individual agricultural enterprise to assess fertilizer needs and organize the farming system, and at the level of an administrative district or region as a whole to analyze the degree of land degradation, develop measures to ensure the reproduction of soil fertility, and predict yields. The range of tasks solved on the basis of this information is extensive. Since the information used to analyze the quality of land resources is spatially distributed, when developing technologies for storing, processing, entering and outputting such information, it is necessary to rely on network geoinformation technologies (GIS technologies). In this area, it is important to develop tools that provide not only information input, but also the calculation of various auxiliary indicators that characterize the dynamics of soil cover changes in individual fields, land users and the area as a whole, display changes in cartographic form, and build ratings reflecting the level of reproduction of soil fertility. At the same time, to ensure high efficiency and accuracy of information on the physico-chemical parameters of soils and the state of plant biomass distributed over large areas, it is necessary to combine information from stationary sensors, space (aviation) sensing systems of the Earth's surface, information obtained in laboratory conditions, with data obtained from UAVs equipped with measuring equipment. The purpose of developing a methodology for collecting and analyzing physico-chemical parameters of soils based on wireless sensor network technology is to identify techniques and methods for practical and theoretical research of the influence of various physico-chemical parameters of soil on crop production, reflecting all stages of research: observation and collection of material, selection of units of analysis and determination of their properties, method of description, methods of analysis and the nature of the interpretation of the phenomenon under study. The task of developing a methodology for collecting and analyzing physico-chemical parameters of soils based on wireless sensor network technology is formulated as follows.

A set of $A=\{a\}$ agro-industrial enterprises engaged in the cultivation of agricultural products using innovative precision farming methods is given. At each enterprise, for each plant crop $K=\{k\}$ and the specified characteristics of soil fertility, there is a set of rational agrotechnical measures $M_k^a \in \{M\}$, leading to the desired value of the selected quality indicator.

It is necessary to find a limited set of physico-chemical parameters $P = \{p_i\}, i = \overline{1,n}$ and decisive rules $\Pi = \{n_j\}, j = \overline{1,J}$, ensuring the development of managerial decisions with a given degree of validity (reliability), such that $\mathrm{optm}_k^a(\mathrm{P},\Pi) = \max \prod_э W_э(U)$. At the same time, the parameters included in the found set must be measurable with the required accuracy $\Delta \mathrm{p}_i \leq \Delta \mathrm{p}_{зад}$ and efficiency $Tp_i = \min\{T_{изм}\}$. Based on the selected data set, it is necessary to develop requirements for $\mathrm{QoS}_i \geq \mathrm{QoS}_{зад}$, including requirements for the performance of the sensor network, its topology and protocols that ensure the functioning of the network in conditions of partial or fully autonomous power supply.

The efficiency of measurements is determined by the time interval from the moment of receiving the measurement results to the moment of the beginning of the implementation of management decisions based on the interpretation of the results obtained. The measurement accuracy is determined by the absolute or relative error, depending on the measured parameter. The nature of the stored data representation is a distributed multi–level knowledge base. End-user service is provided upon request. At the same time, the cost of using data should not lead to a decrease in the profitability of production.

The decomposition of the scientific task consists in the implementation of the following specific tasks: – analysis of the physico-chemical parameters of soils essential for assessing the state of biomass by the main types of crops; – justification of the requirements for the characteristics of a wireless sensor network; – development of methods for collecting and analyzing physico-chemical parameters of soils based on wireless sensor network technology.

2 ANALYSIS OF PHYSICO-CHEMICAL PARAMETERS OF SOILS ESSENTIAL FOR REALIZING THE BIOLOGICAL POTENTIAL OF PLANTS

The selection of a set of physico-chemical parameters to be collected and analyzed is based on the classical agrotechnical analysis of agricultural soils (Ladoni et al., 2009; Montgomery, ; Nield et al., 2007). The main objectives of agrochemical soil analysis of agricultural lands are:

1. assessment of the degree of crop availability of basic nutrients;
2. calculation of fertilizer doses for the planned increase in yield;
3. assessment of the really possible crop yields;
4. assessment of the degree of contamination by heavy metals, radionuclides of soils;
5. assessment of the degree of land degradation;
6. rating assessment of the level of soil fertility of agricultural lands and its dynamics.

The qualitative characteristics of the soils of a particular land plot include the following indicators:

1. the availability of nutrients (micro - and macronutrients);
2. the degree of soil exposure to degradation processes (water and wind erosion, acid-base balance, degree of salinity, etc.);
3. the degree of contamination with heavy metals and radionuclides;
4. information on agricultural technologies for cultivating crops (crop rotation, soil bonus, fertilizer application standards for each field to increase yields);
5. zoning of soils according to various indicators, for example, electrical conductivity.

The requirements for the mineral composition, humidity, thickness of the humus layer, granulometric composition and salinity of the soil vary significantly depending on the type of crops grown, the selected variety, as well as the stage of their life cycle. To identify common patterns, it is necessary to analyze the requirements for fertility factors for all leading field crops grown in the Stavropol Territory, such as winter wheat, sunflower, corn, sugar beet and potatoes. To determine the range of acceptable values of key parameters affecting yield, and to evaluate the possibility of measuring them, with a given accuracy and in a timely manner.

The biological features of the vegetation of basic crops differ significantly, which makes it possible to choose the most productive crops in each specific case of soil and climatic conditions. This requires accurate predictive models of climatic conditions and constant monitoring of the physico-chemical parameters of soils and the state of plant biomass.

The process of plant nutrition significantly depends on the soil moisture in the area of the root system of plants, since the main form of mineral consumption is a solution. Soil moisture is the most variable factor, based on its measurements at almost every stage of the plant life cycle, it is possible to choose the best parameters and timing of agrotechnical measures. At the same time, the soil horizons in which measurements need to be made are located from the groundwater level to the surface.

Taking into account soil and climatic conditions, based on measuring the physico-chemical parameters of the soil, meteorological parameters and processing statistical data on climatic conditions, makes it possible to ensure the completeness of the implementation of plant biopotential due to the rational choice of parameters and timing of agrotechnical measures.

The total number of parameters, divided into three groups: climatic conditions of cultivation, physical parameters of the soil and chemical elements of plant nutrition, is more than a hundred for each of the crops considered. Most of the parameters are similar, the difference lies in the amount of elements required for the normal development of plants.

In accordance with existing regulatory documents, the measurement of parameters that determine the growing conditions of agricultural plants is divided into two types: primary agrochemical soil survey, conducted once every five years; annual survey during operational monitoring of soil fertility during plant vegetation. When conducting state soil monitoring, agrochemical indicators of soils are mandatory for agricultural producers:

1. the content of organic matter in the arable horizon;
2. pH of the aqueous and saline suspension;
3. the content of mobile phosphorus and exchangeable potassium.

The controlled parameters vary significantly in the frequency of measurements and the accuracy of the representation. The results of the analysis of the physico-chemical parameters of soils and climatic conditions are summarized in a summary table, the contents of which allow calculations based on the volume of transmitted traffic and the speed of the wireless sensor network channels.

Table 1. Summary table of parameters to be monitored during the growing season

Winter wheat	1,2-2,0/ 0,6	+++ long day	70-75	30	6.0-7.5	0,6-0,8	6,0	- Chernozems - Dark brown hair - Slightly podzolic - Medium loamy - Grey forest 2,0-2,5
Sunflower seeds	2,0-4,0/ 0,5	+++ short day	60-70	95-125	6.0-6.8	0,3-0,6	5,2	- Sandy loam - Loamy - Brown hair - Grey forest - Podzolic 5.0-5.5
Corn (for grain)	3,0-4,0/ 0,3-0,6	+++ short day	75-80	70-80	5.5-7.5	0,6-0,8	1,7	- Medium and light loamy soils - Sandy loam - Sandy at least 1.8
Sugar beet	2,0-2,5/ 0,6-0,8	+++ long day	30-40	65-75	6.5-7.5	1,2-1,6	7,0	- Chernozems - Brown hair - Grey forest - Dark grey forest - Loamy 2,0-2,5
Potato	0,3-0,7/ 0,5	++ short day	70-80	92-118 62-96	5.5-6.5	0,15-0,25	1,7	- Chernozems - Sandy loam - Loamy - Grey forest 3,5-4,0

Culture Root zone depth/consumption depth, m Illumination The lowest moisture capacity, % Productive moisture reserve, mm pH

Salinity, % Electrical conductivity of water-saturated soil, ECmin, dSm/m Soil type/ humus content at a depth of 0-0.3 m,%

Table 2. Summary table of parameters to be monitored during the growing season

Culture	Soil density, g/cm³	Optimal content of mobile forms of nutrients in the soil, mg/kg in a soil layer of 0-60 cm * (according to Machigin)	Temperature Physiological minimum/optimal range/critical temperature, deg.C				
			Germination	Sprouting	Tillering/ stem growth	Earings/ blooms	Maturation
inter wheat	1,05-1,3	N >50 P >60 K>600 Ca! Mg! Mn! B! Cu! S Mo! Zn! Fe!	1-2/ 12-16/ -25,+25	1-2/ 12-16/ -	8-10/ 13-18/ -	10-24/ 18-20/ -3	10-24/ 22-25/ -3,+35
Sunflower seeds	1,1-1,3	N 30-40 P 45-60 K 400-600 Ca Mg! Mn! B! Cu! Co! S! Mo! Zn! Fe!	4-6/ 20-25/ -6	8-10/ 20-25/ -6	14-16/ 20-25/ -6	14-18/ 20-25/ -2	16-18/ 20-25/ -2,+30
Corn (for grain)	1,1-1,3	N 30-40 P 45-60 K 400-600 Ca Mg! Zn! Mn B! Cu! S! Mo	4-6/ 15-20/ -2	6-8/ 15-20/-6	8-10/ 25-30/ 3	10-12/ 20-25/ -2	6-8/ 20-25/ -2
Sugar beet	1,0-1,3	N >50 P >60 K>600 Ca! Mg! Mn! B! Cu S! Mo!	3-4/ 15-18/ -3	8-10/ 15-23/ -6	16-20/ 25-30/ -2	16-20/ 25-30/ -2	8-10/ 25-30/ -
Potato	0,9-1,2	N 40-50 P 55-60 K 600-700 Ca Mg! Mn B! Cu S! Mo	3-4/ 20-22/ -6	5-6/ 20-22/ -2	7-12/ 17-22/ -1,+50	18-20/ 17-22/ -1,+50	10-16/ 16-20/ +30

3 CALCULATION OF REQUIRED WIRELESS SENSOR NETWORK CHARACTERISTICS

The use of wireless sensor networks (BSN) is one of the most actively developing methods of data collection and transmission. Sensor networks are already finding applications in agriculture in the field of wireless monitoring. The existing BSS hardware and software complexes are mainly designed to solve highly specialized tasks, which complicates the process of adapting systems to changing conditions. Moreover, even if small modifications are required, it is necessary to go through all the design stages anew. At the moment, there are no unified BSS design tools. Each company uses its own proprietary software to analyze development characteristics. For this reason, solutions focused on one specific task cannot be used in another subject area. A wireless sensor network is a distributed, self–organizing set of multiple sensors and actuators connected via a radio channel. The coverage area of such a network can range from several meters to several kilometers due to the relay of messages from one element to another (Balanis, 2005; Kao-Cheng & Edwards, 2008; LoRaWAN Alliance, 2017). To implement a BSS-based data collection system, it is necessary to develop requirements for the basic characteristics of the network.

3.1 Calculation of the required sensor network performance

The sensor network performance is determined by the amount of data being transmitted, the data transfer rate, and the maximum allowable delay, determined by the loss time of the transmitted data. The amount of data transmitted from one stationary sensor in bits can be determined using the expression:

$$V_d = m_i N_i n \tag{1}$$

where N_i- is the number of parameters measured by the sensor; m_i- the number of measuring instruments of the same type located on different soil horizons; n- representation of parameters.

The gateway traffic volume is determined by the expression:

$$V_s = \mathrm{M}\mathrm{V}_d \tag{2}$$

where M – is the number of serviced sensors.

3.2 Calculation of energy consumption and operating time

Energy consumption is one of the key issues for sensor networks, as the end devices (sensors) are usually powered by autonomous sources (mainly batteries and accumulators). The energy consumption is estimated according to three modes of operation of the terminal device: data collection by the microcontroller and data preparation for transmission; data transmission using one of the wireless communication methods; waiting in low-power mode for the next data collection or transmission session (sleep mode).

The choice of wireless sensor network parameters is determined by the parameters of the radio path and the power consumption of the end device. The quality of the radio channel is assessed to ensure high service quality indicators of the BSS.

In general, the evaluation of the radio channel quality is performed in 2 stages: parameter monitoring and metric calculation. The radio channel parameters can be monitored in active, passive and hybrid modes. Active mode involves monitoring connections with other nodes by transmitting test packets at a certain frequency. In passive mode, the traffic existing on the network is analyzed, while packets that can be received by the node are considered, even if they are not addressed to it. Hybrid monitoring is a combination of the two previous options.

The main requirements for the radio channel quality assessment requirements are energy efficiency, accuracy and reactivity. In LoRaWAN networks, the receiving and transmitting parameters are selected based on the amount of information transmitted by the sensor. In this case, the main parameter is the energy efficiency of the sensor.

Thus, the choice of the parameters of an energy-efficient BSS will be based on the developed hardware solution.

The maximum payload size depends on the length of the service fields of the frame. The maximum length of a LoRa packet specified by the LoRa Alliance standard is 256 bytes. If the frame has the most filled fields, the payload length can be no more than 223 bytes. With a minimum field content, the payload length can be no more than 242 bytes.

The standard defines channel speeds of 300 mbit/s, 1200 bit/s, 2400 bit/s, 4800 bit/s, 9600 bit/s, 19200 bit/s, 38400 bit/s, 62500 bit/s. The choice of channel speed depends on the bitrate, which is calculated using the following formula

$$n_{\text{symbol}} = (8 + 4.25) + 8 + \max\left(ceil\left(\frac{8 \cdot 14 - 4 \cdot 8 + 28 + 16 - 20 \cdot 0}{4(8 - 2 \cdot 0)}\right)(1 + 4), 0\right) = 40 symbol \quad (3)$$

From the bitrate calculation, it can be concluded that the optimal channel speed will be 4.8 kbit/s.

The time it takes to transfer data depends on the number of characters sent in the LoRa frame and the time of a single character.

$$T_{\text{ToA}} = n_{\text{symbol}} \cdot T_{\text{symbol}}$$

where n_{symbol} – the number of characters present in the LoRa frame;

$$T_{\text{symbol}} = \frac{2^{\text{SF}}}{\text{Baddwidth}}$$

the time of one character.

n_{symbol} – it depends on many parameters of LoRa and can be expressed by the following formula:

$$n_{\text{symbol}} = \left(n_{\text{preampel}} + 4.25\right) + 8 + \max\left(\text{ceil}\left(\frac{8 \cdot \text{Payload} - 4 \cdot \text{SF} + 28 + 16 - 20 \cdot H}{4(\text{SF} - 2 \cdot \text{DE})}\right)(\text{CR} + 4), 0\right) \quad (4)$$

where Payload– this is the LoRa PHY payload; SF– expansion coefficient;

$H = 1$, when the header is enabled, otherwise $H = 0$;
DE = 1, when low data transfer rate optimization is enabled, otherwise 0;
CR – the encoding speed is from 1 to 4.

The calculation results are summarized in Table 3.

Table 3. Calculation results

Parameter, unit of measurement (designations)	**Values**
n_{symbol}	8
Payload	14
SF	8
H	0
DE	0
CR	1

$$n_{\text{symbol}} = (8 + 4.25) + 8 + \max\left(ceil\left(\frac{8 \cdot 14 - 4 \cdot 8 + 28 + 16 - 20 \cdot 0}{4(8 - 2 \cdot 0)}\right)(1 + 4), 0\right) = 40 symbol$$

$$T_{\text{symbol}} = \frac{2^{\text{SF}}}{\text{Bandwidth}} = \frac{2^8}{125 \cdot 10^3} = 2.048\text{мс}$$

$$T_{\text{ToA}} = n_{\text{symbol}} \cdot T_{\text{symbol}} = 40.25 \cdot 2.048 \cdot 10^{-3} = 82.432\text{мс}$$

The average power of the device during the entire message transmission procedure will be equal to:

$$P_f = \frac{\left[(I_a \cdot U_p) \cdot t_{\text{wait}} + (I_{\text{TX}} \cdot U_p) \cdot t_{\text{CCa}} + (I_{\text{tx}} \cdot U_p) \cdot t_{\text{data}}\right]}{t_f}, \quad (5)$$

$$t_f = t_{\text{wait}} + t_{\text{CCA}} + t_{\text{data}}, \quad (6)$$

where I_a – current in active mode, mA; U_p – supply voltage, V; I_{rx} – current in reception mode, mA; I_{tx} – current in transmission mode, mA; t_{wait} – time to switch between operating modes, ms; t_{CCA} – WOR mode operation time, ms; t_{data} – message transmission time, ms; t_f – total operating time of the transceiver, ms.

The calculation results are summarized in Table 4.

Table 4. Typical device parameter values

Parameter, unit of measurement (designations)	Values
Processor frequency Arduino, MHz	8
Current in reception mode, mA (I_{rx})	15
Supply voltage, V (U_p)	4.2
Current in transmission mode, mA (I_{tx})	650
Current in active mode, mA (I_a)	15
Current in sleep mode, mA (I_s)	3
Time to switch between operating modes, ms (t_{wait})	1500
Operating time of the WOR mode, ms (t_{CCA})	1000
On-air time, ms (T_{ToA})	82.432

Let's calculate the theoretical values according to the formula (5).

$$P_f = \frac{[(15 \cdot 4.2) \cdot 1500 + (15 \cdot 4.2) \cdot 1000 + (650 \cdot 4.2) \cdot 82.432]}{1500 + 1000 + 82.432} \approx 115\text{мВ}$$

Device lifetime:

$$T_l = \frac{E_0}{P_{ed}} \quad (7)$$

where E_0 – battery capacity; P_{ed} – the average power consumed by the device.

$$P_{ed} = \frac{P_f t_{real} + (I_a \cdot U_p) \cdot t_a + (I_s \cdot U_p)(t_c + t_{read} + t_a)}{t_c} \quad (8)$$

$$P_{ed} = \frac{115 \cdot 3.98 + (15 \cdot 4.2) \cdot 2582.432 + (3 \cdot 4.2)(86400000 + 3.98 + 2582.432)}{86400000} = 12.602\text{мВт}$$

From here,

$$T_l = \frac{E_0}{P_{ed}} = \frac{1500}{12.602} = 119.02\text{дней}$$

Table 5. Calculation results

Parameter, unit of measurement (designations)	**Values**
Number of characters transmitted (nsimbol)	40
Transmission time of one character, ms (Tsimbol)	2.048
On-air time, ms (T_{ToA})	82.432
Average device power, W (P_f)	0.65
Total operating time of the transceiver, ms (t_f)	85.432
Arduino code execution time, s (t_a)	1
Device lifetime, days (T_l)	119.02
Average power consumed by the device, milliWatts (*Ped*)	12.602

To assess the correctness of the calculations performed, the energy consumption of the developed device was analyzed using the INA219 device.

INA219 is a voltage, current and power sensor manufactured by Texas Instruments. This microcircuit allows you to measure voltage in the range of 0...26 volts and current with high accuracy. The connection diagram is shown in Figure1.

Figure 1. INA219 connection diagram for measuring energy consumption parameters

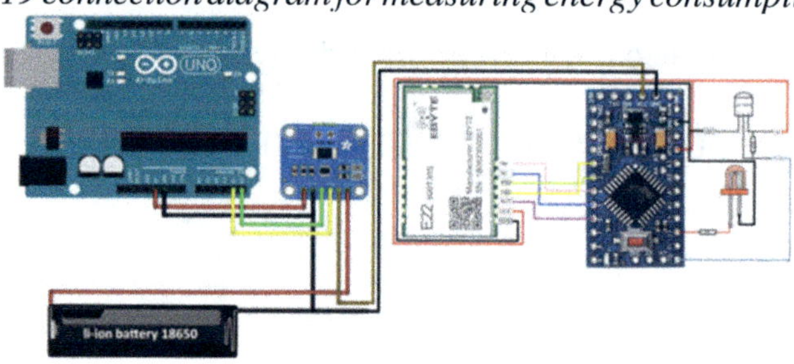

The graph of the energy consumption of the earned sensor when taking readings from the temperature sensor (see Figure 2) and transmitting data to the gateway is shown in Figure3.

Figure 2. Measurement of the actual energy consumption of the developed sensor

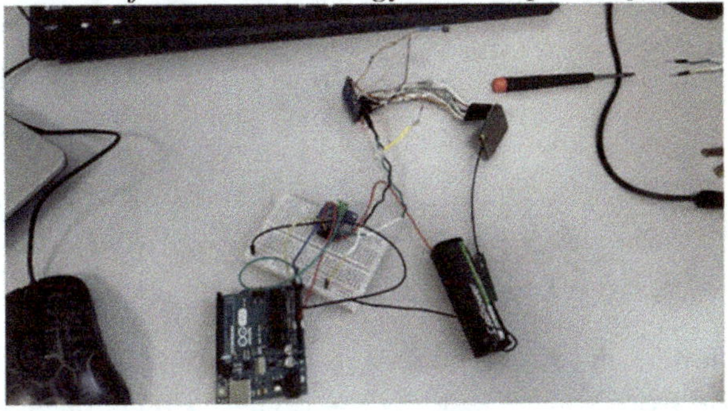

Figure 3. Graph of the energy consumption of the earned sensor

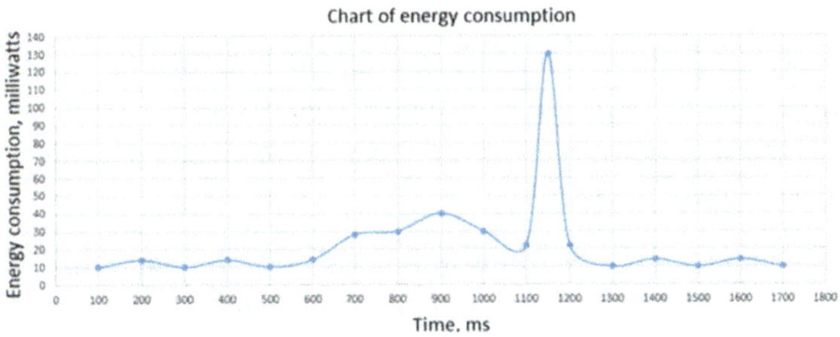

4. CONCLUSIONS

Various standards of low-power wireless networks were considered. The most promising is the IEEE 802.15.4-2006 standard. Based on the specifications of this standard, the effective data transfer rate, current consumption and operating time of the devices were determined at the declared and calculated data transfer rates.

The developed methodology for collecting and analyzing physico-chemical parameters of soils based on wireless sensor network technology, designed to measure physico-chemical parameters of the soil that are critical for the realization of the biopotential of seeds, to collect and analyze the measured data in order to make recommendations on the parameters and timing of agrotechnical measures.,

Together with the methodology for assessing the state of plant biomass based on the integration of multispectral optical and radio band sensors, it is used in the operation of an information management system in order to ensure the completeness of the implementation of plant biopotential.

REFERENCES

Balanis, C. A. (2005). *Antenna theory: Analysis and design* (3rd ed.). Wiley-Interscience.

Kao-Cheng, H., & Edwards, D. J. (2008). *Millimetre wave antennas for gigabit wireless communications: A practical guide to design and analysis in a system context*. John Wiley & Sons Ltd.

Ladoni, M., Bahrami, H. A., Alavipanah, S. K., & Norouzi, A. A. (2009). Estimating soil organic carbon from soil reflectance: A review. *Precision Agriculture, 11*(1), 82–99. DOI: 10.1007/s11119-009-9123-3

LoRaWAN Alliance. (2017). *LoRaWAN® specification v1.1*. https://hz137b.p3cdn1.secureserver.net/wp-content/uploads/2020/11/lorawantm_specification_-v1.1.pdf?time=1687464361

Montgomery, O. L. (n.d.). *An investigation of the relationship between spectral reflectance and the chemical, physical and genetic characteristics of soils* (PhD dissertation, Department of Agronomy, Purdue University). https://docs.lib.purdue.edu/dissertations/AAI7701747/

Nield, S. J., Boettinger, J. L., & Ramsey, R. D. (2007). Mapping gypsic and natric soil areas using Landsat ETM data. *Soil Science Society of America Journal, 71*, 245–252. DOI: 10.1007/978-1-4020-8592-5_16

Compilation of References

Abidi, M., Al-Ahmari, A., Ahmad, A., Ameen, W., & Alkhalefah, H. (2019). Assessment of Virtual Reality-Based Manufacturing Assembly Training System. *International Journal of Advanced Manufacturing Technology*, *105*(9), 3743–3759. DOI: 10.1007/s00170-019-03801-3

Abrashkina, E. D., Agirbov, Y. I., Andreev, O. P., ... (2021). Agroindustrial Sector of Russia: Agriculture 4.0: Monograph in 2 volumes. Moscow: IPR Media, 379.

Adebayo, A. S., Chukwurah, N., & Ajayi, O. O. (2025). Artificial intelligence and machine learning algorithms for advanced threat detection and cybersecurity risk mitigation strategies. *Engineering and Technology Journal*, *10*(3), 4080–4094. DOI: 10.47191/etj/v10i03.18

Afşar Doğrusöz, L., Yazici, S., & Gür Omay, E. G. (2022). Organizational Resilience in Healthcare Organizations: A Research in the Public and Private Sector. *Afet ve Risk Dergisi*, *5*(1), 318–329. DOI: 10.35341/afet.1083886

Agibalov, A. V., Zaporozhtseva, L. A., & Tkacheva, Yu. V. (2019). Scenario approach to setting strategy for rural sustainable development. *Vestnik of Voronezh State Agrarian University*, *12*(3(62)), 94–102. DOI: 10.17238/issn2071-2243.2019.3.94

Agibalov, A. V., Zaporozhtseva, L. A., & Tkacheva, Yu. V. (2020). Formation of a methodology for assessing the quality of sustainability of rural development. *International Agricultural Journal*, *63*(1), 6. https://doi.org/DOI: 10.24411/2588-0209-2020-10130

Ahmad Mir, T., & Singh, M. (2022). Indian Healthcare Sector and the Sustainable Development. *International Journal of Current Research and Review*, *14*(03), 43–47. DOI: 10.31782/IJCRR.2022.14308

Akca, A., Kara, I., & Aydos, M. (2020, July). Privacy, Security and Legal Aspects of Autonomous Vehicles. In *International Conference on Materials Science, Mechanical and Automotive Engineerings and Technology (IMSMATEC'20)* (p. 20).

Aker, J. C., & Mbiti, I. M. (2010). Mobile phones and economic development in Africa. *The Journal of Economic Perspectives*, *24*(3), 207–232. DOI: 10.1257/jep.24.3.207

Aksenova, E. I., Kamynina, N. N., & Turzin, P. S. (2023). Urban stress of a megalopolis resident: Stress factors and correction. *Probl Sotsialnoi Gig Zdravookhranenniia i Istor Med, 31*(Special Issue 2), 1087–1091. DOI: 10.32687/0869-866X-2023-31-s2-1087-1091

Akubue, J. (2019). Challenges of mega cities. *Modern Economy*, *10*(9), 2031–2050. DOI: 10.4236/me.2019.109128

Alali, M., Almogren, A., Hassan, M. M., Rassan, I. A., & Bhuiyan, M. Z. A. (2018). Improving risk assessment model of cyber security using fuzzy logic inference system. *Computers & Security*, *74*, 323–339. DOI: 10.1016/j.cose.2017.09.011

Alberstadt, R. (2014). Drones under international law. *Open Journal of Political Science*, *4*(04), 221–232. DOI: 10.4236/ojps.2014.44023

Alfasi, N., & Margalit, T. (2021). Toward the sustainable metropolis: The challenge of planning regulation. *Sustainability*, *13*(15), 1–21. DOI: 10.3390/su13158189 PMID: 34804601

Al-Hawamleh, A. (2024). Cyber resilience framework: Strengthening defenses and enhancing continuity in business security. *International Journal of Computing and Digital Systems*, *15*(1), 1315–1331. DOI: 10.12785/ijcds/150193

Ali, S. S., Khan, S., Fatma, N., Ozel, C., & Hussain, A. (2024). Utilisation of drones in achieving various applications in smart warehouse management. *Benchmarking*, *31*(3), 920–954. DOI: 10.1108/BIJ-01-2023-0039

Alkhwaldi, A. F., Al-Qudah, A. A., Al-Hattami, H. M., Al-Okaily, M., Al-Adwan, A. S., & Abu-Salih, B. (2024). Uncertainty avoidance and acceptance of the digital payment systems: A partial least squares-structural equation modeling (PLS-SEM) approach. *Global Knowledge, Memory and Communication*, *73*(8/9), 1119–1139. DOI: 10.1108/GKMC-07-2022-0161

Alliance for Financial Inclusion. (2021). Digital Financial Literacy. Retrieved from https://www.afi-global.org/publications/digital-financial-literacy/

Almuhairat, A., Alti, A., & Alswailim, M. (2024). Improving cultural awareness and trust towards m-banking apps in Jordan. *Bulletin of Electrical Engineering and Informatics*, *13*(5), 3695–3704. DOI: 10.11591/eei.v13i5.7824

Alper, A., Pinar, O., & Marin Espinosa, A. G. (2020). Electronic Payments Acceptance Incentives–Literature Review and Country Examples. World Bank Group. https://documents.worldbank.org/en/publication/documents-reports/documentdetail/793041588860799013/Electronic-Payments-Acceptance-Incentives-Literature-Review-and-Country-Examples

Alzoubi, R. H., Shnaikat, B., Abo Murad, M., Al Kloub, R., Alhmesat, M., & Alkharabsheh, O. H. M. (2024). Rethinking growth and survival of hospitality industry in Jordan: The role of responsible and resilient leadership styles in managing crisis. *Cogent Business & Management*, *11*(1), 2402080. Advance online publication. DOI: 10.1080/23311975.2024.2402080

Amaro, P., Anabela, C., & Rui, M. (2019). Lean Thinking: A Transversal and Global Management Philosophy to Achieve Sustainability Benefits. In Alves, A., Kahlen, F. J., Flumerfelt, S., & Siriban-Manalang, A. (Eds.), *Lean Engineering for Global Development* (pp. 1–31). Springer. DOI: 10.1007/978-3-030-13515-7_1

Amiri, Z., Heidari, A., Zavvar, M., Navimipour, N. J., & Esmaeilpour, M. (2024). The applications of nature-inspired algorithms in Internet of Things-based healthcare service: A systematic literature review. *Transactions on Emerging Telecommunications Technologies*, *35*(6), e4969. DOI: 10.1002/ett.4969

Anderson, J. (2020). *U.S. Patent No. 10,815,080*. Washington, DC: U.S. Patent and Trademark Office.

Angeler, D. G., Eyre, H. A., Berk, M., Allen, C. R., Hynes, W., & Linkov, I. (2022). Adaptation, Transformation and Resilience in Healthcare Comment on "Government Actions and Their Relation to Resilience in Healthcare During the COVID-19 Pandemic in New South Wales, Australia and Ontario, Canada.". *International Journal of Health Policy and Management*. Advance online publication. DOI: 10.34172/ijhpm.2022.7043 PMID: 35247939

Ang, J., Goh, C., Saldivar, A., & Li, Y. (2017). Energy-Efficient Through-Life Smart Design, Manufacturing and Operation of Ships in an Industry 4.0 Environment. *Energies*, *10*(5), 610. DOI: 10.3390/en10050610

Anwar, S., Alburaiki, G., Ibrahimi, A., & Qadri, L. (2024)... *Digital Banking Issues Experienced by Older People*, *1–5*, 1–5. Advance online publication. DOI: 10.1109/ICOA62581.2024.10754510

Anyangwe, T., Vanroose, A., & Fanta, A. (2022). Determinants of financial inclusion: Does culture matter? *Cogent Economics & Finance*, *10*(1), 2073656. DOI: 10.1080/23322039.2022.2073656

APQC. 2019. *Process Classification Framework*. Houston, TX: American Productivity & Quality Center. Accessed June 16, 2019. https://www.apqc.org/resource-library/resource-listing/apqc-process-classification-framework-pcf-cross-industry-excel-7

Apruzzese, G., Andreolini, M., Ferretti, L., Marchetti, M., & Colajanni, M. (2022). Modeling realistic adversarial attacks against network intrusion detection systems. *Digital Threats : Research and Practice*, *25*(4), 46. Advance online publication. DOI: 10.1145/3469659

Ardolino, M., Rapaccini, M., Saccani, N., Gaiardelli, P., Crespi, G., & Ruggeri, C. (2018). The Role of Digital Technologies for the Service Transformation of Industrial Companies. *International Journal of Production Research*, *56*(6), 2116–2132. DOI: 10.1080/00207543.2017.1324224

Armendáriz, B., & Morduch, J. (2010). *The Economics of Microfinance*. MIT Press.

Arner, D. W., Auer, R., & Frost, J. (2020). Stablecoins: risks, potential and regulation. SSRN *Electronic Journal*. https://dx.doi.org/DOI: 10.2139/ssrn.3979495

Arner, D. W., Barberis, J., & Buckley, R. P. (2015). The evolution of fintech: A new post-crisis paradigm. *SSRN*, *47*(4), 1271–1319. DOI: 10.2139/ssrn.2676553

Asongu, S. A., Nwachukwu, J. C., Orim, S. M., & Pyke, C. (2019). Crime and social media. *Information Technology & People*, *32*(5), 1215–1233. DOI: 10.1108/ITP-06-2018-0280

Asuamah Yeboah, S. (2024). Navigating the Nexus: Overcoming Challenges in Public-Private Partnerships for Sustainable Building Initiatives in Developing Countries. Munich Personal RePEc Archive. https://mpra.ub.uni-muenchen.de/122667/

Aurazo, J., & Gasmi, F. (2024). Digital payment systems in emerging economies: Lessons from Kenya, India, Brazil, and Peru. *Information Economics and Policy*, *69*, 101113. DOI: 10.1016/j.infoecopol.2024.101113

Avalle, G., De Pace, F., Fornaro, C., Manuri, F., & Sanna, A. (2019). An Augmented Reality System to Support Fault Visualization in Industrial Robotic Tasks. *IEEE Access : Practical Innovations, Open Solutions*, *7*, 132343–132359. DOI: 10.1109/ACCESS.2019.2940887

Awah, L. S., Belle, J. A., Nyam, Y. S., & Orimoloye, I. R. (2024). A systematic analysis of systems approach and flood risk management research: Trends, gaps, and opportunities. *International Journal of Disaster Risk Science*, *15*(1), 45–57. DOI: 10.1007/s13753-024-00544-y

Aysan, A. F., Ozturk, O., & Selim, N. H. (2025). Local power, global reach: The rise and impact of domestic payment networks in the financial landscape. *Journal of Science and Technology Policy Management, 16*(1), 143–162. DOI: 10.1108/JSTPM-12-2023-0227

Badilla, G. L., & Juan, M. T. G. (2021). "The Electronics Industry and Their Chemical Effects in the Environment of Arid Regions of Northwest of Mexico", Indian Journal of Advanced Chemistry (IJAC) ISSN: 2582-8975 (Online), Volume-1 Issue-2, October 2021.

Badilla, G. L., María, M. A. G., Samaniego, E. R., & Sandra, L. T. P. (2014). Microcorrosion and their effect in the operation in of industrial equipments in the electronics industry of Mexicali. In Aliofkhazraei, M. (Ed.), *Book Chapter INTECH, Full Book: Development of Corrosion Protection*.

Bajwa, I. A., Ahmad, S., Mahmud, M., & Bajwa, F. A. (2023). The impact of cyber-attacks awareness on customers' trust and commitment: An empirical evidence from the Pakistani banking sector. *Information and Computer Security, 31*(5), 635–654. DOI: 10.1108/ICS-11-2022-0179

Bakar, S. A., & Dorasamy, M. (2023). From adoption to sustainability: A journey of large-scale agile implementation. *International Journal of Technology, 14*(6), 1367–1379. DOI: 10.14716/ijtech.v14i6.6645

Balanis, C. A. (2005). *Antenna theory: Analysis and design* (3rd ed.). Wiley-Interscience.

Bandura, R., & Ramanujam, S. R. (21 September 2021). *Developing inclusive digital payment systems*. Center for Strategic & International Studies. Retrieved from https://www.csis.org/analysis/developing-inclusive-digital-payment-systems

Bandura, R., McLean, M., & Girgvliani, S. (12 March 2024). *Advancing digital transformation and digital public infrastructure: The role of the private sector*. Center for Strategic & International Studies. Retrieved from https://www.csis.org/analysis/advancing-digital-transformation-and-digital-public-infrastructure-role-private-sector

Bandura, A., & Wessels, S. (1997). *Self-efficacy*. Cambridge University Press.

Barnes, P. (2002). Approaches to community safety: Risk perception and social meaning. *Australian Journal of Emergency Management, 17*(1), 15–23. https://search.informit.org/doi/10.3316/informit.370758269308829

Baron, L., Rouleau, V., Grégoire, S., & Baron, C. (2018). Mindfulness and leadership flexibility. *Journal of Management Development, 37*(2), 165–177. DOI: 10.1108/JMD-06-2017-0213

Barreno, M., Nelson, B., Joseph, A. D., & Tygar, J. D. (2010). The security of machine learning. *Machine Learning, 81*(2), 121–148. DOI: 10.1007/s10994-010-5188-5

Batjargal, T., & Zhang, M. (2021). Review of key challenges in public-private partnership implementation. *Journal of Infrastructure. Policy and Development, 5*(2), 1378. DOI: 10.24294/jipd.v5i2.1378

Battistelli, F., & Galantino, M. G. (2019). Dangers, risks and threats: An alternative conceptualization to the catch-all concept of risk. *Current Sociology, 67*(1), 64–78. DOI: 10.1177/0011392118793675

Batty, M. (1995). The computable city. In *Fourth International Conference on Computers in Urban Planning and Urban Management* (July 11–14). www.acturban.org/biennial/doc_planners/computable_city.htm

Bellis, P., Cunial, M., & Trabucchi, D. (2024). Mastering hybrid worlds through digital leadership: The role of agility in fostering innovation. *Business Horizons, 67*(4), 369–380. DOI: 10.1016/j.bushor.2024.04.002

Bench-Capon, T. J. (2020). Ethical approaches and autonomous systems. *Artificial Intelligence, 281*, 103239. DOI: 10.1016/j.artint.2020.103239

Berghofer, G. (26 September 2023). *How clients' culture influences their financial behavior: Tips and tricks to improve your advisory*. Behavior quant. Retrieved from https://behaviorquant.com/en/understanding-how-clients-culture-influences-their-financial-behavior-tips-and-tricks-to-improve-your-advisory/

Berke, A., Ding, G., Chin, C., Gopalakrishnan, K., Larson, K., Balakrishnan, H., & Li, M. Z. (2023). Drone delivery and the value of customer privacy: A discrete choice experiment with US consumers. *Transportation Research Part C, Emerging Technologies, 157*, 104391. DOI: 10.1016/j.trc.2023.104391

Bernatovic, I., Slavec Gomezel, A., & Černe, M. (2022). Mapping the knowledge-hiding field and its future prospects: A bibliometric co-citation, co-word and coupling analysis. *Knowledge Management Research and Practice, 20*(3), 394–409. DOI: 10.1080/14778238.2021.1945963

Besshaposhny, M. N., Evgrafova, L. V., Lazar, V. V., Pichuzhkin, N. A., & Grachev, A. B. (2021). Innovative mechanism to increase the efficiency of indirect employment of the rural population. In *The Challenge of Sustainability in Agricultural Systems* (pp. 907–913). Springer International Publishing., DOI: 10.1007/978-3-030-73097-0_101

Beth, F. (2023). The Importance of Professional Communities and Collaborations. *Journal of the Advanced Practitioner in Oncology*, *14*(2), 115–116. DOI: 10.6004/jadpro.2023.14.2.1 PMID: 37009410

Bettencourt, L. M. A. (2019). The digital transformation of metropolises. *Metropolis Observatory, 8*, 1–16. https://www.metropolis.org/sites/default/files/resources/Observatory_Issue-Paper-08_Digital-transformation-metropolises.pdf

Bettencourt, L. M. A. (2014). The uses of big data in cities. *Big Data*, *2*(1), 12–22. DOI: 10.1089/big.2013.0042 PMID: 27447307

Better Than Cash Alliance. (2024). *UN Principles for Responsible Digital Payments*. Retrieved from https://responsiblepayments.org/

Better Than Cash Alliance. (30 November 2016). *Responsible Digital Payment Guidelines*. Retrieved from https://www.betterthancash.org/explore-resources/responsible-digital-payments-guidelines

Bhatia-Kalluri, A., & Caraway, B. R. (2023). Transformation of the digital payment ecosystem in India: A case study of Paytm. *Social Inclusion (Lisboa)*, *11*(3), 320–331. DOI: 10.17645/si.v11i3.6687

Bienhaus, F., & Abubaker, H. (2018). Procurement 4.0: Factors Influencing the Digitisation of Procurement and Supply Chains. *Business Process Management Journal*, *24*(4), 965–984. DOI: 10.1108/BPMJ-06-2017-0139

Bine, L. M., Boukerche, A., Ruiz, L. B., & Loureiro, A. A. (2024). Connecting Internet of Drones and urban computing: Methods, protocols and applications. *Computer Networks*, *239*, 110136. DOI: 10.1016/j.comnet.2023.110136

Bitar, M., & Tarazi, A. (2022). Individualism, formal institutional environments, and bank capital decisions. *Journal of Corporate Finance*, *76*, 102244. DOI: 10.1016/j.jcorpfin.2022.102244

Blinova, E. A. (n.d.). Steganographic method based on the line-shift coding method on non-displayed symbols of the electronic text document. *CyberLeninka.* https://cyberleninka.ru/article/n/steganograficheskiy-metod-na-osnove-izmeneniya-mezhdustrochnogo-rasstoyaniya-neotobrazhaemyh-simvolov-strok-elektronnogo-tekstovogo/viewer

Blinova, E. A., & Suschenia, A. A. (n.d.). Several steganographic methods using for embedding of hidden data in electronic text documents. *CyberLeninka.* https://cyberleninka.ru/article/n/primenenie-neskolkih-steganograficheskih-metodov-dlya-osazhdeniya-skrytyh-dannyh-v-elektronnyh-tekstovyh-dokumentah/viewer

Bodenheimer, L. (2018). Leadership reflections: Resiliency in library organizations. *Journal of Library Administration*, *58*(4), 364–374. DOI: 10.1080/01930826.2018.1448651

Bogdanov, A. Y. (2005). *Quantum algorithms and their impact on the security of modern classical cryptographic systems/A. Yu. Bogdanov, IS Kizhvatov*. RGGU.

Bokolo, A. (2020). Managing digital transformation of smart cities through enterprise architecture: A review and research agenda. *Enterprise Information Systems*, *15*(3), 299–331. DOI: 10.1080/17517575.2020.1812006

Bolivar, M. P. R., & Muñoz, L. A. (2019). *E-Participation in Smart Cities: Technologies and Models of Governance for Citizen Engagement*. Springer. DOI: 10.1007/978-3-319-89474-4

Borisenko, B. (2007). Digital watermarks in images. *Information Security/ Информационная безопасность, (4).* https://lib.itsec.ru/articles2/Oborandteh/cifrov_vodyan_znaki_v_izobrazheniyah#:~:text=Основные%20отличия%20цифровых%20водяных%20знаков,наоборот%2C%20в%20полном%20его%20уничтожении

Bormann, R., de Brito, B. F., Lindermayr, J., Omainska, M., & Patel, M. (2019). Towards automated order picking robots for warehouses and retail. In *Computer Vision Systems: 12th International Conference, ICVS 2019, Thessaloniki, Greece, September 23–25, 2019* [Springer International Publishing.]. *Proceedings*, *12*, 185–198.

Börner, D., & Zohmann, A. (2022). VUCA. *Hands on - Manuelle Und Physikalische Therapien in Der Tiermedizin, 4*(04), 145–145. DOI: 10.1055/a-1942-6128

Bose, I. (November 6, 2024). *Regional divide: Paytm and PhonePe dominate north India, Google Pay leads in the south.* Storyboard18. Retrieved from https://www.storyboard18.com/how-it-works/regional-divide-paytm-and-phonepe-dominate-north-india-google-pay-leads-in-the-south-46712.htm

Bounfour, A. (2016). *Digital Futures, Digital Transformation*. Springer Cham Heidelberg., DOI: 10.1007/978-3-319-23279-9

Brassard, G., Hoyer, P., Mosca, M., & Tapp, A. (2000). Quantum amplitude amplification and estimation. arXiv preprint quant-ph/0005055.

Brien, M., & Alues, A. (2018). “Existing Standards as the Framework to Qualify Additive Manufacturing of Metals. In Proceedings of the 2018”, *IEEE Aerospace Conference*, Big Sky, MT, USA, 3–10 March 2018; pp. 1–10.

Bronk, K. C., Reichard, R. J., & Qi Li, J. (2023). A co-citation analysis of purpose: Trends and (potential) troubles in the foundation of purpose scholarship. *The Journal of Positive Psychology, 18*(6), 1012–1026. DOI: 10.1080/17439760.2023.2168563

Brown, M., Henchoz, C., & Spycher, T. (2018). Culture and financial literacy: Evidence from a within-country language border. *Journal of Economic Behavior & Organization, 150*, 62–85. DOI: 10.1016/j.jebo.2018.03.011

Bruno, P., Denecker, O., & Niederkorn, M. (2019). *Global payments report 2019: amid sustained growth, accelerating challenges demand bold actions*. McKinsey Global Banking Practice, September. Retrieved from https://www.mckinsey.com/~/media/mckinsey/industries/financial%20services/our%20insights/tracking%20the%20sources%20of%20robust%20payments%20growth%20mckinsey%20global%20payments%20map/global-payments-report-2019-amid-sustained-growth-vf.ashx

Buchanan, A., & Keohane, R. O. (2015). Toward a drone accountability regime. *Ethics & International Affairs, 29*(1), 15–37. DOI: 10.1017/S0892679414000732

Butt, A. H., Ahmad, H., Muzaffar, A., Irshad, W., Mumtaz, M. U., & Zubair Ahmad Khan, T. (2024). WeChat gamification: Mobile payment impact on word of mouth and customer loyalty. *Spanish Journal of Marketing - ESIC*. Advance online publication. DOI: 10.1108/SJME-01-2023-0021

Cabral, T. S. (2020). Liability and artificial intelligence in the EU: Assessing the adequacy of the current Product Liability Directive. *Maastricht Journal of European and Comparative Law, 27*(5), 615–635. DOI: 10.1177/1023263X20948689

Cabusora, H., Cordova, E., Kirk, A., & Watson, I. (18 January 2022). *Can Fintech Resolve Public Benefit Obstacles?* Change-Machine. https://change-machine.org/can-fintech-resolve-public-benefit-obstacles/

Cai, K., & Yu, Z. (2024). Research on the antecedents of data privacy concern toward intelligent connected vehicles. *International Journal of Human-Computer Interaction*, •••, 1–14.

Candrasa, L., Cahyadi, L., Cahyadi, W., & Cen, C. C. (2024). Change management strategies: Building organizational resilience in the digital era. *Journal of Ecohumanism, 3*(7), 4125–4135. DOI: 10.62754/joe.v3i7.4534

Carter, E. D., & Cordero, M. L. (2022). Salir adelante: Social capital and resilience during the Covid-19 pandemic in Argentina. *Health & Place*, *77*, 102870. DOI: 10.1016/j.healthplace.2022.102870

Carter, K. M., Idika, N., & Streilein, W. W. (2014). Probabilistic threat propagation for network security. *IEEE Transactions on Information Forensics and Security*, *9*(9), 1394–1405. DOI: 10.1109/TIFS.2014.2334272

Chakraborty, D. (2019). Versatile Performance in Vuca World: A Case Study. *Ushus -. Journal of Business and Management*, *18*(4), 1–8. DOI: 10.12725/ujbm.49.1

Chakravorty, I. (2021). Key Factors Fuelling India's Second COVID Surge. *The Physician*, *7*(1), 1–6. DOI: 10.38192/1.7.1.9

Chang, J. (5 August 2024). How cultural differences influence security and convenience in digital payment experiences. UX Across Cultures. Retrieved from https://www.uxmatters.com/mt/archives/2024/08/how-cultural-differences-influence-security-and-convenience-in-digital-payment-experiences.php

Charteris, J., Smardon, D., & Kemmis, S. (2022). Leadership in the built spaces of innovative learning environments: Leading change in people and practices in the perfectly self-managing society. *Studies in Continuing Education*, *44*(2), 212–231. DOI: 10.1080/0158037X.2021.1928051

Chaudhri, V., Armbrust, N., & Nguyen, S. (2024). Permanently on standby: Practitioner perspectives on the complexities of crisis planning. *Journal of Contingencies and Crisis Management*, *32*(3), e12600. Advance online publication. DOI: 10.1111/1468-5973.12600

Chen, C., & Song, M. (2019). Visualizing a field of research: A methodology of systematic scientometric reviews. *PLoS One*, *14*(10), e0223994. DOI: 10.1371/journal.pone.0223994

Chen, H., Hu, Z., & Solak, S. (2021). Improved delivery policies for future drone-based delivery systems. *European Journal of Operational Research*, *294*(3), 1181–1201. DOI: 10.1016/j.ejor.2021.02.039

Cheruiyot, R., & Venter, R. (2024). Complex systems and sustainable leadership: Enhancing resilience and sustainability of community-based social enterprises in Soweto, South Africa. *Sustainability (Basel)*, *16*(19), 8555. DOI: 10.3390/su16198555

Chidziwisano, G. H., Wyche, S., & Kisyula, R. (2020). *SmartChama: Designing Collaborative Digital Financial Systems for Chama Groups in Kenya*. Information and Communication Technologies and Development., DOI: 10.1145/3392561.3394641

Chingwena, T., & Scheepers, C. B. (2022). Dramatic social change (COVID-19) moderating complexity leadership and organisational adaptability in Zimbabwean SMEs. *European Business Review, 34*(6), 749–775. DOI: 10.1108/EBR-01-2022-0015

Chockalingam, S., Hadziosmanovic, D., Pieters, W., Teixeira, A., & van Gelder, P. (2017). Integrated safety and security risk assessment methods: A survey of key characteristics and applications. *arXiv*. https://arxiv.org/abs/1707.02140 DOI: 10.1007/978-3-319-71368-7_5

Clarke, A., English, V., Harris, H., & Wells, F. (2001). Ethical considerations. *International Journal of Pharmaceutical Medicine*, *15*(2), 89–94. DOI: 10.2165/00124363-200104000-00012

Clark, K. L., & Kowalski, S. P. (2012). Harnessing the power of patent information to accelerate innovation. *Wiley Interdisciplinary Reviews. Data Mining and Knowledge Discovery*, *2*(5), 427–435. DOI: 10.1002/widm.1066

Cogan, N., Kennedy, C., Beck, Z., McInnes, L., MacIntyre, G., Morton, L., Tanner, G., & Kolacz, J. (2022). ENACT study: What has helped health and social care workers maintain their mental well-being during the COVID-19 pandemic? *Health & Social Care in the Community*, *30*(6), e6656–e6673. DOI: 10.1111/hsc.13992

Cornell, A., Mahan, S., & Riedel, R. (2023). Commercial drone deliveries are demonstrating continued momentum in 2023. *URL:* https://www. mckinsey. com/ industries/aerospace-and-defense/our-insights/future-air-mobility-blog/commercial-drone-deliveries

Costa, I., Machado, C. S., & Conceição, O. (2024). Automation and the Importance of Reskilling Workers: A Case Study in the Automotive Industry. In *Proceedings of The 19th European Conference on Management, Leadership and Governance*. Academic Conferences International. DOI: 10.34190/ecmlg.20.1.2923

Coykendall, J., Cotteleer, M., Holdowsky, J., & Mahto, M. (2014). *3D Opportunity in Aerospace and Defense: Additive Manufacturing Takes Flight* (Vol. 1). Deloitte University Press.

Creese, S., Dutton, W. H., & Esteve-González, P. (2021). The social and cultural shaping of cybersecurity capacity building: A comparative study of nations and regions. *Personal and Ubiquitous Computing*, *25*(5), 941–955. DOI: 10.1007/s00779-021-01569-6 PMID: 33994905

Das, S., & Nene, M. J. (2017). A survey on types of machine learning techniques in intrusion prevention systems. *2017 2nd International Conference on Wireless Systems and Networks (WiSPNET)*, 1–5. DOI: 10.1109/WiSPNET.2017.8300169

Dasarraju, V. K. (2024). The Evolution of Cross-Border Payments and Remittances: Integrating QR Codes, Blockchain, and AI. *Journal of Artificial Intelligence & Cloud Computing*, *3*(1), 1–3. DOI: 10.47363/JAICC/2024(3)364

de Diego Ruiz, E., Almodóvar, P., & Birkinshaw, J. (2024). The effects of a firm's internationalization, age, and environmental turbulence on the capabilities that comprise strategic agility. *The International Entrepreneurship and Management Journal*, *20*(3), 1935–1961. DOI: 10.1007/s11365-024-00971-9

de Graaff, N., ten Brink, T., & Parmar, I. (2020). China's rise in a liberal world order in transition–introduction to the FORUM. *Review of International Political Economy*, *27*(2), 191–207. DOI: 10.1080/09692290.2019.1709880

de Lange, M., Synnes, K., & Leindecker, G. (2019). Smart Citizens in the Hackable City: On the Datafication, Playfulness, and Making of Urban Public Spaces Through Digital Art. In. Lecture Notes in Computer Science: Vol. 11380. *Smaniotto Costa, C., et al. CyberParks – The Interface Between People, Places and Technology* (pp. 157–166). Springer., DOI: 10.1007/978-3-030-13417-4_13

De Sisto, M., Shearing, C., Heffernan, T., & Sanderson, D. (2024). Reshaping disaster management: An integrated community-led approach. *Australian Journal of Public Administration*, •••, 1–28. DOI: 10.1111/1467-8500.12668

Dean, H. (2015). *Building innovation capability in an intergovernmental aid organization: A case study of UNICEF s Innovation Unit* (Master's thesis).

Debbarma, S. (2023). COVID-19 Pandemic and Challenges Faced by Healthcare Professionals in India. *Cureus*. Advance online publication. DOI: 10.7759/cureus.38278 PMID: 37255890

Debellis, F., De Massis, A., Messeni Petruzzelli, A., Frattini, F., & Del Giudice, M. (2021). Strategic agility and international joint ventures: The willingness-ability paradox of family firms. *Journal of International Management*, *27*(1), 100739. DOI: 10.1016/j.intman.2020.100739

DebRoy, T., Mukherjee, T., Milewski, J., Elmer, J., Ribic, B., Blecher, J., & Zhang, W. (2019). Scientific, technological and economic issues in metal printing and their solutions. *Nature Materials*, *18*(10), 1026–1032. DOI: 10.1038/s41563-019-0408-2 PMID: 31263223

Demirgüç-Kunt, A., Klapper, L., Singer, D., Ansar, S., & Hess, J. (2017). The Global Findex Database 2017: Measuring financial inclusion and the fintech revolution. *World Bank Group*. https://openknowledge.worldbank.org/server/api/core/bitstreams/4c546107-6b3e-5069-b8d1-b2cc25ea2aa6/content

Denny, B. (2024). Impact of Soundbox by Third Party Application Providers (UPI) on Retailers, with Special Reference to Greater Chennai Corporation. *Journal of Corporate Finance Management and Banking System*, *4*(43), 16–26. DOI: 10.55529/jcfmbs.43.16.26

Derossi, A., Di Palma, E., Moses, J. A., Santhoshkumar, P., Caporizzi, R., & Severini, C. (2023). Avenues for non-conventional robotics technology applications in the food industry. *Food Research International*, *173*, 113265. DOI: 10.1016/j.foodres.2023.113265

Desrues, T. (2020). Authoritarian resilience and democratic representation in Morocco: Royal interference and political parties' leaderships since the 2016 elections. *Mediterranean Politics*, *25*(2), 254–262. DOI: 10.1080/13629395.2018.1543038

Deufel, P., Kemper, J., & Brettel, M. (2019). Pay now or pay later: A cross-cultural perspective on online payments. *Journal of Electronic Commerce Research*, *20*(3), 141–154. http://jecr.org/sites/default/files/2019vol20no3_Paper1.pdf

Dhandapani, S., & Kaur, K. (2024). Instructional leadership: The key to addressing psychological issues post-covid-19. *Journal of Interdisciplinary Studies in Education*, *13*(S1), 16–33. DOI: 10.32674/p547wx36

Ding, Y., Jin, M., Li, S., & Feng, D. (2021). Smart logistics based on the internet of things technology: An overview. *International Journal of Logistics*, *24*(4), 323–345. DOI: 10.1080/13675567.2020.1757053

Dolata, M., & Schwabe, G. (2023). Moving beyond privacy and airspace safety: Guidelines for just drones in policing. *Government Information Quarterly*, *40*(4), 101874. DOI: 10.1016/j.giq.2023.101874

Domingo Galindo, L. (2016). *The challenges of logistics 4.0 for the supply chain management and the information technology* (Master's thesis, NTNU).

Döner, E., & EfeoGlu, İ. E. (2023). Being affected by VUCA factors? Developing the 'Perceived VUCA Exposure' Scale. *GAB Akademi,* 3(2), 28–53.

Dong, L., Duarte, F., Duranton, G., Santi, P., Barthelemy, M., Batty, M., Bettencourt, L., Goodchild, M., Hack, G., Liu, Y., Pumain, D., Shi, W., Verbavatz, V., West, G. B., Yeh, A. G. O., & Ratti, C. (2024). Defining a city — Delineating urban areas using cell-phone data. *Nature Cities*, *1*(2), 117–125. Advance online publication. DOI: 10.1038/s44284-023-00019-z

Dulina, N. V. (2005). The state of the modern Russian city and the need to develop a strategy for its development. *Regional Economics: Theory and Practice*, *6*, 36–37.

Dulina, N. V., & Tokarev, V. V. (2005). Monitoring as a means of information support for social management of a modern city. *Regional Economics: Theory and Practice*, *6*, 43.

Dwivedi, S., & Gupta, A. (2024). Strategically addressing skill gaps and imbalances among health employees. In *Contemporary Challenges in Social Science Management: Skills Gaps and Shortages in the Labour Market* (pp. 17-33). Emerald Publishing Limited. DOI: 10.1108/S1569-37592024000112A015

Dwivedi, S., Poddar, A., Gupta, A., Memoria, M., & Kumar, R. (2022, December). Analyzing the effectiveness of a health care waste reduction education and training interventions: A step towards green healthcare. In *2022 2nd International Conference on Innovative Sustainable Computational Technologies (CISCT)* (pp. 1-5). IEEE. DOI: 10.1109/CISCT55310.2022.10046456

Dwivedi, R., Nerur, S., & Balijepally, V. (2023). Exploring artificial intelligence and big data scholarship in information systems: A citation, bibliographic coupling, and, co-word analysis. *International Journal of Information Management Data Insights*, *3*(2), 100185. DOI: 10.1016/j.jjimei.2023.100185

Dwivedi, Y. K., Hughes, L., Ismagilova, E., Aarts, G., Coombs, C., Crick, T., Duan, Y., Dwivedi, R., Edwards, J., Eirug, A., Galanos, V., Ilavarasan, P. V., Janssen, M., Jones, P., Kar, A. K., Kizgin, H., Kronemann, B., Lal, B., Lucini, B., & Williams, M. D. (2021). Artificial Intelligence (AI): Multidisciplinary perspectives on emerging challenges, opportunities, and agenda for research, practice and policy. *International Journal of Information Management*, *57*, 101994. DOI: 10.1016/j.ijinfomgt.2019.08.002

Eberz, S., Lang, S., Breitenmoser, P., & Niebert, K. (2023). Taking the lead into sustainability: Decision makers' competencies for a greener future. *Sustainability (Basel)*, *15*(6), 4986. DOI: 10.3390/su15064986

Edwards, B., Stallings, H., & Cheng, I. (30 July, 2014). *Using human-centered design for e-payment systems in Indonesia*. CGAP. Retrieved from https://www.cgap.org/blog/using-human-centered-design-for-e-payment-systems-in-indonesia

Eelu, S., & Nakakawa, A. (2018). Framework towards enhancing adoption of electronic payment in a developing economy: A case of Uganda. *The African Journal of Information Systems*, *10*(3), 222–245. https://digitalcommons.kennesaw.edu/ajis/vol10/iss3/5

ElSalamouny, E. (2011). *Probabilistic trust models in network security* (Doctoral dissertation, University of Southampton).

Eng, T.-Y., Mohsen, K., & Wu, L.-C. (2023). Wireless information technology competency and transformational leadership in supply chain management: Implications for innovative capability. *Information Technology & People*, *36*(3), 969–995. DOI: 10.1108/ITP-06-2021-0489

ENISA. (2006). *Risk management: Implementation principles and inventories for risk management/risk assessment methods and tools.* https://www.enisa.europa.eu/publications/risk-management-principles-and-inventories

Euromonitor International. (25 April 2024). Asia-Pacific accounts for 52% of global paperless transactions: Euromonitor International. Retrieved from https://www.euromonitor.com/press/press-releases/april-2024/asia-pacific-accounts-for-52-of-global-paperless-transactions-euromonitor-international#:~:text=SINGAPORE%2C%20Singapore%20%E2%80%93%20Asia%2DPacific,to%20findings%20from%20Euromonitor%20International

European Investment Fund (EIF). (2024). EIF working paper 2024/97, *Addressing financial and digital illiteracy: Insights from microfinance institutions and fintech organisations.* https://www.eif.org/news_centre/publications/EIF_Working_Paper_2024_97.htm

Evans, N. D. (2017). *Mastering digital business: How powerful combinations of disruptive technologies are enabling the next wave of digital transformation.* BCS - The Chartered Institute for IT.

Evans, S., & Bahrami, H. (2020). Super-flexibility in practice: Insights from a crisis. *Global Journal of Flexible Systems Managment*, *21*(3), 207–214. DOI: 10.1007/s40171-020-00246-6

Evgrafova, L. V. (2015). Analysis of the efficiency of the dairy cattle breeding. *International Agricultural Journal*, *2*, 62–64.

Evgrafova, L. V., & Ershova, V. L. (2020). Methodical aspects of formation of agrotouristic cluster structure in the economy of the region. *International Scientific Journal*, *2*(2), 30–36. DOI: 10.34286/1995-4638-2020-71-2-30-36

Evgrafova, L. V., Sergeeva, N. A., Ismailova, I. Z., Eremin, V. I., & Sherstyuk, M. V. (2022). Assessing the functional efficiency of tourist and recreational clusters in the regions of Russia. *AIP Conference Proceedings*, *2467*, 070011. DOI: 10.1063/5.0092777

Evseeva, S., Evseeva, O., & Rawat, P. (2022). Employee development and digitalization in BANI world. *Communications in Computer and Information Science,* 1619(CCIS), 253–264. DOI: 10.1007/978-3-031-14985-6_18

Falch, M., & Henten, A. (2010). Public private partnerships as a tool for stimulating investments in broadband. *Telecommunications Policy*, *34*(9), 496–504. DOI: 10.1016/j.telpol.2010.07.010

Faling, M., & Biesbroek, R. (2019). Cross-boundary policy entrepreneurship for climate-smart agriculture in Kenya. *Policy Sciences*, *52*(4), 525–547. DOI: 10.1007/s11077-019-09355-1

Familiar, B., & Barnes, J. (2017). *Business in Real-Time Using Azure IoT and Cortana Intelligence Suite: Driving Your Digital Transformation*. Apress., DOI: 10.1007/978-1-4842-2650-6

Federal Reserve Bank of Atlanta. (June 2023). *Shifting the focus: Digital payments and the path to financial inclusion*. https://www.atlantafed.org/-/media/documents/promoting-safer-payments-innovation/publications/2023/06/13/SCOPI_Report.pdf

Feng, N., & Li, M. (2011). An information systems security risk assessment model under uncertain environment. *Applied Soft Computing*, *11*(7), 4332–4340. DOI: 10.1016/j.asoc.2010.06.005

Fenz, S., Heurix, J., Neubauer, T., & Pechstein, F. (2014). Current challenges in information security risk management. *Computers & Security*, *45*, 1–17. DOI: 10.1016/j.cose.2014.01.005

Fernandes, V., Wong, W., & Noonan, M. (2023). Developing adaptability and agility in leadership amidst the COVID-19 crisis: Experiences of early-career school principals. *International Journal of Educational Management*, *37*(2), 483–506. DOI: 10.1108/IJEM-02-2022-0076

Fernández-Pérez de la Lastra, S., & Sánchez-Gardey, G. (2024). Organizational ambidexterity: A reconceptualization and research agenda for the VUCA international context. *Journal of Contingencies and Crisis Management*, *32*(2), e12565. Advance online publication. DOI: 10.1111/1468-5973.12565

Ferreira, J. J. M., Fernandes, C. I., & Ratten, V. (2016). A co-citation bibliometric analysis of strategic management research. *Scientometrics*, *109*(1), 1–32. DOI: 10.1007/s11192-016-2008-0

Feyen, E., Frost, J., Gambacorta, L., Natarajan, H., & Saal, M. (2021). *Fintech and the digital transformation of financial services: implications for market structure and public policy. BIS papers*, Bank for International Settlements, number 117. https://www.bis.org/publ/bppdf/bispap117.pdf

Figueira, P. T., Bravo, C. L., & López, J. L. R. (2020). Improving information security risk analysis by including threat-occurrence predictive models. *Computers & Security*, *88*, 101609. DOI: 10.1016/j.cose.2019.101609

Filippi, E., Banno, M., & Trento, S. (2023). Automation technologies and their impact on employment: A review, synthesis and future research agenda. *Technological Forecasting and Social Change*, *191*, 122448. DOI: 10.1016/j.techfore.2023.122448

Finn, D., Chandrasekhar, D., & Xiao, Y. (2023). A region recovers: Planning for resilience after superstorm Sandy. *Journal of Planning Education and Research*, *43*(1), 136–149. DOI: 10.1177/0739456X19864145

Florek-Paszkowska, A., Ujwary-Gil, A., & Godlewska-Dzioboń, B. (2021). Business innovation and critical success factors in the era of digital transformation and turbulent times. *Journal of Entrepreneurship. Management and Innovation*, *17*(4), 7–28. DOI: 10.7341/20211741

Fomina, A. (2023). Comfortable urban environment of the megapolis and digitalization. *AIP Conference Proceedings*, *2560*(1), 040005. Advance online publication. DOI: 10.1063/5.0134578

Ford, J. B. (1965). The challenge of the megalopolis. *The International Migration Digest*, *2*(1), 41–51. DOI: 10.2307/3002979

Ford, S., & Despeisse, M. (2016). Additive manufacturing and sustainability: An exploratory study of the advantages and challenges. *Journal of Cleaner Production*, *13*, 1573–1587. DOI: 10.1016/j.jclepro.2016.04.150

Fox, S. J. (2022). Drones: Foreseeing a'risky'business? Policing the challenge that flies above. *Technology in Society*, *71*, 102089. DOI: 10.1016/j.techsoc.2022.102089

Fridgeirsson, T. V., Kristjansdottir, B. H., & Ingason, H. T. (2021). An Alternative Risk Assessment Routine for Decision Making; Towards a VUCA Meter to Assess the Volatility, Uncertainty, Complexity and Ambiguity of Complex Projects. In *Research on Project* (pp. 41–54). Programme and Portfolio Management., DOI: 10.1007/978-3-030-60139-3_4

Fuentealba, D., Flores-Fernández, C., & Carrasco, R. (2023). Análisis bibliométrico y de contenido sobre VUCA. *Revista Española de Documentación Científica, 46*(2), e354. DOI: 10.3989/redc.2023.2.1968

Fung, B. S. C., & Halaburda, H. (2016). *Central bank digital currencies: a framework for assessing why and how* (No. 2016-22). Bank of Canada Staff Discussion Paper. https://dx.doi.org/DOI: 10.2139/ssrn.2994052

Gagliardi, J. P., Renaud, J., & Ruiz, A. (2012). Models for automated storage and retrieval systems: A literature review. *International Journal of Production Research, 50*(24), 7110–7125. DOI: 10.1080/00207543.2011.633234

Gantman, E. R., & Dabós, M. P. (2018). Research output and impact of the fields of management, economics, and Sociology in spain and france: An analysis using google scholar and scopus. *Journal of the Association for Information Science and Technology, 69*(8), 1054–1066. DOI: 10.1002/asi.24020

Ganzhur, M. A., Dzyuba, Y. V., & Panchenko, V. A. (n.d.). Features of digital steganography as a method of ensuring data concealment. *CyberLeninka.* https://cyberleninka.ru/article/n/osobennosti-tsifrovoy-steganografii-kak-metoda-obespecheniya-sokrytiya-dannyh/viewer

Gao, M., Hugenholtz, C. H., Fox, T. A., Kucharczyk, M., Barchyn, T. E., & Nesbit, P. R. (2021). Weather constraints on global drone flyability. *Scientific Reports, 11*(1), 12092. DOI: 10.1038/s41598-021-91325-w

Gao, S., Meng, F., Gu, Z., Liu, Z., & Farrukh, M. (2021). Mapping and clustering analysis on environmental, social and Governance field a bibliometric analysis using scopus. *Sustainability (Basel), 13*(13), 7304. DOI: 10.3390/su13137304

Garg, V., Niranjan, S., Prybutok, V., Pohlen, T., & Gligor, D. (2023). Drones in last-mile delivery: A systematic review on Efficiency, Accessibility, and Sustainability. *Transportation Research Part D, Transport and Environment, 123*, 103831. DOI: 10.1016/j.trd.2023.103831

Gedam, V., Raut, R., Inamdar, Z., Narkhede, B., Dharaskar, S., & Narvane, V. (2022). <scp>COVID</scp> -19 critical success factors in Indian healthcare industry—A <scp>DEMATEL</scp> approach. *Journal of Multi-Criteria Decision Analysis, 29*(1–2), 135–149. DOI: 10.1002/mcda.1763 PMID: 40477520

Gefen, D., Karahanna, E., & Straub, D. W. (2003). Trust and TAM in online shopping: An integrated model. *Management Information Systems Quarterly, 27*(1), 51–90. DOI: 10.2307/30036519

George, C. (2017). *Specification for Control and Qualification of Laser Powder Bed Fusion Metallurgical Processes*. NASA Marshall Space Flight Center.

Gerling, E. Yu., & Ahrameeva, K. A. (n.d.). Method of linguistic steganography based on keyword word. *CyberLeninka.* https://cyberleninka.ru/article/n/metod-lingvisticheskoy-steganografii-osnovannyy-na-opornom-slove/viewer

Goel, N., & Nath, V. (2020, March). An exploratory study on digital payment systems and its impact on trust and continuance intention in newly remonetized and digitized era. In *Proceedings of the International Conference on Innovative Computing & Communications (ICICC)*. https://doi.org/DOI: 10.2139/ssrn.3562948

GOI. (February 2024). Impact of DIGIDGAN mission on India's digital payment ecosystem. Retrieved from https://www.meity.gov.in/writereaddata/files/Impact-Assessment-of-DigiDhan-Mission.pdf

Goodwin Groen, R., & Klapper, L. (July 2020). Advancing Women's Digital Financial Inclusion. Better Than Cash Alliance. *Women's World Banking, and the World Bank Group*. https://www.gpfi.org/sites/gpfi/files/sites/default/files/saudig20_women.pdf

Govers, M., & van Amelsvoort, P. (2023). A theoretical essay on socio-technical systems design thinking in the era of Digital transformation. *Gruppe. Interaktion. Organisation. Zeitschrift Für Angewandte Organisationspsychologie*, *54*(1), 27–40. DOI: 10.1007/s11612-023-00675-8

Grames, E. M., Stillman, A. N., Tingley, M. W., & Elphick, C. S. (2019). An automated approach to identifying search Terms for systematic reviews using keyword co-occurrence networks. *Methods in Ecology and Evolution*, *10*(10), 1645–1654. DOI: 10.1111/2041-210X.13268

Grand View Research. (2023). *Digital Payment Market Size, Share & Growth Report, 2030. Retrieved from* https://www.grandviewresearch.com/industry-analysis/digital-payment-solutions-market

Grau, G. I. (2023). An Appraisal of the Proposal for a Directive on Liability for Defective Products. *Journal of European Consumer and Market Law*, *12*(5).

Grobler, A., & Flotman, A.-P. (2021). Servant leadership, team-based learning and hope and optimism: A sectoral Comparative study. *South African Journal of Business Management*, *52*(1), a2444. DOI: 10.4102/sajbm.v52i1.2444

Grote, M., Oakey, A., Pilko, A., Krol, J., Blakesley, A., Cherrett, T., Scanlan, J., Anvari, B., & Martinez-Sykora, A. (2024). The effects of costs on drone uptake in multi-modal logistics systems within a healthcare setting. *Transport Economics and Management*, *2*, 58–75. DOI: 10.1016/j.team.2024.03.001

Grover, L. K. (1996, July). A fast quantum mechanical algorithm for database search. In *Proceedings of the twenty-eighth annual ACM symposium on Theory of computing* (pp. 212-219).

Grutters, J. P. C., van Asselt, M. B. A., Chalkidou, K., & Joore, M. A. (2015). Healthy Decisions: Towards Uncertainty Tolerance in Healthcare Policy. *PharmacoEconomics*, *33*(1), 1–4. DOI: 10.1007/s40273-014-0201-7 PMID: 25145801

GSMA. (2022). *State of the Industry Report on Mobile Money.* GSMA Intelligence. Retrieved from https://www.gsma.com/sotir/wp-content/uploads/2022/03/GSMA_State_of_the_Industry_2022_English.pdf

GSMA. (23 March 2023). *The digital financial literacy toolkit: Addressing the gap in low and middle-income countries*. Retrieved from https://www.gsma.com/solutions-and-impact/connectivity-for-good/mobile-for-development/gsma_resources/the-digital-financial-literacy-toolkit-addressing-the-gap-in-low-and-middle-income-countries/

Gunasekaran, A., Subramanian, N., & Papadopoulos, T. (2017). Information technology for competitive advantage within logistics and supply chains: A review. *Transportation Research Part E, Logistics and Transportation Review*, *99*, 14–33. DOI: 10.1016/j.tre.2016.12.008

Gupta, A., Chaithra, N., Jha, J., Sayal, A., Gupta, V., & Memoria, M. (2023, May). Machine learning algorithms for disease diagnosis using medical records: a comparative analysis. In *2023 4th international conference on intelligent engineering and management (ICIEM)* (pp. 1-6). IEEE. DOI: 10.1109/ICIEM59379.2023.10165850

Gupta, O., Joshi, P., & Gupta, A. (2023, May). Prioritizing the enablers for adoption of AI based health care wearables among elderly people. In *2023 4th International Conference on Intelligent Engineering and Management (ICIEM)* (pp. 1-4). IEEE. DOI: 10.1109/ICIEM59379.2023.10165740

Gu, Q., & Day, C. (2013). Challenges to teacher resilience: Conditions count. *British Educational Research Journal*, *39*(1), 22–44. DOI: 10.1080/01411926.2011.623152

Haddaway, N. R., Page, M. J., Pritchard, C. C., & McGuinness, L. A. (2022). PRISMA2020: An R package and Shiny app For producing PRISMA 2020-compliant flow diagrams, with interactivity for optimised digital transparency and Open Synthesis. *Campbell Systematic Reviews*, *18*(2), e1230. DOI: 10.1002/cl2.1230

Hamsal, M., Dwidienawati, D., Ichsan, M., Syamil, A., & Trigunarsyah, B. (2022). Multi-perspective approach to building Team resilience in project management—A case study in Indonesia. *Sustainability (Basel), 14*(20), 13137. DOI: 10.3390/su142013137

Hancock, P. A., Nourbakhsh, I., & Stewart, J. (2019). On the future of transportation in an era of automated and autonomous vehicles. *Proceedings of the National Academy of Sciences of the United States of America, 116*(16), 7684–7691. DOI: 10.1073/pnas.1805770115

Hansen, L. M. P., & Pistelli, M. (9 July, 2024). *Effective public-private partnerships to advance financial inclusion and financial well-being of forcibly displaced persons and their hosts- Aligning the business cases.* e-MFP. Retrieved from https://www.e-mfp.eu/post/effective-public-private-partnerships-advance-financial-inclusion-and-financial-well-being

Harrison, R. T. (2017). Leadership leadership development and all that jazz. *Leadership, 13*(1), 81–99. DOI: 10.1177/1742715016681120

Hass, B., Pooley, J., Feuring, M., Suvarna, V., & Harrington, A. (2012). Health technology assessment and its role in the future development of the Indian healthcare sector. *Perspectives in Clinical Research, 3*(2), 66. DOI: 10.4103/2229-3485.96449 PMID: 22701823

Hegde, S., Larsen, E., Torbett, O., Ponnala, S., Pohl, E., Sze, R., & Schaeubinger, M. M. (2023). A proactive learning Approach toward building adaptive capacity during COVID-19: A radiology case study. *Applied Ergonomics, 110*, 104009. DOI: 10.1016/j.apergo.2023.104009

Helmrich, A., & Chester, M. (2022). Navigating exploitative and explorative leadership in support of infrastructure resilience. *Frontiers in Sustainable Cities, 4*, 791474. DOI: 10.3389/frsc.2022.791474

Helsper, E., & Eynon, R. (2013). Distinct skill pathways to digital engagement. *European Journal of Communication, 15*(4), 739–758. DOI: 10.1177/0267323113499113

Hermans, J., Vercauteren, F., & Preneel, B. (2010, March). Speed records for NTRU. In *Cryptographers' track at the RSA conference* (pp. 73–88). Springer Berlin Heidelberg.

Hickman, X. (2023). A challenge for higher education leadership: The multi-faceted and irreversible impact of the COVID-19 pandemic. *Journal of Higher Education Policy and Leadership Studies, 4*(4), 147–158. DOI: 10.61186/johepal.4.4.147

Hilbert, M. (2016). The bad news is that the digital access divide is here to stay: Domestically installed bandwidths among 172 countries. *Telecommunications Policy, 40*(6), 567–581. DOI: 10.1016/j.telpol.2016.01.006

Hill, M., & Hoss, M. A. K. (2018). Reclaiming American Indian women leadership: Indigenous pathway to leadership. *Open Journal of Leadership, 7*(3), 225–236. DOI: 10.4236/ojl.2018.73013

Hoff, R., Helmrich, A., Dirks, A., Kim, Y., Li, R., & Chester, M. (2023). Dynamic criticality for infrastructure prioritization In complex environments. *Environmental Research: Infrastructure and Sustainability, 3*(1), 015011. DOI: 10.1088/2634-4505/acbe15

Hofstede, G. (1980). Culture and organizations. *International Studies of Management & Organization, 10*(4), 15–41. DOI: 10.1080/00208825.1980.11656300

Hofstede, G. (1984). *Culture's consequences: International differences in work-related values*. Sage Publications.

Hofstede, G. (2001). *Culture's Consequences: Comparing Values, Behaviors, Institutions, and Organizations Across Nations*. Sage Publications.

Holley, D., Coulson, K., Buckley, C., & Corradini, E. (2022). Wellbeing in the workplace: Exploring the VUCA approach. *Journal of Learning Development in Higher Education, 25*(25). Advance online publication. DOI: 10.47408/jldhe.vi25.954

Hong, J., Cross Francis, D., Wang, Q., Lewis, L., Parsons, A., Neill, C., & Meek, D. (2020). The role of trust: Teacher capacity during school leadership transition. *Frontiers in Education, 5*(108), 1–9. DOI: 10.3389/feduc.2020.00108

Huang, Y., Wang, X., & Wang, X. (2020). Mobile payment in China: Practice and its effects. *Asian Economic Papers, 19*(3), 1–18. DOI: 10.1162/asep_a_00779

Hube, B., Stockport, G., & Soutar, G. (2022). A cogwheel model of dynamic capabilities: Evidence from an Australian university. *Australian Journal of Public Administration, 81*(4), 569–588. DOI: 10.1111/1467-8500.12554

Huo, W., Xiohui, W., Zulfiqar, M., Chand, A., & Ullah, M. R. (2024). Communication dynamics: Fintech's role in promoting sustainable cashless transactions. *Humanities & Social Sciences Communications, 11*(1), 1–15. DOI: 10.1057/s41599-024-03729-4

Hussaini, W., & Hussaini, S. G. (2023). Risks of a megacity. *Journal of Survey in Fisheries Sciences, 10*(1), 1576–1588. DOI: 10.53555/sfs.v10i1.2414

Illanes, P., Lund, S., Mourshed, M., Rutherford, S., & Tyreman, M. (2018). Retraining and reskilling workers in the age of automation. *McKinsey Global Institute, 8*(1), 1–8.

Ilyushechkin, E. A. (n.d.). Research of robustness of digital watermarking methods to distortions. *CyberLeninka.* https://cyberleninka.ru/article/n/issledovanie-ustoychivosti-metodov-vstraivaniya-tsifrovyh-vodyanyh-znakov-k-iskazhayuschim-vozdeystviyam/viewer

Indrianti, Y., Sasmoko, S., Abdinagoro, S. B., & Rahim, R. K. (2024). Building bridges to entrepreneurial resilience: Exploring the mediating role of business model innovation capacity in ambidextrous leadership and entrepreneurial mindfulness. *Uncertain Supply Chain Management, 12*(4), 2245–2252. DOI: 10.5267/j.uscm.2024.6.007

ISO/IEC. (2022). *ISO/IEC 27005:2022 – Information security, cybersecurity and privacy protection: Guidance on managing information security risks.* https://www.iso.org/standard/80585.html

Istishev, V. (2021). Data protection with digital watermarks. *Habr.* https://habr.com/ru/companies/droider/articles/568776/

Ivanenko, V. G., & Ushakov, N. V. (n.d.). Digital watermarks in electronic document management. *ResearchGate.* https://www.researchgate.net/publication/318831123_Digital_watermarks_in_electronic_document_circulation

Jack, W., & Suri, T. (2011). *Mobile Money: The Economics of M-Pesa. NBER Working Paper Series*, 16721. https://doi.org/DOI: 10.3386/w16721

Jacome, V. (2023). Limits of energy access research on unreliability. *Journal of International Development, 35*(8), 2626–2644. Advance online publication. DOI: 10.1002/jid.3790

Jain, N., Gambhir, A., & Pandey, M. (2024). Unmanned Aerial Networks—UAVs and AI. In *Recent Trends in Artificial Intelligence Towards a Smart World: Applications in Industries and Sectors* (pp. 321–351). Springer Nature Singapore. DOI: 10.1007/978-981-97-6790-8_12

Javaid, M., Haleem, A., Singh, R. P., & Suman, R. (2021). Substantial capabilities of robotics in enhancing industry 4.0 implementation. *Cognitive Robotics, 1*, 58–75. DOI: 10.1016/j.cogr.2021.06.001

Jha, W. C. U. (2014). *Drone Wars: Ethical, Legal and Strategic Implications: Ethical, Legal and Strategic Implications*. KW Publishers Pvt Ltd.

Johnson, Lee 2020 News Consumption across Media Platforms and Content https://www.researchgate.net/publication/344619753

Jones, J. (2006). *An introduction to factor analysis of information risk (FAIR)*. https://en.wikipedia.org/wiki/Factor_analysis_of_information_risk

Juncos, A. E. (2018). Resilience in peacebuilding: Contesting uncertainty, ambiguity, and complexity. *Contemporary Security Policy*, *39*(4), 559–574. DOI: 10.1080/13523260.2018.1491742

Jun, L., & Weilin, W. (2024). A bibliometric analysis of the evolving mechanisms of shadow education research. *European Journal of Education*, *59*(3), 1–18. DOI: 10.1111/ejed.12646

Kabanov, Yu. A., & Chugunov, A. V. (2018). Conceptualization of concepts used in "smart city" and "e-governance" studies: An experience of scientometric analysis. *International Journal of Open Information Technologies*, *6*(11), 54–58.

Kachniewska, M., & Para, A. (2023). Feminine vs. masculine: Expectations of leadership styles in hotels during the COVID-19 pandemic. *Sustainability (Basel)*, *15*(13), 10602. DOI: 10.3390/su151310602

Kaivo-oja, J. R. L., & Lauraeus, I. T. (2018). The VUCA approach as a solution concept to corporate foresight challenges and global technological disruption. *Foresight*, *20*(1), 27–49. DOI: 10.1108/FS-06-2017-0022

Kalanje, C. M. (2006). *Role of intellectual property in innovation and new product development*. World Intellectual Property Organization.

Kao-Cheng, H., & Edwards, D. J. (2008). *Millimetre wave antennas for gigabit wireless communications: A practical guide to design and analysis in a system context*. John Wiley & Sons Ltd.

Karabacak, B., & Sogukpinar, I. (2005). ISRAM: Information security risk analysis method. *Computers & Security*, *24*(6), 428–438. DOI: 10.1016/j.cose.2004.07.004

Kargapolova, E. V., & Davydova, Yu. A. (2022). The well-being of the population of the Russian region in conditions of socio-economic constraints (on the example of the Moscow agglomeration). *CITISE*, *3*, 301–316.

Karjaluoto, H., Shaikh, A. A., Leppäniemi, M., & Luomala, R. (2020). Examining consumers' usage intention of contactless payment systems. *International Journal of Bank Marketing*, *38*(2), 332–351. DOI: 10.1108/IJBM-04-2019-0155

Karneli, O. (2022). Application of the Concept of Strategic Human Resources Management in Improving Organizational Performance in the Vuca Era. *Eduvest - Journal Of Universal Studies*, *2*(4), 704–709. DOI: 10.36418/edv.v2i4.381

Kaya, Y. (2023). Agile leadership from the perspective of dynamic capabilities and creating value. *Sustainability (Basel), 15*(21), 15253. DOI: 10.3390/su152115253

Keck, M., & Etzold, B. (2013). Risks and resilience in Asian megacities: Guest editorial. *Erdkunde, 67*(1), 1–3. DOI: 10.3112/erdkunde.2013.01.01

Kelly, S. E., Moher, D., & Clifford, T. J. (2016). Quality of conduct and reporting in rapid reviews: An exploration of Compliance with PRISMA and AMSTAR guidelines. *Systematic Reviews, 5*(1), 79. DOI: 10.1186/s13643-016-0258-9

Khando, K., Islam, M. S., & Gao, S. (2022). The emerging technologies of digital payments and associated challenges: A systematic literature review. *Future Internet, 15*(1), 21. DOI: 10.3390/fi15010021

Khan, Z., & Lew, Y. K. (2018). Post-entry survival of developing economy international new ventures: A dynamic capability perspective. *International Business Review, 27*(1), 149–160. DOI: 10.1016/j.ibusrev.2017.06.001

Khogali, H. O., & Mekid, S. (2023). The blended future of automation and AI: Examining some long-term societal and ethical impact features. *Technology in Society, 73*, 102232. DOI: 10.1016/j.techsoc.2023.102232

Kim, J., Moon, H., & Jung, H. (2020). Drone-based parcel delivery using the rooftops of city buildings: Model and solution. *Applied Sciences (Basel, Switzerland), 10*(12), 4362. DOI: 10.3390/app10124362

Kim, P. S. (2009). Enhancing public accountability for developing countries: Major constraints and strategies. *Australian Journal of Public Administration, 68*(s1), S89–S100. DOI: 10.1111/j.1467-8500.2009.00626.x

Kirby, A. (2012). Current research on cities and its contribution to urban studies. *Cities (London, England), 29*(1), 3–8. DOI: 10.1016/j.cities.2011.12.004

Kirvan, P., & Irei, A. (2023). Using the FAIR model to quantify cyber-risk. *TechTarget*. https://www.techtarget.com/searchsecurity/tip/Using-the-FAIR-model-to-quantify-cyber-risk

Kisiała, W., & Rącka, I. (2021). Spatial and statistical analysis of urban poverty for sustainable city development. *Sustainability (Basel), 13*(2), 858. DOI: 10.3390/su13020858

Klapper, L., El-Zoghbi, M., & Hess, J. (2016). *Achieving the Sustainable Development Goals: The role of financial inclusion. CGAP*. Retrieved from https://www.cgap.org/sites/default/files/Working-Paper-Achieving-Sustainable-Development-Goals-Apr-2016_0.pdf

Kligyte, G., & Barrie, S. (2014). Collegiality: Leading us into fantasy-the paradoxical resilience of collegiality in academic leadership. *Higher Education Research & Development*, *33*(1), 157–169. DOI: 10.1080/07294360.2013.864613

Klumpp, M., & Ruiner, C. (2018). Regulation for artificial intelligence and robotics in transportation, logistics, and supply chain management: Background and developments. *Network Industries Quarterly*, *20*(2), 3–7.

Kogai, E. A., & Dremova, L. A. (2021). The state of civil society in the Kursk region in 2020. In *Effectiveness of national projects: Socio-economic processes, civil society institutions, education* (pp. 11–16). Kursk Institute of Management, Economics and Business.

Koh, L. Y., Lee, J. Y., Wang, X., & Yuen, K. F. (2023). Urban drone adoption: Addressing technological, privacy and task–technology fit concerns. *Technology in Society*, *72*, 102203. DOI: 10.1016/j.techsoc.2023.102203

Koiwanit, J. (2018). Analysis of environmental impacts of drone delivery on an online shopping system. *Advances in Climate Change Research*, *9*(3), 201–207. DOI: 10.1016/j.accre.2018.09.001

Komoju (January 17, 2024). Southeast Asia's payment methods: The local payment culture by country. Retrieved from https://en.komoju.com/blog/payment-method/southeast-asia/

Kondakci, S. (2007). A new assessment and improvement model of risk propagation in information security. *International Journal of Information and Computer Security*, *1*(3), 341–366. DOI: 10.1504/IJICS.2007.013959

Kondakci, S. (2015). Analysis of information security reliability: A tutorial. *Reliability Engineering & System Safety*, *133*, 275–299. DOI: 10.1016/j.ress.2014.09.021

Kotarba, M. (2018). Digital Transformation of Business Models. *Foundations of Management.*, *10*(1), 123–142. DOI: 10.2478/fman-2018-0011

Kouty, M. (2024). Boosting the intra-African digital trade in the AfCFTA context: Does regulatory framework matter? *Digital Economy and Sustainable Development*, 2(1), 9. Advance online publication. DOI: 10.1007/s44265-024-00036-3

Kraus, K. (2023). *The nature of digital transformation: Energy of the economy. How to feel, understand and use it.* Baltija Publishing., DOI: 10.30525/978-9934-26-287-6

Kreier, F. (2022). Drones bearing parcels deliver big carbon savings. *Nature*. Advance online publication. DOI: 10.1038/d41586-022-02101-3

Krupinski, E. A. (2019). Writing systematic reviews of the literature—It really is a systematic process. *Journal of Digital Imaging*, *32*(2), 199–200. DOI: 10.1007/s10278-018-00176-x

Kumar, K., Chaudhary, S., Anandaram, H., Kumar, R., Gupta, A., & Joshi, K. (2023, February). Industry 4.0 and health care system with special reference to mental health. In *2023 1st International Conference on Intelligent Computing and Research Trends (ICRT)* (pp. 1-5). IEEE.

Kumar, S., & Ranga, V. (2015). A novel approach of zero watermarking for text documents. *International Journal of Advanced Research in Computer Science.* https://scholar.archive.org/work/cvvyhadj4fadrohwpc3ycavyha/access/wayback/http://ijarf.com/wp-content/uploads/2015/07/IJARF-18-21.pdf

Kumar, A. (2023). The Transformation of The Indian Healthcare System. *Cureus.* Advance online publication. DOI: 10.7759/cureus.39079 PMID: 37378105

Kumar, T. (2020). Community mental health services in India: The pandemic and beyond. *Indian Journal of Social Psychiatry*, *36*(5), 168. DOI: 10.4103/ijsp.ijsp_126_20

Kutsepal, S., Zinchenko, N., & Perebyynis, Y. (2020). Specificity of social existence in megalopolis. *SHS Web of Conferences, 85*, 01003. https://doi.org/DOI: 10.1051/shsconf/20208501003

Ladoni, M., Bahrami, H. A., Alavipanah, S. K., & Norouzi, A. A. (2009). Estimating soil organic carbon from soil reflectance: A review. *Precision Agriculture*, *11*(1), 82–99. DOI: 10.1007/s11119-009-9123-3

Lane, M. (2021). The impact of AI on the labour market: is this time different. *OECD. AI, Labour Market Economist,* https://oecd. ai/en/wonk/impact-ai-on-the-labour-market-is-this-timedifferent, *erişim, 25*, 2021.

Lane, P. J., Clay-Williams, R., Johnson, A., Garde, V., & Barrett-Beck, L. (2021). Creating a healthcare variant CYNEFIN Framework to improve leadership and urgent decision-making in times of crisis. *Leadership in Health Services, ahead-of-print(ahead-of-print),* 454–461. DOI: 10.1108/LHS-03-2021-0013

Lapin, N. I., & Belyaeva, L. A. (Eds.). (2009). *Regions in Russia: Socio-cultural portraits of regions in the all-Russian context.* Academia.

Ledesma, J. (2014). Conceptual frameworks and research models on resilience in leadership. *SAGE Open*, *4*(3), 1–8. DOI: 10.1177/2158244014545464

Lee, S. W. (2011, June). Probabilistic risk assessment for security requirements: A preliminary study. In *2011 Fifth International Conference on Secure Software Integration and Reliability Improvement* (pp. 11-20). IEEE. DOI: 10.1109/SSIRI.2011.12

Li, D., Ma, J., Cheng, T., van Genderen, J. L., & Shao, Z. (2018). Challenges and opportunities for the development of megacities. *International Journal of Digital Earth*, *12*(12), 1382–1395. DOI: 10.1080/17538947.2018.1512662

Lisdiono, P., Said, J., Yusoff, H., & Hermawan, A. A. (2022). examining leadership capabilities, risk management practices, and organizational resilience: The case of state-owned enterprises in Indonesia. *Sustainability (Basel)*, *14*(10), 6268. DOI: 10.3390/su14106268

Lisha, C., Goh, C. F., Yifan, S., & Rasli, A. (2017). Integrating guanxi into technology acceptance: An empirical investigation of WeChat. *Telematics and Informatics*, *34*(7), 1125–1142. DOI: 10.1016/j.tele.2017.05.003

Liu, L., Ju, J., Feng, Y., & Hu, Q. 2019. Impact of Governance Structure Characteristics of Public-private Partnerships on Smart City Project Success: Evidence from a Multi-case Study in China // Proceedings of the 52nd Hawaii International Conference on System Sciences. DOI: DOI: 10.24251/HICSS.2019.397

Liu, Y. (2019). An optimization-driven dynamic vehicle routing algorithm for on-demand meal delivery using drones. *Computers & Operations Research*, *111*, 1–20. DOI: 10.1016/j.cor.2019.05.024

Livingstone, O., & Anthony, B. P. (2023). Documenting fisheries co-management in action: Case of the Grand Cape Mount County co-management Association, Liberia. *Marine Policy*, *157*, 105844. DOI: 10.1016/j.marpol.2023.105844

Long, N. D. B., Ooi, P. T., Le, T. V., Thiet, L. T., Ai, T. V., An, L. Q., Hudson, A., Tan, K. S., & Van, N. T. L. (2022). Leading in the age of the fourth industrial revolution–a perspective from Vietnam. *International Journal of Technology*, *13*(5), 949–957. DOI: 10.14716/ijtech.v13i5.5839

LoRaWAN Alliance. (2017). *LoRaWAN® specification v1.1.* https://hz137b.p3cdn1.secureserver.net/wp-content/uploads/2020/11/lorawantm_specification_-v1.1.pdf?time=1687464361

Lozano, S., Calzada-Infante, L., Adenso-Díaz, B., & García, S. (2019). Complex network analysis of keywords co-occurrence in the recent efficiency analysis literature. *Scientometrics*, *120*(2), 609–629. DOI: 10.1007/s11192-019-03132-w

Luo, X., Li, H., Zhang, J., & Shim, J. P. (2010). Examining multi-dimensional trust and multi-faceted risk in initial acceptance of emerging technologies: An empirical study of mobile banking services. *Decision Support Systems*, *49*(2), 222–234. DOI: 10.1016/j.dss.2010.02.008

Lu, Y., Yang, S., Chau, P. Y. K., & Cao, Y. (2011). Dynamics between the trust transfer process and intention to use mobile payment services: A cross-environment perspective. *Information & Management*, *48*(8), 393–403. DOI: 10.1016/j.im.2011.09.006

Ly, P., & Mason, G. (2012). Individual preferences over development projects: Evidence from microlending on Kiva. *Voluntas*, *23*(4), 1036–1059. DOI: 10.1007/s11266-011-9255-8

Lyu, P., Liu, X., & Yao, T. (2023). A bibliometric analysis of literature on bibliometrics in recent half-century. *Journal of Information Science*, 01655515231191233. Advance online publication. DOI: 10.1177/01655515231191233

Magwegwe, F. (2021). Ayanda Mbatha: Resilience in action during the COVID-19 pandemic (A). *Emerald Emerging Markets Case Studies*, *11*(4), 1–12. DOI: 10.1108/EEMCS-06-2021-0201

Mahajan, R., Lim, W. M., Kumar, S., & Sareen, M. (2023). COVID-19 and management education: From pandemic to endemic. *International Journal of Management Education*, *21*(2), 100801. DOI: 10.1016/j.ijme.2023.100801

Maheshwari, A. (2019). *Digital Transformation: Building Intelligent Enterprises*. Wiley.

Majnoor, N., & Vinayagam, K. (2023). The ascendency of the paradigm shift from organizational change management to change agility. *International Journal of Professional Business Review*, *8*(4), e01151. DOI: 10.26668/businessreview/2023.v8i4.1151

Mäkinen, E. I. (2018). Complexity leadership theory and the leaders of transdisciplinary science. *Informing Science*, *21*, 133–155. DOI: 10.28945/4009

Mariani, M. M., Machado, I., Magrelli, V., & Dwivedi, Y. K. (2023). Artificial intelligence in innovation research: A systematic review, conceptual framework and future research directions. *Technovation*, *122*, 102623. DOI: 10.1016/j.technovation.2022.102623

Martsinkevich, V. I., Larionova, G. S., Tereshchenko, V. V., Sitnikova, K. A., & Gorlushkina, N. N. (n.d.). Analysis of the possibilities of parsing electronic text documents for the automation of standard control. *CyberLeninka.* https://cyberleninka.ru/article/n/analiz-vozmozhnostey-parsinga-elektronnyh-tekstovyh-dokumentov-dlya-avtomatizatsii-normokontrolya/viewer

Martynova, M. D., & Bochkova, I. A. (2023). Transformation of moral values of youth in the field of digital communication. *Primo Aspectu, 3*(55), 55–64. DOI: 10.35211/2500-2635-2023-3-55-55-64

Mas, I., & Radcliffe, D. (2010). *Mobile payments go viral: M-PESA in Kenya.* World bank Document. https://documents.worldbank.org/curated/en/638851468048259219

Mathew, J., Gupta, N., & Jagose, J. (2023). *Innovations In the Indian Healthcare Industry to Brave the Vuca Times*. Asia Pacific Journal of Health Management., DOI: 10.24083/apjhm.v18i1.1703

Maupin, C. K., Mohan, G., Choudhury, A., Deepak, P., & Jin, F. (2024). Network-based approaches to leadership: An organizing framework review, and recommendations. *The Leadership Quarterly, 35*(1), 101753. DOI: 10.1016/j.leaqua.2023.101753

Mbiti, I., & Weil, D. N. (2015). Mobile banking: The impact of M-Pesa in Kenya. In *African successes, Volume III: Modernization and development* (pp. 247-293). University of Chicago Press. https://doi.org/DOI: 10.7208/9780226315867-009

McBride, N., & Liyala, S. (2021). Memoirs from Bukhalalire: A poetic inquiry into the lived experience of M-PESA mobile money usage in rural Kenya. *European Journal of Information Systems*, •••, 1–22. DOI: 10.1080/0960085X.2021.1924088

Mehta, P. (2017). Framework of Indian Healthcare System and its. *Challenges*, •••, 247–271. DOI: 10.4018/978-1-5225-0640-9.ch011

Melián-Alzola, L., Domínguez-Falcón, C., & Martín-Santana, J. D. (2020). The role of the human dimension in organizational agility: An empirical study in intensive care units. *Personnel Review, 49*(9), 1945–1964. DOI: 10.1108/PR-08-2019-0456

Mennella, C., Maniscalco, U., De Pietro, G., & Esposito, M. (2024). Ethical and regulatory challenges of AI technologies in healthcare: A narrative review. *Heliyon, 10*(4), e26297. DOI: 10.1016/j.heliyon.2024.e26297

Mervyn, K., Amoo, N., & Malby, R. (2019). Challenges and insights in inter-organizational collaborative healthcare networks: An empirical case study of a place-based network. *The International Journal of Organizational Analysis, 27*(4), 875–902. DOI: 10.1108/IJOA-05-2018-1415

Meyerowitz, D., Lew, C., & Svensson, G. (2018). Scenario-planning in strategic decision-making: Requirements, benefits and inhibitors. *foresight*, 20(6), 602–621. DOI: 10.1108/FS-04-2018-0036

Mijwil, M. M., Faieq, A. K., & Al-Mistarehi, A.-H. (2022). The Significance of Digitalisation and Artificial Intelligence in The Healthcare Sector: A Review. *Asian Journal of Pharmacy. Nursing and Medical Sciences*, *10*(3). Advance online publication. DOI: 10.24203/ajpnms.v10i3.7065

Millings, M., Burke, L., & Robinson, G. (2019). Lost in transition? The personal and professional challenges for probation leaders engaged in delivering public sector reform. *Probation Journal*, *66*(1), 60–76. DOI: 10.1177/0264550518820120

Misaros, M., Stan, O. P., Donca, I. C., & Miclea, L. C. (2023). Autonomous Robots for Services—State of the Art, Challenges, and Research Areas. *Sensors (Basel)*, *23*(10), 4962. DOI: 10.3390/s23104962

Mishra, J., & Sharma, S. (2024). A Study on impact of adaptability and accessibility of digital payment systems and its impact on growth and development of economy. *Academy of Marketing Studies Journal*, *28*(1), 1–5. https://www.abacademies.org/articles/a-study-on-impact-of-adaptability-and-accessibility-of-digital-payment-systems-and-its-impact-on-growth-and-development-of-economy-16173.html

Mitchell, M. (2016). Authentic leadership in illness blogs: What we can learn from jaw surgery bloggers. *Cogent Social Sciences*, *2*(1), 1271270. DOI: 10.1080/23311886.2016.1271270

Mitchell, S., Steinbach, J., Flanagan, T., Ghabezi, P., Harrison, N., O'reilly, S., Killian, S., & Finnegan, W. (2023). Evaluating the sustainability of lightweight drones for delivery: Towards a suitable methodology for assessment. *Functional Composite Materials*, *4*(1), 4. DOI: 10.1186/s42252-023-00040-4

Mittal, H., & Singh, A. (2024). Crypto in India: Time for SEBI to take the lead. DNLUSLJ, 1, 20-25. https://dnluslj.in/wp-content/uploads/2024/04/DNLU-SLJ-SPECIAL-VOLUME-4.pdf

Miyamoto, K., Sato, Y., & Kitazume, K. (2005). Private-sector participation in infrastructure projects and value for money: Economic and financial impacts. *Transportation Research Record: Journal of the Transportation Research Board*, *1932*(1), 16–22. DOI: 10.1177/0361198105193200103

Mohamed Shaffril, H. A., Samsuddin, S. F., & Abu Samah, A. (2021). The ABC of systematic literature review: The basic Methodological guidance for beginners. *Quality & Quantity*, *55*(4), 1319–1346. DOI: 10.1007/s11135-020-01059-6

Mohammad, W. A., Nazih Diab, Y., Elomri, A., & Triki, C. (2023, April). Innovative solutions in last mile delivery: Concepts, practices, challenges, and future directions. In *Supply Chain Forum* [). Taylor & Francis.]. *International Journal (Toronto, Ont.)*, *24*(2), 151–169.

Mohsan, S. A. H., Othman, N. Q. H., Li, Y., Alsharif, M. H., & Khan, M. A. (2023). Unmanned aerial vehicles (UAVs): Practical aspects, applications, open challenges, security issues, and future trends. *Intelligent Service Robotics*, *16*(1), 109–137. DOI: 10.1007/s11370-022-00452-4

Mohsan, S. A. H., Zahra, Q. U. A., Khan, M. A., Alsharif, M. H., Elhaty, I. A., & Jahid, A. (2022). Role of drone technology helping in alleviating the COVID-19 pandemic. *Micromachines*, *13*(10), 1593. DOI: 10.3390/mi13101593

Montasser, D., Prijadi, R., & Balqiah, T. E. (2023). The mediating effect of IT-enabled dynamic capabilities and organizational readiness on the relationship between transformational leadership and digital business model innovation: Evidence from Indonesia incumbent firms. *SAGE Open*, *13*(2), 21582440231181588. DOI: 10.1177/21582440231181588

Montgomery, O. L. (n.d.). *An investigation of the relationship between spectral reflectance and the chemical, physical and genetic characteristics of soils* (PhD dissertation, Department of Agronomy, Purdue University). https://docs.lib.purdue.edu/dissertations/AAI7701747/

Morales, S. N., Martínez, L. R., Gómez, J. A. H., López, R. R., & Torres-Argüelles, V. (2019). Predictors of organizational resilience by factorial analysis. *International Journal of Engineering Business Management*, *11*, 1–13. DOI: 10.1177/1847979019837046

Mo, S. Y. K., Beling, P. A., & Crowther, K. G. (2009, April). Quantitative assessment of cyber security risk using Bayesian Network-based model. In *2009 Systems and Information Engineering Design Symposium* (pp. 183-187). IEEE. DOI: 10.1109/SIEDS.2009.5166177

Munshi, M., Ritchie, A., & Nair, A. (2007). *Community-managed microfinance: a new model from Sri Lanka*. World Bank. Retrieved from https://www.semanticscholar.org/paper/Community-managed-Microfinance-A-New-Model-from-Sri-Munshi-Ritchie/38b23fd5a76703e00b621f180a99e19c71c29483

Mustak, M., Salminen, J., Plé, L., & Wirtz, J. (2021). Artificial intelligence in marketing: Topic modeling, scientometric analysis, and research agenda. *Journal of Business Research*, *124*, 389–404. DOI: 10.1016/j.jbusres.2020.10.044

Mustofa, A., & Tjahjono, H. K. (2024). Improving organizational agility through ethical leadership, employee job satisfaction and Islamic OCB. *Asian Journal of Business Research, 14*(2), 129–150. DOI: 10.14707/ajbr.240176

Naik, N., Hameed, B. Z., Shetty, D. K., Swain, D., Shah, M., Paul, R., Aggarwal, K., Ibrahim, S., Patil, V., Smriti, K., Shetty, S., Rai, B. P., Chlosta, P., & Somani, B. K. (2022). Legal and ethical consideration in artificial intelligence in healthcare: Who takes responsibility? *Frontiers in Surgery*, *9*, 862322. DOI: 10.3389/fsurg.2022.862322

Najmon, J., Raeisi, S., & Tovar, A. (2019). In Froes, F., & Boyer, R. (Eds.), *Review of additive manufacturing technologies and applications in the aerospace industry. In Additive Manufacturing for the Aerospace Industry* (pp. 7–31). Elsevier Inc.

Nakamura, H., & Kajikawa, Y. (2018). Regulation and innovation: How should small unmanned aerial vehicles be regulated? *Technological Forecasting and Social Change*, *128*, 262–274. DOI: 10.1016/j.techfore.2017.06.015

Nayak, B., Bhattacharyya, S. S., & Krishnamoorthy, B. (2023). Integrating the dialectic perspectives of resource-based view and industrial organization theory for competitive advantage–a review and research agenda. *Journal of Business and Industrial Marketing*, *38*(3), 656–679. DOI: 10.1108/JBIM-06-2021-0306

Ndung'u, N. (2021). *A digital financial services revolution in Kenya: The M-Pesa case study*. African Economic Research Consortium., Retrieved from https://www.researchgate.net/publication/349548752_A_Digital_Financial_Services_Revolution_in_Kenya_The_M-Pesa_Case_Study

Nesti, G. (2019). Mainstreaming gender equality in smart cities: Theoretical, methodological and empirical challenges. *Information Polity*, *24*(3), 289–304. DOI: 10.3233/IP-190134

Ngo, T., Kashani, A., Imbalzano, G., Nguyen, K., & Hui, D. (2018). Additive manufacturing (3D printing): A review of materials, methods, applications and challenges. *Composites. Part B, Engineering*, *143*, 172–196. DOI: 10.1016/j.compositesb.2018.02.012

Nguyen, T., Le, C. V., Nguyen, M., Nguyen, G., Lien, T.T.H., & Nguyen, O. (2024). The organisational impact of agility: A systematic literature review. *Management Review Quarterly,* 1-49. DOI: 10.1007/s11301-024-00446-9

Nicol, D. M., Sanders, W. H., & Trivedi, K. S. (2004). Model-based evaluation: From dependability to security. *IEEE Transactions on Dependable and Secure Computing*, *1*(1), 48–65. DOI: 10.1109/TDSC.2004.11

Nicoletti, B. (2018). Agile Procurement: Volume II: Designing and Implementing a Digital Transformation. Cham, Switzerland. *Palgrave Macmillan., 281p*. Advance online publication. DOI: 10.1007/978-3-319-61085-6

Nield, S. J., Boettinger, J. L., & Ramsey, R. D. (2007). Mapping gypsic and natric soil areas using Landsat ETM data. *Soil Science Society of America Journal, 71*, 245–252. DOI: 10.1007/978-1-4020-8592-5_16

Niemeyer-Rens, R. F., Sullivan, G., & Self, S. W. (2022). Leadership adaptability within higher education. *Journal of Higher Education Theory and Practice*, *22*(9), 1–15. DOI: 10.33423/jhetp.v22i9.5357

Nkurunziza, G., Munene, J., Ntayi, J., & Kaberuka, W. (2019). Business process re-engineering in developing economies: Lessons from microfinance institutions (MFIs) in Uganda. *Innovation & Management Review*, *16*(2), 118–142. DOI: 10.1108/INMR-03-2018-0010

Nugraha, R. A., Jeyakodi, D., & Mahem, T. (2016). Urgency for legal framework on drones: Lessons for Indonesia, India, and Thailand. *Indon. L. Rev.*, *6*(2), 137. DOI: 10.15742/ilrev.v6n2.229

Nurjihadi, M., Ibrahim, O., & Pasaribu, S. M. (2024). University education and innovation resistance: A multigroup analysis of digital payment barriers among Indonesian farmers. *International Journal of Academic Research in Progressive Education and Development*, *13*(4), 2199–2216. DOI: 10.6007/IJARPED/v13-i4/23812

Øby, E. (2024). Assessing transparency and methodological precision in variable measurement within organizational research: Implications for validity. *Quality & Quantity*, •••, 1–18. DOI: 10.1007/s11135-024-01991-x

Ochnev, D. V., & Chirkin, E. S. (n.d.). Digital watermarks as a method of protection of text printed documents. *CyberLeninka*. https://cyberleninka.ru/article/n/tsifrovye-vodyanye-znaki-kak-metod-zaschity-tekstovyh-pechatnyh-dokumentov/viewer

Odunola, A. B., Olawumi, A. A., & Adesola, A. G. (2015). Performance prediction model for network security risk management. *Communications on Applied Electronics*, *2*(8), 1–7. DOI: 10.5120/cae2015651816

Olaes, T. (2025). Using the FAIR model for cyber risk quantification. *Balbix*. https://www.balbix.com/insights/fair-model-for-risk-quantification-pros-and-cons/

Omoniwa, B. (2021). Nearly cashless: Digital transformation or cultural transmission? EconStor. Retrieved from https://www.econstor.eu/bitstream/10419/289862/1/1886219311.pdf

Ongsulee, P. (2017, November). Artificial intelligence, machine learning and deep learning. In *2017 15th international conference on ICT and knowledge engineering (ICT&KE)* (pp. 1-6). IEEE. DOI: 10.1109/ICTKE.2017.8259629

Opoku-Okuampa, I. Y. (2024). The Influence of Cultural Factors on Financial Decision-Making in Ghana. *International Journal of Finance and Banking Research, 10*(6), 118–125. http://www.intjfobr.com/article/10.11648/j.ijfbr.20241006.12. DOI: 10.11648/j.ijfbr.20241006.12

Oreshkina, E. I., & Favorskaya, M. N. (n.d.). Classification of embedding digital watermarking methods. *CyberLeninka.* https://cyberleninka.ru/article/n/klassifikatsiya-metodov-naneseniya-tsifrovyh-vodyanyh-znakov/viewer

Osasona, F., Amoo, O. O., Atadoga, A., Abrahams, T. O., Farayola, O. A., & Ayinla, B. S. (2024). Reviewing the ethical implications of AI in decision making processes. *International Journal of Management & Entrepreneurship Research, 6*(2), 322–335. DOI: 10.51594/ijmer.v6i2.773

Padee, V. (31 October, 2023). Overcoming cultural barriers in payment systems. Apexx Global. Retrieved from https://apexx.global/blog/overcoming-cultural-barriers-in-payment-systems/

Panda, S., & Sharma, R. (2020). Impact of patent rights on innovation: A meta-analysis. *Journal of Public Affairs, 20*(2), e2000. DOI: 10.1002/pa.2000

Pandey, D. (2023). Empowering rural communities through digital literacy initiatives in India. [IARE]. *International Assulta of Research and Engagement, 1*(1), 1–12. DOI: 10.5281/iare.v1i1.31

Pandit, M. (2021). Critical factors for successful management of VUCA times. *BMJ Leader, 5*(2), 121–123. DOI: 10.1136/leader-2020-000305 PMID: 36170532

Pappas, I. O., Mikalef, P., Giannakos, M. N., Krogstie, J., & Lekakos, G. (2018). Big data and business analytics ecosystems: Paving the way towards digital transformation and sustainable societies. *Information Systems and e-Business Management, 16*(3), 479–491. DOI: 10.1007/s10257-018-0377-z

Paracha, A., Arshad, J., Farah, M., & Ismail, K. (2024). Machine learning security and privacy: A review of threats and countermeasures. *EURASIP Journal on Information Security, 10*(1), 10. Advance online publication. DOI: 10.1186/s13635-024-00158-3

Parenti, R. (2020). Regulatory sandboxes and innovation hubs for FinTech. *Study for the Committee on Economic and Monetary Affairs, Policy Department for Economic, Scientific and Quality of Life Policies, European Parliament, Luxembourg, 65*. Retrieved from https://www.europarl.europa.eu/supporting-analyses

Pascal, I. N., & Edith, O. (2023). Cross-cultural approaches to managing the adoption of cashless policies in the workplace. *International Journal of Management, Social Sciences. Peace and Conflict Studies*, *6*(2), 345–353. https://ijmsspcs.com/index.php/IJMSSPCS/article/view/538/587

Pasmore, W., Winby, S., Mohrman, S. A., & Vanasse, R. (2019). Reflections: Sociotechnical systems design and organization change. *Journal of Change Management*, *19*(2), 67–85. DOI: 10.1080/14697017.2018.1553761

Patel, A. S., Rao, V. K., & Radhakrishnan, M. K. (2023). Impact of Mobile Banking Platforms Paytm and Google Pay on Financial Inclusion in Rural and Semi-Urban Areas in India. *Journal of Finance and Accounting*, *7*(5), 113–122. DOI: 10.53819/81018102t4205

Patil, P. P., Dwivedi, Y. K., & Rana, N. P. (2017). Digital payments adoption: an analysis of literature. In *Digital Nations–Smart Cities, Innovation, and Sustainability: 16th IFIP WG 6.11 Conference on E-Business, E-Services, and E-Society, I3E 2017, Delhi, India, November 21–23, 2017, Proceedings 16* (pp. 61-70). Springer International Publishing. https://doi.org/DOI: 10.1007/978-3-319-68557-1_7

Patil, P. P., Rana, N. P., & Dwivedi, Y. K. (2018). Digital payments adoption research: A review of factors influencing consumer's attitude, intention and usage. In *Challenges and Opportunities in the Digital Era: 17th IFIP WG 6.11 Conference on e-Business, e-Services, and e-Society, I3E 2018, Kuwait City, Kuwait, October 30–November 1, 2018, Proceedings 17* (pp. 45-52). Springer International Publishing. https://doi.org/DOI: 10.1007/978-3-030-02131-3_6

Paul, S., Ramaprasad, A., & Wickramasinghe, N. (2013). Introduction to Technology Mediated Collaborations in Healthcare Minitrack. *2013 46th Hawaii International Conference on System Sciences*, 791–791. DOI: 10.1109/HICSS.2013.508

Pauner, C., Kamara, I., & Viguri, J. (2015, December). Drones. Current challenges and standardisation solutions in the field of privacy and data protection. In *2015 ITU Kaleidoscope: Trust in the Information Society (K-2015)* (pp. 1-7). IEEE.

Peltier, T. R. (2005). *Information security risk analysis*. CRC Press. https://www.crcpress.com/Information-Security-Risk-Analysis/Peltier/p/book/9780849333466

Poikonen, S., & Campbell, J. F. (2021). Future directions in drone routing research. *Networks*, *77*(1), 116–126. DOI: 10.1002/net.21982

Politi, E., Panagiotopoulos, I. E., Varlamis, I., & Dimitrakopoulos, G. (2021, April). A Survey of UAS Technologies to Enable Beyond Visual Line Of Sight (BVLOS) Operations. In *VEHITS* (pp. 505-512).

Popov, A. E. (2024). Methods of steganographic information hiding in images. *Science and Technology, (5)*. https://na-journal.ru/5-2024-informacionnye-tekhnologii/11724-metody-steganograficheskogo-skrytiya-informacii-v-izobrajeniyah

Poszler, F., Geisslinger, M., Betz, J., & Lütge, C. (2023). Applying ethical theories to the decision-making of self-driving vehicles: A systematic review and integration of the literature. *Technology in Society*, *75*, 102350. DOI: 10.1016/j.techsoc.2023.102350

Qadri, E. N. (2023). Critical Factors that Affect the Adoption of Mobile Payment Services in Developed and Developing Countries. Master's Degree Project. Retrieved from https://uu.diva-portal.org/smash/get/diva2:1800492/FULLTEXT01.pdf

Rahman, M. M., Kshetri, N., Sayeed, S. A., & Rana, M. M. (2024). AssessITS: Integrating procedural guidelines and practical evaluation metrics for organizational IT and cybersecurity risk assessment. *arXiv*. https://arxiv.org/abs/2410.01750

Raja, A. S. (2021). Business Research in the VUCA World. *Ushus Journal of Business Management*, *20*(1), v–xvi. DOI: 10.12725/ujbm.54.0

Ramadan, M., Bou Zakhem, N., Baydoun, H., Daouk, A., Youssef, S., El Fawal, A., Elia, J., & Ashaal, A. (2023). Toward digital transformation and business model innovation: The Nexus between leadership, organizational agility, and knowledge transfer. *Administrative Sciences*, *13*(8), 185. DOI: 10.3390/admsci13080185

Ramli, K., Spigel, B., Williams, N., Mawson, S., & Jack, S. (2023). Managing through a crisis: Emotional leadership strategies of high-growth entrepreneurs during the COVID-19 pandemic. *Entrepreneurship and Regional Development*, *35*(1-2), 24–48. DOI: 10.1080/08985626.2022.2143905

Raney, A. F. (2014). Agility in adversity: Integrating mindfulness and principles of adaptive leadership in the administration of a community mental health center. *Clinical Social Work Journal*, *42*(3), 312–320. DOI: 10.1007/s10615-014-0487-0

Rani, I. H., & Widyowati, L. A. (2021). The role of visionary leader competency in bridging company's levers of controls and dynamic capabilities. *Journal of Logistics. Informatics and Service Science*, *8*(2), 1–18. DOI: 10.33168/LISS.2021.0201

Raposo, V. L. (2024). When facial recognition does not 'recognise': Erroneous identifications and resulting liabilities. *AI & Society*, *39*(4), 1857–1869. DOI: 10.1007/s00146-023-01634-z

Ratti, K., & Claudel, M. (2017). *Cities of Tomorrow: Sensors, Networks, Hackers and the Future of Urban Life*. Gaidar Institute Publishing House.

Reis, A. C. S. (September 2022). The impact of culture on risk aversion in financial decisions: evidence from a TV show. Retrieved from https://repositorio-aberto.up.pt/bitstream/10216/146593/2/597064.pdf

Rejeb, A., Rejeb, K., Simske, S. J., & Treiblmaier, H. (2023). Drones for supply chain management and logistics: A review and research agenda. *International Journal of Logistics*, *26*(6), 708–731. DOI: 10.1080/13675567.2021.1981273

Resnik, D. B., & Elliott, K. C. (2019). Using drones to study human beings: Ethical and regulatory issues. *Science and Engineering Ethics*, *25*(3), 707–718. DOI: 10.1007/s11948-018-0032-6

Rethlefsen, M. L., Kirtley, S., Waffenschmidt, S., Ayala, A. P., Moher, D., Page, M. J., Koffel, J. B., Blunt, H., Brigham, T., Chang, S., Clark, J., Conway, A., Couban, R., de Kock, S., Farrah, K., Fehrmann, P., Foster, M., Fowler, S. A., Glanville, J., & Young, S. (2021). PRISMA-S: An extension to the PRISMA Statement for Reporting Literature Searches in Systematic Reviews. *Systematic Reviews*, *10*(1), 1–19. DOI: 10.1186/s13643-020-01542-z

Reyes, D. L., Bisbey, T., Day, D., & Salas, E. (2021). Translating 6key insights from research on leadership and management in times of crisis. *BMJ Leader*, *5*(4), 291–294. DOI: 10.1136/leader-2020-000349

Ritchie, A. (2007). *Community-based financial organizations: a solution to access in remote rural areas?* World Bank, ARD. Retrieved from https://www.rfilc.org/wp-content/uploads/2020/08/1175091210909_community_based_FO_rural_solution.pdf

Roblek, V., Dimovski, V., Jovanov Oblak, K., Meško, M., & Peterlin, J. (2024). Leadership and managerial challenges to ensure agile management as a method to enable business success: A Delphi study of the Slovenian health organisations. *Measuring Business Excellence*, *28*(1), 39–51. DOI: 10.1108/MBE-09-2023-0122

Rodina, L., Harris, L., Ziervogel, G., & Wilson, J. (2024). Resilience countercurrents: Water infrastructures, informality, and inequities in Cape Town, South Africa. *World Development*, *180*, 106619. DOI: 10.1016/j.worlddev.2024.106619

Rodwin, L. (1981). Problems of the metropolis. In *Cities and city planning. Environment, development, and public policy: Cities and development* (pp. 41–51). Springer., DOI: 10.1007/978-1-4684-1089-1_6

Romashkina, G. F., & Khudyakova, M. V. (2020). Sociological analysis of human capital factors and resources. *Economic and Social Changes: Facts, Trends. Forecast*, *13*(6), 232–251.

Romero-Lankao, P., Hughes, S., Qin, H., Hardoy, J., Rosas-Huerta, A., Borquez, R., & Lampis, A. (2014). Scale, urban risk and adaptation capacity in neighborhoods of Latin American cities. *Habitat International*, *42*, 224–235. DOI: 10.1016/j.habitatint.2013.12.008

Rosa, F. D. F., Bonacin, R., & Jino, M. (2017). The security assessment domain: A survey of taxonomies and ontologies. *arXiv*. https://arxiv.org/abs/1706.09772

Rowe, S. W., Arghode, V., & Bhattacharyya, S. S. (2024). A study on adaptive performance, work-related psychological health and demographics in Episcopal Church bishops. *Journal of Work-Applied Management*, *16*(1), 31–47. DOI: 10.1108/JWAM-02-2023-0015

Rudrappa, S., Agarkhed, D. V., & Vaidya, S. S. (2019). Healthcare Systems: India. In *Quality Spine Care* (pp. 211–224). Springer International Publishing. DOI: 10.1007/978-3-319-97990-8_13

Ryazanova, L. V. (2020). *2020 Digital transformation of state and society: collection of articles*. Rusains.

Sadaf, M., Iqbal, Z., Javed, A. R., Saba, I., Krichen, M., Majeed, S., & Raza, A. (2023). Connected and automated vehicles: Infrastructure, applications, security, critical challenges, and future aspects. *Technologies*, *11*(5), 117. DOI: 10.3390/technologies11050117

Safitra, M. F., Lubis, M., & Fakhrurroja, H. (2023). Counterattacking cyber-threats: A framework for the future of cyber-security. *Sustainability (Basel)*, *15*(18), 13369. DOI: 10.3390/su151813369

Said, M., Aeschliman, S., & Stathopoulos, A. (2023). Robots at your doorstep: Acceptance of near-future technologies for automated parcel delivery. *Scientific Reports*, *13*(1), 18556. DOI: 10.1038/s41598-023-45371-1

Saiyid, M. (2023). Pix: Brazil's Successful Instant Payment System. *IMF Staff Country Reports*, 53-61. Retrieved from https://www.elibrary.imf.org/view/journals/002/2023/289/article-A004-en.xml

Sames, W., List, F., Pannala, S., Dehoff, R., & Babu, S. (2016). The metallurgy and processing science of metal additive manufacturing. *International Materials Reviews*, *61*(5), 315–360. DOI: 10.1080/09506608.2015.1116649

Sampson, M., McGowan, J., Cogo, E., & Horsley, T. (2006). Managing database overlap in systematic reviews using batch citation matcher: Case studies using Scopus. *Journal of the Medical Library Association*, 94(4), 461–463, e219. +E.

Sanusi, Sh. (2019). Eyeing fear and anxiety: Postcolonial modernity and cultural identity in the urban space of The Eye. *Otherness: Essays and Studies*, *7*(1), 70–99.

Saputra, N., Putera, R. E., Zetra, A., Valentina, T. R., & Mulia, R. A. (2024). Capacity building for organizational performance: A systematic review, conceptual framework, and future research directions. *Cogent Business & Management*, *11*(1), 2434966. Advance online publication. DOI: 10.1080/23311975.2024.2434966

Saravanan, S., & Chithra, M. (2024). Adaptation of digital payment system among rural consumers in Enhancing the performance of the MFI. *International Journal of Arts. Science and Humanities*, *11*(1), 59–65. DOI: 10.34293/sijash.v11iS1-June.7791

Sawik, B. (2024). Optimizing Last-Mile Delivery: A Multi-Criteria Approach with Automated Smart Lockers, Capillary Distribution and Crowdshipping. *Logistics (Basel)*, *8*(2), 52. DOI: 10.3390/logistics8020052

Sayal, A., Gupta, A., Jha, J., Chaithra, N., & Nithin, A. (2025). Application of Artificial Intelligence in Healthcare Sector: A Multidimensional Perspective. In *Driving Global Health and Sustainable Development Goals With Smart Technology* (pp. 115–138). IGI Global Scientific Publishing.

Schäffer, B., Pieren, R., Heutschi, K., Wunderli, J. M., & Becker, S. (2021). Drone noise emission characteristics and noise effects on humans—A systematic review. *International Journal of Environmental Research and Public Health*, *18*(11), 5940. DOI: 10.3390/ijerph18115940

Scheepers, C. B., Mathibe, M. S., & Verachia, A. (2023). Localisation of PPE manufacturing during COVID-19 (local manufacturing partnership and Kingsgate Clothing South Africa). *Emerald Emerging Markets Case Studies*, *13*(1), 1–13. DOI: 10.1108/EEMCS-10-2022-0352

Schell, J. A. (2023). Designing for academic resilience in hands-on courses in times of crisis: Two models for supporting hands-on online learning drawn from the COVID-19 Pandemic. *The American Behavioral Scientist*, *67*(13), 1632–1654. DOI: 10.1177/00027642221118292

Schneier, B. (2000). *Secrets and lies: Digital security in a networked world*. Wiley. https://www.schneier.com/books/secrets-and-lies/

Schoemaker, P. J. H., Heaton, S., & Teece, D. (2018). Innovation, dynamic capabilities, and leadership. *California Management Review*, *61*(1), 1–28. DOI: 10.1177/0008125618790246

Schulze, J. H., & Pinkow, F. (2020). Leadership for organisational adaptability: How enabling leaders create adaptive space. *Administrative Sciences*, *10*(3), 37. DOI: 10.3390/admsci10030037

Sehrawat, V. (2018). Liability issue of domestic drones. *Santa Clara High-Technology Law Journal*, *35*, 110.

Selsky, J. W., & Parker, B. (2005). Cross-sector partnerships to address social issues: Challenges to theory and practice. *Journal of Management*, *31*(6), 849–873. DOI: 10.1177/0149206305279601

Sergeev, V. F., & Preobrazhenskaya, K. V. (2023). Megapolis as a risk space: Destructive factors of personal identity and methodologies for their description and research. *Bulletin of the Russian Christian Humanitarian Academy*, *24*(1), 259–265.

Sergi, I., Montanaro, T., Benvenuto, F. L., & Patrono, L. (2021). A smart and secure logistics system based on IoT and cloud technologies. *Sensors (Basel)*, *21*(6), 2231. DOI: 10.3390/s21062231

Shaikh, E., Al-Ali, A. R., Muhammad, S., Mohammad, N., & Aloul, F. (2023). Security analysis of a digital twin framework using probabilistic model checking. *IEEE Access : Practical Innovations, Open Solutions*, *11*, 26358–26374. DOI: 10.1109/ACCESS.2023.3257171

Shameli-Sendi, A., Aghababaei-Barzegar, R., & Cheriet, M. (2016). Taxonomy of information security risk assessment (ISRA). *Computers & Security*, *59*, 32–62. DOI: 10.1016/j.cose.2015.11.001

Shapiro, J. M. (2008). Smart cities: Quality of life, productivity, and the growth effects of human capital. *The Review of Economics and Statistics*, *88*(2), 324–335. DOI: 10.1162/rest.88.2.324

Sharafaldin, I., Lashkari, A. H., & Ghorbani, A. (2018). Toward generating a new intrusion detection dataset and intrusion traffic characterization. *Proceedings of the 10th International Conference on Information Systems Security and Privacy*, 108–116. DOI: 10.5220/0006639801080116

Shende, A., & Kathiriya, S. (2024). Consumer Protection and Payment Compliance in the Digital Age: Challenges, Innovations, and Regulatory Responses. 9. Retrieved from https://www.ijnrd.org/papers/IJNRD2401114.pdf

Shetty, A., Yu, M., Kurzhanskiy, A., Grembek, O., Tavafoghi, H., & Varaiya, P. (2021). Safety challenges for autonomous vehicles in the absence of connectivity. *Transportation Research Part C, Emerging Technologies*, *128*, 103133. DOI: 10.1016/j.trc.2021.103133

Shetty, K. P. (2022). A Study on the Transformation of the Indian Healthcare Sector during the COVID-19 Pandemic. *International Journal for Research in Applied Science and Engineering Technology, 10*(1), 1057–1064. DOI: 10.22214/ijraset.2022.39986

Shilling, J. D. (2007). *The Nexus between infrastructure and environment* (Vol. 1). Evaluation Cooperation Group.

Shin, D.-H. (2010). The effects of trust, security, and privacy in social networking: A security-based approach to understand the pattern of adoption. *Interacting with Computers, 22*(5), 428–438. DOI: 10.1016/j.intcom.2010.05.001

Shor, P. W. (1994). Algorithms for quantum computers: Discrete logarithms and factoring. In *Proc. 35th Annual Symposium on Foundations of Computer Science*, IEEE Computer Society Press, Los Alamitos, CA (pp. 124-134).

Sibley, J. (16 July 2024). *Redefining community finance: Practical strategies for payment leaders to drive financial inclusion*. The Payments Association. Retrieved from https://thepaymentsassociation.org/article/redefining-community-finance-practical-strategies-for-payment-leaders-to-drive-financial-inclusion/

Sihag, V., Choudhary, G., Choudhary, P., & Dragoni, N. (2023). Cyber4drone: A systematic review of cyber security and forensics in next-generation drones. *Drones (Basel), 7*(7), 430. DOI: 10.3390/drones7070430

Silalahi, S. A. F., Junaidi, A., Nasution, L. Z., Sutoto, A., Hutomo, A. S., Sutrisno, J., Lukas, A., Ginting, J., Suripto, S., & Lestari, R. (2023). Grit and business survivability among small business during the COVID-19 pandemic: The moderating role of servant leadership. *Cogent Business & Management, 10*(3), 2284442. Advance online publication. DOI: 10.1080/23311975.2023.2284442

Sim, K., Hwang, W.-S., & Choi, M.-R. (2019). Building a New Smart City: Integrating Local Culture and Technology. *Journal of Digital Convergence., 17*(9), 193–198. DOI: 10.14400/JDC.2019.17.9.193

Simpson, C., Ataii, E., Kemp, E., & Zhang, Y. (2019). *Mobility 2030: Transforming the mobility landscape*. KPMG International.

Sindila, A., Foss, N. J., & Zhan, X. (2023). Building Resilience for Surviving and Thriving in a VUCA Context. *AIB Insights, 23*(3). Advance online publication. DOI: 10.46697/001c.73812

Sinha, D., & Sinha, S. (2020). Managing in a VUCA World: Possibilities and Pitfalls. *Journal of Technology Management for Growing Economies, 11*(1), 17–21. DOI: 10.15415/jtmge.2020.111003

Siregar, A. A., Afiff, A. Z., & Halim, R. E. (2023). Linking agile leadership and business sustainability through the mediation of political and social capabilities. *Journal of Open Innovation*, *9*(4), 100153. DOI: 10.1016/j.joitmc.2023.100153

Slater, S. F., Mohr, J. J., & Sengupta, S. (2014). Radical product innovation capability: Literature review, synthesis, and illustrative research propositions. *Journal of Product Innovation Management*, *31*(3), 552–566. DOI: 10.1111/jpim.12113

Sleiman, K. A. A., Juanli, L., Lei, H., Liu, R., Ouyang, Y., & Rong, W. (2021). User trust levels and adoption of mobile payment systems in China: An empirical analysis. *SAGE Open*, *11*(4), 21582440211056599. DOI: 10.1177/21582440211056599

Sodiya, E. O., Umoga, U. J., Amoo, O. O., & Atadoga, A. (2024). AI-driven warehouse automation: A comprehensive review of systems. *GSC Advanced Research and Reviews*, *18*(2), 272–282. DOI: 10.30574/gscarr.2024.18.2.0063

Sokrovolska, N., Prodius, O., Zhuk, O., Yamnenko, G., & Fedyna, V. (2022). Public-private partnership in microfinance. *Financial & Credit Activity: Problems of Theory & Practice*, *4*(45), 93–103. Advance online publication. DOI: 10.55643/fcaptp.4.45.2022.3842

Sommestad, T., Ekstedt, M., & Johnson, P. (2010). A probabilistic relational model for security risk analysis. *Computers & Security*, *29*(6), 659–679. DOI: 10.1016/j.cose.2010.02.002

Song, H., Srinivasan, R., Sookoor, T., & Jeschke, S. (2017). *Smart Cities: Foundations, Principles, and Applications*. Wiley., DOI: 10.1002/9781119226444

Song, Y., Lei, L., Wu, L., & Chen, S. (2023). Studying domain structure: A comparative analysis of bibliographic coupling analysis and co-citation analysis considering all authors. *Online Information Review*, *47*(1), 123–137. DOI: 10.1108/OIR-12-2020-0540

Song, Y., Sun, X., & Duan, W. (2023). Co-word analysis of first-authored publications written by top cited scholars of social work. *Research on Social Work Practice*, *33*(8), 929–943. DOI: 10.1177/10497315221148661

Sorooshian, S., Khademi Sharifabad, S., Parsaee, M., & Afshari, A. R. (2022). Toward a modern last-mile delivery: Consequences and obstacles of intelligent technology. *Applied System Innovation*, *5*(4), 82. DOI: 10.3390/asi5040082

Soyer, B., & Tettenborn, A. (2022). Artificial intelligence and civil liability—Do we need a new regime? *International Journal of Law and Information Technology*, *30*(4), 385–397. DOI: 10.1093/ijlit/eaad001

Spante, M., Hashemi, S. S., Lundin, M., & Algers, A. (2018). Digital competence and digital literacy in higher education research: Systematic review of concept use. *Cogent Education*, *5*(1), 1519143. DOI: 10.1080/2331186X.2018.1519143

Spillers, F., & Gavine, A. (2018). Localization of mobile payment systems: cultural and temporal rhythms in user adoption. *Nordic Conference on Human-Computer Interaction*, 862–867. https://doi.org/DOI: 10.1145/3240167.3240276

Stamp, M. (2022). *Introduction to machine learning with applications in information security* (2nd ed.). Chapman and Hall/CRC., DOI: 10.1201/9781003264873

Steffen, B., Braun von Reinersdorff, A., & Rasche, C. (2023). IT-Based Decision Support for Holistic Healthcare Management in Times of VUCA, Disorder, and Disruption. *Applied Sciences (Basel, Switzerland)*, *13*(10), 6008. DOI: 10.3390/app13106008

Stoneburner, G., Goguen, A., & Feringa, A. (2002). *Risk management guide for information technology systems* (NIST SP 800-30). https://csrc.nist.gov/publications/detail/sp/800-30/rev-1/final

Strebkov, T. (28 October, 2024). The presence and future of fintech UX & product design: Survey results. Pixels and Sense. Retrieved from https://www.pixelsandsense.com/insights/the-presence-and-future-of-fintech-ux-product-design-survey-results

Sultania, A. (2024). Are We Underestimating the Success of Self Help Groups in India? A Systematic Review. *Open Journal of Business and Management*, *12*(4), 2640–2661. DOI: 10.4236/ojbm.2024.124137

Sundararaj, J., & Meera, R. (2024). Adoption and User Behaviour Towards Digital Payments by Millennial. *South Eastern European Journal of Public Health*, •••, 822–828. DOI: 10.70135/seejph.vi.2126

Sunil, S. V., & Nalwaya, N. (2023). Digital Payment Systems–An Overview of Categories and Extant Opportunities and Challenges. [IJRASET]. *International Journal for Research in Applied Science and Engineering Technology*, *11*(4), 2321–9653. DOI: 10.22214/ijraset.2023.50019

Sun, N., Ding, M., Jiang, J., Xu, W., Mo, X., Tai, Y., & Zhang, J. (2023). Cyber threat intelligence mining for proactive cybersecurity defense: A survey and new perspectives. *IEEE Communications Surveys and Tutorials*, *25*(4), 3273282. DOI: 10.1109/COMST.2023.3273282

Sun, Y., Bashir, A. K., Tariq, U., & Xiao, F. (2021). Effective malware detection scheme based on classified behavior graph in IIoT. *Ad Hoc Networks*, *120*, 102558. DOI: 10.1016/j.adhoc.2021.102558

Szolomicki, J., & Golasz-Szolomicka, H. (2019). Technological advances and trends in modern high-rise buildings. *Buildings (Basel, Switzerland)*, *9*(9), 193. DOI: 10.3390/buildings9090193

Tarr, A. A., Tarr, J. A., Thompson, M., & Ellis, J. (Eds.). (2021). *Drone law and policy: Global development, risks, regulation and insurance*. Routledge. DOI: 10.4324/9781003028031

The Open Group. (2013). *Open FAIR™ risk analysis (O-RA) standard*. https://publications.opengroup.org/c13g

Thompsett, L. (4 October, 2024). The future of finance: Emerging technologies deep-dive. FinTech Magazine. Retrieved from https://fintechmagazine.com/articles/the-future-of-finance-emerging-technologies-deep-dive

Thude, B. R., Juhl, A. G., Stenager, E., von Plessen, C., & Hollnagel, E. (2019). Staff acting resiliently at two hospital wards. *Leadership in Health Services (Bradford, England)*, *32*(3), 445–457. DOI: 10.1108/LHS-09-2018-0042

Torre, D., & Xu, Q. (2020). Digital payments in China: adoption and interactions among applications. *Revue d'économie industrielle*, (172), 55-82. Retrieved from https://shs.hal.science/halshs-03313693/file/torre-xu-def.pdf

Townsend, A. M. (2013). *Smart Cities: Big Data, Civic Hackers, and the Quest for a New Utopia*. W. W. Norton & Company, Inc.

Tran, C., & Ross, T. (2021). Quantitative risk assessment with the FAIR model. *LogicGate*. https://www.logicgate.com/blog/the-fair-model-an-objective-approach-to-risk-measurement/

Triandis, H. C. (1995). *Individualism and collectivism*. Westview Press. Triandis, H. C. https://psycnet.apa.org/record/1995-97791-000

Trujillo, C. M., & Long, T. M. (2018). Document co-citation analysis to enhance transdisciplinary research. *Science Advances*, *4*(1), e1701130. DOI: 10.1126/sciadv.1701130

Trukhachev, A., Ivolga, A., & Varivoda, V. (2018). Analysis of agricultural tourism and study of its development potential. *Research Journal of Pharmaceutical, Biological and Chemical Sciences*, *9*(3), 792–801.

Tsiamis, N., Efthymiou, L., & Tsagarakis, K. P. (2019). A comparative analysis of the legislation evolution for drone use in OECD countries. *Drones (Basel)*, *3*(4), 75. DOI: 10.3390/drones3040075

Tubis, A. A., Poturaj, H., Dereń, K., & Żurek, A. (2024). Risks of Drone Use in Light of Literature Studies. *Sensors (Basel), 24*(4), 1205. DOI: 10.3390/s24041205

Umlauf, R., & Burchardt, M. (2022). Infrastructure-as-a-service: Empty skies, bad roads, and the rise of cargo drones. *Environment and Planning A. Economy and Space*, *54*(8), 1489–1509.

Urs, N., Roja, A., & Nisioi, I. (2024). Digital transformation in organizational management: A bibliometric analysis. *NISPAcee Journal of Public Administration and Policy*, *17*(1), 198–227. DOI: 10.2478/nispa-2024-0009

Urso, G. (2020). Metropolisation and the challenge of rural-urban dichotomies. *Urban Geography*, *42*(1), 37–57. DOI: 10.1080/02723638.2020.1760536

Utomo, F. C., & Latukismo, T. H. (2022). Trends and patterns in workforce agility literature: A scopus-based bibliometric analysis. *Journal of Wireless Mobile Networks, Ubiquitous Computing and Dependable Applications*, *13*(4), 211–224. DOI: 10.58346/JOWUA.2022.I4.014

Uzorka, A., & Kalabuki, K. (2024). The transformative impact of technological advancements in educational leadership on student experiences and outcomes. *Education and Information Technologies*, *29*(16), 1–22. DOI: 10.1007/s10639-024-12714-y

van Dijk, J. (2020). *The Digital Divide*. Polity Press., DOI: 10.1002/asi.24355

Vargas, J., Alsweiss, S., Toker, O., Razdan, R., & Santos, J. (2021). An overview of autonomous vehicles sensors and their vulnerability to weather conditions. *Sensors (Basel)*, *21*(16), 5397. DOI: 10.3390/s21165397

Vatsa, P., Ma, W., & Zheng, H. (2024). Mobile payment adoption in China: Do demographic and socioeconomic factors matter? *MDE. Managerial and Decision Economics*, *45*(3), 1428–1434. DOI: 10.1002/mde.4086

Velner, T., Rubinstein, Z., & Adini, B. (2023). Organizational Resilience Among Health Organizations in Israel. *Prehospital and Disaster Medicine*, *38*(S1), s143–s143. DOI: 10.1017/S1049023X23003758

Viano, S., Shahrokhi, F., & Hunter, S. B. (2024). Improvement science and school leadership: The precarious path to dynamic school improvement. *Frontiers in Education*, *9*, 1371664. Advance online publication. DOI: 10.3389/feduc.2024.1371664

Vidjikant, S. (31 July, 2024). User experience in digital payments – A guide through best practices. Softjourn. Retrieved from https://softjourn.com/insights/user-experience-in-digital-payments-a-guide-through-best-practices

Vinichenko, M. (2021). Megapolis residents' readiness to participate in megaprojects under the conditions of the digitalization of the economy and the use of artificial intelligence. *Revista Gestão Inovação e Tecnologias, 11*(4), 2926–2940. https://www.academia.edu/103501261/Megapolis_Residents_Readiness_to_Participate_in_Megaprojects_Under_the_Conditions_of_the_Digitalization_of_the_Economy_and_the_Use_of_Artificial_Intelligence

Voigt, P., & Von dem Bussche, A. (2017). The eu general data protection regulation (gdpr). *A Practical Guide, 1st Ed., Cham: Springer International Publishing, 10*(3152676), 10-5555. https://link.springer.com/book/10.1007/978-3-319-57959-7

Vorobyova, I. V., & Zhdanova, N. E. Personal predictors of students in the perception of stress factors of the real and virtual urban environment // Pedagogical education in Russia. 2022. No. 1. P. 129-136. DOI: DOI: 10.26170/2079-8717_2022_01_15

Walch, C. (2019). Adaptive governance in the developing world: Disaster risk reduction in the State of Odisha, India. *Climate and Development, 11*(3), 238–252. DOI: 10.1080/17565529.2018.1442794

Wang, A. J. A. (2005, March). Information security models and metrics. In *Proceedings of the 43rd annual ACM Southeast Conference-Volume 2* (pp. 178-184).

Wang, D., Yang, Y., Yang, L., & Yang, H. (2023). Bibliometric analysis and visualization of endocrine therapy for breast cancer research in the last two decade. *Frontiers in Endocrinology (Lausanne), 14*, 1287101. DOI: 10.3389/fendo.2023.1287101

Wangen, G., Hallstensen, C., & Snekkenes, E. (2017). A framework for estimating information security risk assessment method completeness. *Journal of Computer Security, 25*(5), 517–544. https://link.springer.com/article/10.1007/s10207-017-0382-0

Wang, J., & Gu, L. (2017). Why is wechat pay so popular? *Issues in Information Systems, 18*(4), 1–8. https://iacis.org/iis/2017/4_iis_2017_1-8.pdf

Wang, L., Zhong, H., Ma, W., Abdel-Aty, M., & Park, J. (2020). How many crashes can connected vehicle and automated vehicle technologies prevent: A meta-analysis. *Accident; Analysis and Prevention, 136*, 105299. DOI: 10.1016/j.aap.2019.105299

Wang, N., Christen, M., & Hunt, M. (2021). Ethical considerations associated with "humanitarian drones": A scoping literature review. *Science and Engineering Ethics, 27*(4), 51. DOI: 10.1007/s11948-021-00327-4

Wang, N., Mutzner, N., & Blanchet, K. (2023). Societal acceptance of urban drones: A scoping literature review. *Technology in Society, 75*, 102377. DOI: 10.1016/j.techsoc.2023.102377

Wang, T., Ramik, D. M., Sabourin, C., & Madani, K. (2012). Intelligent systems for industrial robotics: Application in logistic field. *The Industrial Robot*, *39*(3), 251–259. DOI: 10.1108/01439911211217071

Wang, X., Reardon, L., & To, L. S. (2022). Visualizing Nepal's electricity supply resilience from a whole-systems perspective: A participatory approach. *Energy Research & Social Science*, *85*, 102409. DOI: 10.1016/j.erss.2021.102409

Wawrla, L., Maghazei, O., & Netland, T. (2019). Applications of drones in warehouse operations. *Whitepaper. ETH Zurich, D-MTEC*, *212*, 1–12.

Weerts, K., Vermeulen, W., & Witjes, S. (2018). On corporate sustainability integration research: Analysing corporate leaders' experiences and academic learnings from an organisational culture perspective. *Journal of Cleaner Production*, *203*, 1201–1215. DOI: 10.1016/j.jclepro.2018.07.173

Weng, H., & Feng, Y. (2022). Discussion of 'cocitation and coauthorship networks of statisticians. *Journal of Business & Economic Statistics*, *40*(2), 486–490. DOI: 10.1080/07350015.2022.2037432

West, J. P., & Bowman, J. S. (2016). The domestic use of drones: An ethical analysis of surveillance issues. *Public Administration Review*, *76*(4), 649–659. DOI: 10.1111/puar.12506

Whyte, J., Naderpajouh, N., Clegg, S., Matous, P., Pollack, J., & Crawford, L. (2022). Project leadership: A research agenda for a changing world. *Project Leadership and Society*, *3*, 100044. DOI: 10.1016/j.plas.2022.100044

Wilson, R. L. (2014, May). Ethical issues with use of drone aircraft. In *2014 IEEE International Symposium on Ethics in Science, Technology and Engineering* (pp. 1-4). IEEE. DOI: 10.1109/ETHICS.2014.6893424

Winkler, P., Kretschmer, J., & Wamprechtsamer, P. (2024). Navigating through digitalization challenges in strategic communication: Introducing the VUCA radar. *Journal of Communication Management (London)*. Advance online publication. DOI: 10.1108/JCOM-11-2023-0119

Wiranda, N., & Sadikin, F. (2021). Machine learning for security and security for machine learning: A literature review. *2021 International Conference on Information and Communications Technology (ICOIACT)*, 1–7. DOI: 10.1109/ICOIACT53268.2021.9563985

Wischnewsky, A. (2024). Nearly Cashless: Digital Transformation or Cultural Transmission? *Available at SSRN* 4794163. https://dx.doi.org/DOI: 10.2139/ssrn.4794163

Wohlers, T., Campbell, R., & Caffrey, T. (2016). *3D Printing and Additive Manufacturing State of the Industry: Annual Worldwide Progress Report*. Wohlers Associates.

Worakul, W. (2006). Community-Based Microfinance: An Empowering Approach Towards poverty Alleviation and Community Self-Reliance. *UNDP Thailand*. Retrieved from https://www.undp.org/sites/g/files/zskgke326/files/migration/th/Microfinancebook.pdf

World Bank. (2022). South Asia's digital opportunity. World Bank Group. Retrieved from https://openknowledge.worldbank.org/server/api/core/bitstreams/778a3dc7-9da1-5722-a128-50b6b2aa38d7/content

Wowk, K., Adams, M., & Martinez, E. (2023). Translating the complexity of disaster resilience with local leaders. *Frontiers in Communication*, *8*, 1100265. DOI: 10.3389/fcomm.2023.1100265

Yamin, M. A. (2021). Investigating the drivers of supply chain resilience in the wake of the COVID-19 pandemic: Empirical evidence from an emerging economy. *Sustainability (Basel)*, *13*(21), 11939. DOI: 10.3390/su132111939

Yang, C. C., Yang, S. Y., & Chang, Y. C. (2023). Predicting older adults' mobile payment adoption: An extended TAM model. *International Journal of Environmental Research and Public Health*, *20*(2), 1391. DOI: 10.3390/ijerph20021391 PMID: 36674145

Yang, L., Han, L., & Liu, N. (2019). A new approach to journal co-citation matrix construction based on the number of co-cited articles in journals. *Scientometrics*, *120*(2), 507–517. DOI: 10.1007/s11192-019-03141-9

Yeomans, L., & Bowman, S. (2021). Internal crisis communication and the social construction of emotion: University leaders' sensegiving discourse during the COVID-19 pandemic. *Journal of Communication Management (London)*, *25*(3), 196–213. DOI: 10.1108/JCOM-11-2020-0130

Yermalovich, P., & Mejri, M. (2020, October). Information security risk assessment based on decomposition probability via bayesian network. In *2020 International Symposium on Networks, Computers and Communications (ISNCC)* (pp. 1-8). IEEE. DOI: 10.1109/ISNCC49221.2020.9297330

Yoon, B., Kim, H., Youn, G., & Rhee, J. (2023). 3d position estimation of objects for inventory management automation using drones. *Applied Sciences (Basel, Switzerland)*, *13*(19), 10830. DOI: 10.3390/app131910830

Yoon, C. (2009). The effects of national culture values on consumer acceptance of e-commerce: Online shoppers in China. *Information & Management, 46*(5), 294–301. DOI: 10.1016/j.im.2009.06.001

Yu, M., Wen, J., Smith, S. M., & Stokes, P. (2022). Building-upHresilience and being effective leaders in the workplace: A systematic review and synthesis model. *Leadership and Organization Development Journal, 43*(7), 1098–1117. DOI: 10.1108/LODJ-09-2021-0437

Zabaniotou, A. (2020). A systemic approach to resilience and ecological sustainability during the COVID-19 pandemic: Human, societal, and ecological health as a system-wide emergent property in the Anthropocene. *Global Transitions, 2*, 116–126. DOI: 10.1016/j.glt.2020.06.002

Zahari, A. I., Abdul Manan, D. I., Mohamed, N., & Said, J. (2023). Impact of dynamic leadership and marketing planning on organizational resilience during Covid-19: Higher learning institutions. *SAGE Open, 13*(4), 1–12. DOI: 10.1177/21582440231212042

Zarif, A., Bandyopadhyay, S., Miller, G., & Malawana, J. (2024). Delivering medical leadership training through the Healthcare Leadership Academy: A four year analysis. *BMC Medical Education, 24*(1), 194. DOI: 10.1186/s12909-024-05031-y

Zhanga, A., Yue, X., & Kong, Y. (2011, June). Exploring culture factors affecting the adoption of mobile payment. In *2011 10th International Conference on Mobile Business* (pp. 263-267). IEEE. https://doi.org/DOI: 10.1109/ICMB.2011.32

Zhang, H., Zhao, B., Yuan, H., Zhao, J., Yan, X., & Li, F. (2019). SQL injection detection based on deep belief network. *ACM Transactions on Internet Technology, 19*(4), 46. Advance online publication. DOI: 10.1145/3331453.3361280

Zhang, J., Liu, Y., Gu, X., & Chai, J. (2024). A bibliometrics and visualization analysis of ropivacaine research from 2000 to 2023. *Frontiers in Medicine, 11*, 1465308. DOI: 10.3389/fmed.2024.1465308

Zhang, Q., Zhang, J., Zaccour, G., & Tang, W. (2018). Strategic technology licensing in a supply chain. *European Journal of Operational Research, 267*(1), 162–175. DOI: 10.1016/j.ejor.2017.11.036

Zhang-Zhang, Y., Rohlfer, S., & Varma, A. (2022). Strategic people management in contemporary highly dynamic VUCA contexts: A knowledge worker perspective. *Journal of Business Research, 144*, 587–598. DOI: 10.1016/j.jbusres.2021.12.069

Zhao, Y., & Pan, Y. H. (2023). A Study of the Impact of Cultural Characteristics on Consumers' Behavioral Intention for Mobile Payments: A Comparison between China and Korea. *Sustainability (Basel), 15*(8), 6956. DOI: 10.3390/su15086956

Zhukov, A. (2015, September). Lightweight cryptography: modern development paradigms. In *Proceedings of the 8th International Conference on Security of Information and Networks* (pp. 7-7).

Zieher, S., Olcay, E., Kefferpütz, K., Salamat, B., Olzem, S., Elsbacher, G., & Meeß, H. (2024). Drones for automated parcel delivery: Use case identification and derivation of technical requirements. *Transportation Research Interdisciplinary Perspectives*, *28*, 101253. DOI: 10.1016/j.trip.2024.101253

Zupic, I., & Čater, T. (2015). Bibliometric methods in management and organization. *Organizational Research Methods*, *18*(3), 429–472. DOI: 10.1177/1094428114562629

About the Contributors

Prakasha G S is an Associate Professor at the School of Education, Christ University, Bangalore, India. He has completed his MEd from the Regional Institute of Education, India. PhD from Bangalore University. He also completed an MSc (Chemistry) and MPhil in Education. He cleared the National Eligibility Test (NET) and was awarded a junior research fellowship (JRF). He holds IB Educator certificates in MYP and DP from DePaul University, Chicago, USA. He is a visiting research fellow, International University, Malaysia. He visited Germany, France, Switzerland, Canada, & USA. He led EdTech Project with Switzerland's Teacher-Ed University (hep, Vaud). He visited Miami University, University of Cincinnati-Ohio, University of Mount Union, Johns Hopkins University-Washington DC, Syracuse University, Binghamton University, Campbellsville University- Kentucky, New York University, & the Catholic University of America. His research areas are EdTech, TeacherEdu, & HigherEdu. He has published 50+ research articles indexed in SCOPUS and WOS databases. He wrote several book chapters, articles, & conference papers. He has presented papers worldwide, to name a few: Bangkok, Switzerland, Russia, University of Oxford-UK, & Binghamton University, USA. He is an active editorial board member for SAGE, SPRINGER NATURE, and Taylor & Francis. He received a research grant from Sophia University, Tokyo, JAPAN. At the Centre for Advanced Research and Training (CART), he has conducted many national-level research data analysis workshops at Christ University. He has trained PhD scholars, students, and faculty members in quantitative data analysis using SPSS. He has delivered lectures nationwide, including ICSSR- New Delhi, XLRI, Mount Carmel, Jyothi Nivas, Christ Nagar, Thiagarajar College, & St. Joseph's University. He conducts research activities that have national and international emphasis. He completed a Major Research PROJECT on Teacher & Teacher Educator Competencies. His PROJECT with NGO-GRAMA contributed a sustainable & replicable "Informal-Education-Model" to uplift rural education in India. He authored 6 books 1) Teaching Internship Models Around the World-

Routledge, 2) Educational Perspectives on Digital Technologies in Modeling & Management- IGI, 3) Digital Technologies- IGI: Insights in Education and Industry, 4) Diagnostic Testing, 5) Attention-deficit-behaviour, 6) Teacher-effectiveness by Lap Lambert, Germany.

Juan J. Leiva-Olivencia is Full Professor in the Department of Didactics and Educational Organisation at the University of Málaga (Spain). His professional career is linked to research projects on educational innovation, ICT, inclusive education, interculturality and teacher training. He combines his teaching and research work with the role of evaluator for various international agencies for the evaluation of quality in higher education.

***

Swati Dwivedi is a research scholar in the Department of Management Studies of Graphic Era Deemed to be University, Dehradun, India. She had published scopus indexed two articles.

Julia Georgievskaya Candidate of Sociological Science, Associate Professor of Philosophy, Sociology and Linguistics at Astrakhan State University of Architecture and Civil Engineering.

Oindrilla Ghosh is a Research Scholar hailing from the School of Law at Christ (Deemed to be University) Bangalore, India, who focuses her research on the health of tribal communities. Her work includes total 25 publications that are indexed in SCOPUS and Web of Sciences conference proceedings. She completed her early education at Loreto Day School, Bowbazar, and achieved distinction in her undergraduate (B.A/LL.B) and postgraduate (LL.M) studies at Calcutta University. Having served as an Assistant Professor of Law at Kingston Law College, Barasat, Kolkata, Ghosh has actively participated in numerous National Conferences, International Conferences, UN Global Summits, webinars, and seminars organized by prestigious academic institutions. Her expertise lies in health perspectives, environmental studies, human behavior, and related areas. Additionally, she has conducted various workshops and lectures in educational institutions, and is engaged with NGO's to help the needy

Ashulekha Gupta is a Professor in the Department of Management Studies, Dehradun, India. She is an exemplary teacher and avid researcher in the diverse field of International Business, Economics and Marketing. She has a distinguished academic record and holds a Doctorate and a Post-Graduation in Economics from HNB Garhwal Central University, Garhwal. She also got a Post Graduate Diploma in

Marketing Management from AIMA, New Delhi. Dr. Gupta has more than 18 years of teaching, research and Industry experience. She was associated with different Academic Institutions and Corporate as Faculty and administrative positions. Dr. Gupta's research interests include International economics, Environmental and behavioral Economics. She has numerous publications in national and international journals of repute ABDC ranking and presented more than 30 papers in national and international conferences contributed 5 Chapters in edited books. Her recent research interest in the area of Environmental Economics is published as a book chapter in edited book of Springer Nature, Singapore. Dr. Gupta is a reviewer of several National and International journals published by Springer and Taylor & Francis.

Jipson Joseph has a PhD in Ethics from DVK Bangalore. Currently, he is a Research Scholar in the School of Law at Christ University, Bangalore. His research interests are in the fields of ethics, gender studies, constitutional law, human rights, human security, food security, environmental justice, developmental policies and global poverty.

Binod Kumar is a graduate student known for his academic excellence, with his total 20 scholarly works indexed in Scopus database on Neuroscience in Organizational Behavior, Neuromarketing and AI Applications on Wildlife Conservation. Hailing from Ranchi, Jharkhand in India, Binod is currently pursuing a Master of Laws (LL.M) at the prestigious Radha Govind University. His passion for technology and innovation along with climate changes is evident through his high academic performance and numerous projects. Binod has a keen interest in artificial intelligence, machine learning, neuropsychology, climate changes and disaster management, fields in which he has undertaken several research projects. He is particularly fascinated by the potential of AI to solve real-world problems and is dedicated to learning and applying new techniques in these areas. He volunteers at local NGOs, teaching underprivileged children basic computer skills. With a strong foundation in law and a commitment to making a positive impact, Binod Kumar is poised to become a future leader in the legal industry.

Nisha K. Kumar holds master's degree in business administration with a specialization in Finance and PhD in management. Presently, an independent researcher. She has a teaching experience of 6 years. She has published research articles in peer reviewed journals.

Gustavo López Badilla Expert in Phsycochemical studies as Electronic, Electrical, Elecotrmecahnical, Mechanical,Corrosion, Materials, Micro and Nano Techniques of Analysis, Ergonomy (Basic, Environmental and Thermical

Environmental), Climatic Change, Lean and Sustainable Manufacturing, Industrial Operations.

Sergey Melnikov In 2016, he graduated from the North Caucasus Federal University with a degree in 230102 "Automated Information Processing and Management Systems". In 2016, he enrolled in postgraduate studies at the Department of Information Communications and began working as a laboratory engineer. In 2020, he successfully defended his thesis for the degree of Candidate of Technical Sciences on the topic "Improving the positioning accuracy of unmanned aerial vehicles in conditions of distortion or suppression of the GPS/GLONASS navigation field." In the 2020-2021 academic year, he worked at the department as an assistant, and since 2021 he has been working at the department as an associate professor at the Department of Information Communications.

Victor Nemchinov is Leading research fellow in the Department of comparative cultural studies, Institute of Oriental studies, Russian Academy of Sciences, e-mail: viators@mail.ru He is the author of more than 300 academic publications on humanitarian issues of analog and electronic cross-cultural communication, historical consciousness, world cultural heritage and civilizations, picture of the world transformation and on advanced dialogical forms of education. He authored Mixed Economy: Controlling National Development (1994), Testing Russian Teenagers in the Time of Transition. in: Youth and History: a Comparative European Survey on Historical Consciousness and Political Attitudes among Adolescents (1997), the chapter on Russian social sciences in UNESCO's World Social Science Report (1999), The Quest for the Formation of a New Identity: A Workshop on the Way of Teaching History in: Worldwide Teaching of History: Present and Future. (2006), Fragmentary Paradigm Shift: a Predicament of Pending Cultural Challenges. In: Deep North. Transmediale parcours 2 (2009), Orientalist and the Book: Reflections on Changing Role of Libraries in the Information Field of the 21st Century. in: Oriental Analytics (2016), compiled and edited Multiculturalism in Russia and countries of the East: problems of globalization, regionalization and identity in the first third of the 21st century (2024). He is member of ICOM, founder and director of the Museum of Historical Consciousness, member of the International society of history didactics, founder and curator of the advanced learning program "Diaversity". Dr. Nemchinov worked as senior visiting fellow in advanced study international institutes including ZiF, Bilefeld University, Collegium Budapest and IFK Vienna. Victor is also known as paintings connoisseur and for his technique of instant graphic sketches on glossy paper.

Ananya Pandey is a Research Scholar in School of Law at Christ University, Bangalore. She has her research interests in the field of Constitutional Law, Environment Law, Human Rights, Food Security, Intellectual Property Law and Jurisprudence.

Meera Sharma is currently working as an Associate Professor with Uttaranchal University Dehradun(Uttarakhand) India. She has also worked as an Associate Professor with JBIT – Dehradun. Previously worked as Head of the Department, and Associate Professor with Roorkee Institute of Technology, Roorkee (India). She has more than eighteen years Academic and Administrative experience. She completed her Ph.D. from Uttarakhand Technical University, Dehradun(UK) in Batch 2010-2014. She has participated in International Conference at World Bank Washington DC, USA Also presented research paper at international conference at American University of Phonom Penh 2024, Cambodia. She could fetch full travel and stay grant from IIM Indore for paper presented in 2010. She has authored several Research papers in National and International Journals and conferences of repute (Scopus, webofsc, etc) and published several patents. She has been awarded with letter of appreciation by IIT Delhi for her contribution in Research in 2010. She received several govt. funds for academic activities. She received Award of Appreciation in Research during AICTE sponsored FDP in 2015. She has also been awarded Excellent & Dedicated Academician Award by ITU San Jose, California, USA under faculty exchange program. Awarded with prize and appreciation certificate for contributing in State Level Brainstorming Session organized by Uttarakhand Council for Science and Technology- Uttarakhand on International Women`s Day 2024. She has visited San Jose, San Francisco, Washington, Chicago, Dallas (Texas), Cambodia, Bangkok and Vietnam for her research work and also received Certificate of Appreciation and Best academician award for her research visit at ITU, California Campus. Her Area of Interest is Organizational Behaviour, Organizational Structure and Personnel Management, Human Resource Management, International Human Resource Management etc

Index

A

aerospace industry 2, 8, 11, 12, 15, 17, 18, 19, 22, 24
Analog and Virtual 158
analysis methods 52, 62, 87
atmospheric agents 1, 2, 12

B

Balancing 61, 74, 75, 77, 196
BANI 97, 98, 99, 100, 101, 102, 103, 104, 105, 106, 107, 108, 109, 110, 111, 114, 128, 158, 168, 190
Bayesian Networks 81, 86, 88, 90, 91, 93, 94

C

climatic conditions 179, 184, 287, 288, 290, 291
Correlation Matrix 72, 73
Cryptography 39, 40, 41, 44, 45, 140
Cyber Risk Quantification 58

D

Data 26, 27, 42, 43, 44, 48, 49, 50, 52, 53, 54, 55, 56, 57, 61, 62, 63, 64, 65, 66, 67, 68, 69, 70, 71, 72, 73, 74, 75, 76, 77, 78, 82, 83, 84, 85, 86, 87, 88, 89, 90, 91, 93, 94, 103, 105, 106, 107, 114, 128, 129, 130, 137, 140, 141, 143, 147, 150, 155, 158, 160, 161, 162, 163, 169, 170, 187, 209, 210, 211, 212, 214, 222, 226, 227, 229, 231, 234, 237, 239, 243, 257, 260, 263, 265, 268, 269, 271, 272, 273, 275, 277, 282, 287, 288, 289, 291, 293, 294, 295, 296, 297, 298, 299, 300
Data-Driven Security 94
Delivery 6, 7, 9, 22, 43, 132, 133, 194, 195, 218, 230, 259, 261, 262, 264, 266, 267, 268, 270, 271, 275, 276, 277, 278, 280, 281, 282, 283, 285
Digital Age 97, 98, 99, 100, 101, 102, 103, 104, 105, 106, 107, 108, 109, 129, 255
Digital Detox 109
DIGITALIZATION 31, 114, 124, 127, 128, 129, 158, 161, 162, 170, 172, 230, 237, 243, 255
Digital Literacy 97, 99, 101, 102, 104, 107, 108, 109, 190, 194, 208, 211, 212, 213, 214, 227, 234, 236, 237, 238, 239, 242, 254, 256
DOCX 142, 143, 144, 145, 147, 148, 152

E

Emotional Resilience 97, 102
Error Coding 42

F

Financial Behaviour 204
Financial Inclusion 191, 192, 193, 194, 195, 196, 198, 200, 203, 214, 216, 223, 226, 230, 232, 233, 235, 236, 237, 238, 239, 241, 242, 245, 246, 249, 250, 251, 252, 254, 256
FinTech 190, 191, 194, 195, 197, 199, 200, 201, 202, 203, 204, 205, 222, 224, 225, 229, 230, 231, 235, 236, 237, 240, 241, 242, 243, 245, 246, 248, 249, 250, 251, 254, 256, 257
FORMATTING CHANGES 140, 142, 143, 144, 145, 149, 150, 153, 154

H

Hash Functions 42

I

Indian Healthcare 25, 27, 28, 30, 33, 34, 35, 36, 37
information assets 51
information security 42, 47, 48, 49, 50, 52, 53, 54, 56, 58, 59, 61, 62, 63, 64, 66, 67, 68, 69, 70, 77, 79, 81, 82, 83, 84, 85, 86, 88, 89, 90, 94, 95, 96, 140,

143, 155
INTERNET RESOURCES 127, 128, 129, 130, 131, 132, 133, 134, 135
ISO 48, 49, 50, 51, 52, 54, 56, 57, 58

L

Lattice Cryptography 41
LINGUISTIC CHANGES 140, 150, 154
low-power wireless networks 288, 299

M

Machine Learning 35, 61, 62, 63, 64, 65, 66, 67, 68, 69, 70, 71, 72, 74, 77, 78, 79, 83, 84, 85, 87, 94, 102, 236, 262, 269, 281
Machine Learning Models 65, 68, 71
Management 23, 26, 30, 34, 35, 36, 37, 47, 48, 49, 50, 51, 52, 53, 54, 55, 56, 57, 58, 59, 67, 70, 81, 82, 84, 88, 94, 95, 104, 109, 112, 113, 114, 115, 116, 117, 118, 119, 120, 121, 122, 123, 124, 125, 128, 129, 136, 137, 155, 161, 169, 170, 175, 206, 215, 230, 231, 243, 247, 249, 250, 251, 252, 254, 256, 258, 260, 262, 263, 264, 268, 276, 277, 278, 279, 280, 281, 282, 283, 285, 288, 289, 299
Manufacturing technologies 1, 4, 5, 20, 24
Megalopolis 157, 158, 160, 166, 167, 169, 170, 171
Mental Health 28, 36, 97, 98, 99, 100, 101, 102, 103, 104, 105, 106, 107, 108, 109, 110, 111, 121
METROPOLIS 127, 128, 129, 130, 169, 171
micro deterioration 1, 2
Mindfulness 97, 98, 101, 103, 104, 109, 112, 117, 121
Multiplier effect 173, 174, 175, 179, 180, 182, 183, 186

O

OCTAVE 50, 51, 52
Online Communities 97, 109

P

PDF 142, 143, 144, 147, 148, 152, 155, 169, 249, 250, 251, 252, 253, 254, 255, 257, 300
physico-chemical parameters 287, 288, 289, 290, 291, 299
POPULATION 33, 49, 128, 129, 130, 131, 136, 160, 161, 162, 166, 170, 179, 180, 181, 188, 193, 208, 223, 232, 240, 241, 254
Privacy 58, 68, 79, 106, 107, 129, 208, 210, 211, 214, 227, 228, 230, 234, 236, 256, 268, 272, 273, 274, 276, 277, 278, 280, 282
Probabilistic Models 81, 86, 87, 88, 90, 93, 94

Q

Quantum Computer 40, 41
Quantum World 41

R

Random Forest 71
Resilience 25, 26, 27, 28, 31, 33, 34, 37, 51, 70, 82, 92, 93, 97, 98, 99, 100, 101, 102, 103, 105, 109, 112, 113, 114, 115, 116, 117, 118, 119, 121, 122, 123, 124, 128, 147, 170, 190, 260
risk assessment 35, 47, 48, 49, 50, 51, 52, 53, 54, 55, 56, 58, 59, 70, 82, 89, 95, 96
risk management 47, 48, 49, 50, 51, 53, 54, 56, 57, 58, 59, 95, 112, 118
RISKS 47, 48, 49, 50, 51, 52, 53, 54, 55, 56, 58, 70, 86, 87, 88, 89, 90, 93, 102, 106, 127, 135, 140, 157, 158, 160, 163, 165, 166, 167, 168, 169, 170, 205, 210, 211, 214, 237, 246, 270, 272, 284
Rural tourism 173, 174, 175, 176, 177, 178, 179, 180, 181, 182, 183, 184, 185, 186, 187

S

Security 31, 32, 40, 42, 44, 45, 47, 48, 49, 50, 52, 53, 54, 55, 56, 58, 59, 61, 62, 63, 64, 65, 66, 67, 68, 69, 70, 71, 77, 79, 81, 82, 83, 84, 85, 86, 88, 89, 90, 91, 94, 95, 96, 106, 112, 117, 121, 129, 140, 143, 151, 155, 163, 165, 166, 167, 179, 183, 191, 192, 193, 199, 200, 201, 205, 206, 208, 209, 210, 211, 212, 214, 215, 219, 222, 227, 228, 234, 236, 239, 240, 244, 247, 248, 256, 268, 272, 273, 274, 276, 281, 282, 283
Security Optimization 84
SEM technique 2, 21, 22
Shor's Algorithm 40
Social Dangers 163, 164, 166
Strategy 32, 48, 64, 98, 159, 170, 188, 197, 202, 211, 221, 226, 227, 228, 236, 240
STRUCTURAL CHANGES 149, 150, 151, 153, 154
Supply Chain 114, 117, 121, 124, 260, 261, 264, 265, 266, 278, 279, 280, 281, 282, 285
Sustainability 9, 23, 24, 25, 27, 33, 112, 113, 114, 115, 116, 117, 118, 121, 122, 123, 124, 169, 170, 175, 188, 255, 258, 275, 278, 281
Sustainable development 27, 33, 34, 37, 129, 175, 186, 187, 188, 230, 252
Symmetric Cryptography 40

T

Technology 5, 23, 26, 27, 30, 31, 32, 35, 36, 37, 39, 40, 44, 52, 59, 62, 79, 80, 97, 98, 100, 101, 102, 103, 105, 106, 107, 108, 109, 110, 111, 112, 114, 115, 118, 129, 136, 137, 140, 141, 156, 162, 169, 189, 191, 193, 194, 196, 198, 199, 201, 208, 209, 211, 212, 213, 214, 215, 218, 219, 220, 221, 223, 224, 225, 226, 227, 229, 231, 233, 234, 237, 239, 240, 241, 243, 244, 247, 248, 252, 254, 256, 257, 261, 262, 264, 265, 267, 268, 271, 273, 274, 275, 276, 277, 278, 279, 280, 281, 282, 283, 285, 287, 288, 289, 299
Threat Modeling 81
threats and vulnerabilities 50, 56, 82, 88, 90
Transportation 15, 181, 253, 261, 262, 264, 265, 266, 269, 274, 275, 276, 278, 279, 280, 283, 285
TXT 142, 151, 152

U

USING METADATA 140, 147, 148, 153, 154

V

VIRTUAL SPACE 163, 165, 166, 167, 168
VUCA 25, 26, 27, 28, 29, 30, 31, 32, 33, 34, 35, 36, 37, 97, 98, 99, 100, 101, 102, 103, 104, 105, 106, 107, 108, 109, 110, 111, 114, 115, 124, 125, 128, 157, 158, 161, 190
Vulnerability Assessment 55

W

Warehouse 262, 263, 264, 268, 276, 283, 284, 285
Well-Being 31, 97, 98, 99, 100, 101, 102, 103, 104, 105, 106, 107, 108, 109, 110, 111, 113, 134, 135, 136, 170, 251, 271
wireless sensor network 287, 288, 289, 291, 293, 294, 299

www.ingramcontent.com/pod-product-compliance
Lightning Source LLC
LaVergne TN
LVHW081536300926
849161LV00006B/1448